Building Valve Amplifiers

Second Edition

Morgan Jones

ELSEVIER

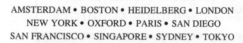

AMSTERDAM • BOSTON • HEIDELBERG • LONDON
NEW YORK • OXFORD • PARIS • SAN DIEGO
SAN FRANCISCO • SINGAPORE • SYDNEY • TOKYO

Newnes is an imprint of Elsevier

Newnes

Newnes is an imprint of Elsevier
32 Jamestown Road, London NW1 7BY, UK
225 Wyman Street, Waltham, MA 02451, USA
525 B Street, Suite 1800, San Diego, CA 92101-4495, USA

First Edition 2004
Second Edition 2014

British Library Cataloguing-in-Publication Data
A catalogue record for this book is available from the British Library

Library of Congress Cataloging-in-Publication Data
A catalog record for this book is available from the Library of Congress

ISBN: 978-0-08-096638-0

For information on all Newnes publications
visit our website at store.elsevier.com

14 15 16 17 18   10 9 8 7 6 5 4 3 2

Building Valve Amplifiers

CONTENTS

Contents

PREFACE

As its title suggests, this book is concerned with the practical problems of **building** a valve amplifier that has already been designed, making it the companion book to *Valve Amplifiers* (fourth edition), which contains far more equations and is concerned with circuit design.

This new edition's word count is more than twice that of the first edition. So where is all the new material? All chapters have been updated, but noting the most commonly asked questions, the metalworking and test principles chapters have doubled in size, and the entirely new practical projects chapter is the second largest. The wiring and performance testing chapters have grown significantly, and an appendix appeared.

Although construction techniques have changed little in the 10 years since the first edition, measurement techniques and data analysis have been transformed by computers. Digital oscilloscopes have ousted analogue, and audio measurement systems based on a recording quality sound card offer an unprecedented performance to price ratio. Analogue circuit modelling via LTspice has spread from the professional world to the amateur, and the author took to it like a duck to juggling. However, perseverance paid off, and he now finds it invaluable when dealing with semiconductors and passive components whose characteristics are tied firmly to fundamental physics. Sadly, valve characteristics are determined by production engineering, making valve models less accurate, and distortion predictions distinctly dubious.

One possible way of building an amplifier is to choose the most expensive components on offer, have a chassis CNC-machined from a single billet of aerospace aluminium alloy (very popular commercial ploy at the moment), then use whichever designer wire is currently fashionable. However, the author assumes that you are clutching this book because you want to know how to build a valve amplifier that is better than you could afford for the same cost ready-made. For that reason, some of the physics that you slept through at school will reappear, but with the valuable bonus that it will allow you to make reasoned choices that improve quality or save money — usually both.

It is undoubtedly easier to do metalwork in a fully equipped machine shop, but even if domestic harmony precludes machine tools requiring an engine hoist to move them, surprisingly good work can be done by a power drill in a stand plus a few carefully selected hand tools. Nevertheless, drill presses, lathes, and bandsaws are all now available at amateur prices (and perhaps more significantly, **sizes**) making them well worth investigating. In addition to the standard techniques, a number of "cheats" will be shown that allow you to produce work of a standard that appears to have come from a precision machine shop. This will enable your creation to be a thing of beauty that can be proudly displayed.

The rules for good audio construction are not complex. It's just that there are rather a lot of them. Once the logic is understood, good layout comes naturally.

Even the most carefully considered designs need a little fettling once built. Electronic test equipment ranges from 3½ digit DVMs through gigahertz oscilloscopes, megahertz impedance analysers and signal generators, back to PC-based virtual instruments. They all cost money, but once you understand the operating principles, you can choose which features are worth paying for, which can be safely ignored, and how to use what you **can** afford to its best advantage.

Startlingly, years of experience don't make the author any less frightened at the instant of first switch-on. Accidents do happen, but there are ways of minimising the quantity of smoke. Sometimes, an amplifier is stubborn, and just doesn't **quite** work properly, requiring genuine faultfinding.

This book is distilled from years of bludgeoning recalcitrant electronics, thumping metal, and sucking teeth at the price of good test equipment.

Acknowledgements

The author would like to thank Euan MacKenzie for proofreading (once again), although he should emphasise that any remaining errors are entirely the author's responsibility.

Especial thanks are due to Susan for sustaining the author during the many hours he was closeted in the laboratory.

Section 1

Construction

CHAPTER 1

PLANNING

In this first chapter, we will plan the mechanical layout of a valve amplifier and find that it is determined by fundamental physics. At this stage, freedom of choice is unlimited, whereas it will later be restricted, so it is important that the choices and compromises made now are the best ones. Whilst good planning will not save a poor design, poor planning can certainly ruin a good one. Any component that requires holes to be drilled or cut in the chassis should undergo thorough electrical testing before a layout is planned — especially if the part is not current production. It is galling to discover that a component is faulty (and worse, irreplaceable) after custom metalwork has been done and wiring completed — details on component testing are given in Chapter 4.

Chassis layout

A valve amplifier uses a number of large components needing relative positioning that minimises the length of connecting wires, yet prevents them and their wiring from interfering with each other. Chassis layout breaks down into the following considerations:

1. Electromagnetic induction: Minimising hum induction from chokes and transformers into each other and into valves.
2. Heat: Output valves etc. are hot and must be cooled. Conversely, capacitors run cool and should be kept that way.

Building Valve Amplifiers. DOI: http://dx.doi.org/10.1016/B978-0-08-096638-0.00001-1

3. Unwanted voltage drops: All wires have resistance, so the wiring must be arranged to minimise any adverse effects of these voltage drops.
4. Electrostatic induction: Minimising hum from AC power wiring is not often a problem, because even thin conductive foil provides perfect electrostatic screening, but paths should be kept as short as possible.
5. Mechanical/safety: Achieving an efficient chassis arrangement that is easily made, maintained, and used.
6. Acoustical: Almost all components are microphonic, but valves are the worst. We should consider which components are most sensitive to vibration, and minimise their exposure.
7. Aesthetic: The highest expression of engineering is indistinguishable from art. If you have a superb chassis layout, it will probably look good. Conversely, if it looks horrible, it is probably a poor layout. . .

We have a seven-dimensional problem. A poor transistor amplifier might be able to hide behind the fence of negative feedback, but amplifiers having an output transformer rarely tolerate more than 25 dB of feedback before their stability becomes distinctly questionable. Consequently, chassis layout is critical to performance.

The large components are generally the mains transformers, output transformers, power supply chokes, power supply capacitors, and valves. The traditional way of deciding their positioning was to cut out pieces of paper of the same size and shuffle them around on a piece of graph paper. Alternatively, the lumps themselves can be arranged and glanced at over a few days until the best layout presents itself to the viewer's subconscious.

Finally, components can be shuffled around and a chassis designed using an engineering drawing package on a computer, with the enormous bonus that a template of the layout can be printed with all the fixing holes precisely positioned, saving errors in marking out. Although it takes time to draw a valve holder or a transformer precisely, it only has to be done once, and you will quickly build up a library of mechanical parts. In consequence, the author has almost forgotten how to perform traditional marking out using a scriber, ruler, and square.

It is vital to make the chassis large enough!

This point cannot be emphasised too strongly. Achieving neat construction on a cramped chassis requires a great deal more skill and patience than on a spacious chassis. There are many considerations that must be taken into account, so it is vital that this stage is not rushed. Each of the following design considerations might not make a great difference in itself, but the sum of their effects is the difference between a winner and an "also ran".

Electromagnetic induction

Almost all of the larger components either radiate a magnetic field or are sensitive to one. Not all of a transformer's primary flux reaches the secondary, so leakage flux might induce currents into grid wiring, thereby developing voltages across associated resistances. Whether or not these currents and voltages are significant depends on the signal level and source impedance at that point, so output valves are less of a problem than the input stage.

Coupling between wound components

Wound components such as transformers and chokes can easily couple into one another, so hum can be produced by a mains transformer inducing current directly into an output transformer. Fortunately, the cure is reasonably simple, and may be summarised by a simple ratio whose value must be minimised:

$$\text{Induction} \propto \frac{\cos \theta}{d^3}$$

The angle θ and distance d are shown in the diagram. See Figure 1.1.

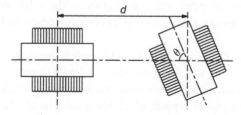

Figure 1.1
Orienting transformers for minimum coupling.

Rotating transformer cores by 90° (cos 90° = 0), so that the coil of one transformer (or choke) is not aligned with the other is very effective, and typically results in an immediate 25 dB of practical improvement. Even better, if one coil is driven from an oscillator whilst the interference developed in the other is monitored (oscilloscope or amplifier/loudspeaker), careful adjustment of relative angles can often gain a further 25 dB.

Because coupling decays with the cube of distance [1], as the distance between offending items is increased, the interference falls away rapidly. However, simply increasing the **gap** between two adjacent transformers from 6 to 25 mm does not materially reduce the interference, because the transformers are typically 75 mm cubes, and the spacing that applies is the distance between centres, which has only changed from 81 to 100 mm, resulting in only 5.5 dB of theoretical improvement. Unfortunately, when large transformers are this close, coupling no longer obeys the inverse cube law because the dipole equation upon which it is based carries the implicit assumption that the separation is much greater than dipole length, so a 3 dB reduction, or less, is more likely.

Another consequence of size is that if other layout considerations force an output transformer and mains transformer to almost touch, we should not only ensure that their coils are at right angles to one another, but also align their centres. When the centres are aligned, each edge of one coil induces significant current into the receiving coil, but because the edge distances are the same, their induced currents are equal and opposite, so cancel. But if one transformer is slid to one side (whilst remaining at right angles), the edge distances differ, complete cancellation no longer occurs, and increased induction results.

Be aware that the previous argument of edge distance equality and consequent cancellation assumes that the transformer manufacturer made each winding fill complete layers. They usually do, because it avoids difficult winding, but a dual chamber bobbin having the mains primary in one chamber and secondaries in the other destroys this fundamental assumption and only testing can determine its optimum orientation.

Although smoothing chokes are gapped, and therefore inevitably leaky, they don't generally have much alternating voltage across them, so their leakage is low, and they can often be used to shield output transformers from the mains transformer. The exception to this rule is the choke input power supply, which has a substantial alternating voltage across its choke, so its leakage field can be significant.

A poorly designed mains transformer's core can easily be saturated by the large current pulses drawn by a large reservoir capacitor in combination with a semiconductor rectifier, producing a particularly ragged leakage flux, and this can be quickly identified using a search coil (see Chapter 4 for details on search coil construction). See Figure 1.2.

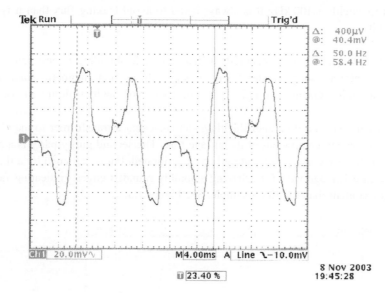

Figure 1.2
Leakage flux caused by a saturating mains transformer.

Shielding

The effectiveness of magnetic shielding is determined by the ratio of the shield's thickness to penetration depth, so we must minimise penetration depth [2]:

$$\delta = \sqrt{\frac{2}{\omega\mu_0\mu_\mathrm{r}\sigma}}$$

where:

δ = penetration depth (m)
ω = angular frequency = $2\pi f$
μ_0 = permeability of free space = $4\pi \times 10^{-7}$ H/m
μ_r = relative permeability (1 for air, >5000 for steel)
σ = electrical conductivity (1/Ωm).

Minimising penetration depth means maximising the denominator, so our first observation is that low frequencies (<100 kHz) have large penetration depths and will therefore be difficult to shield; <100 kHz it is always better to avoid leaking flux than it is to attempt to shield it.

Unsurprisingly, the equation also tells us that magnetic shielding is best achieved using a magnetic material (such as steel), but it also tells us that magnetic shielding can be obtained at high frequencies simply by a good conductor, such as aluminium or copper.

Thus, we can expect a thin steel shielding can enclosing a transformer to mainly attenuate the high frequency component of its leakage flux. The second guilty party is a 50-year-old mains transformer whose core material has deteriorated, but is screened by a thin steel can resulting in a leakage waveform having smoothly rounded edges, indicating far less high frequency content than the first example. See Figure 1.3.

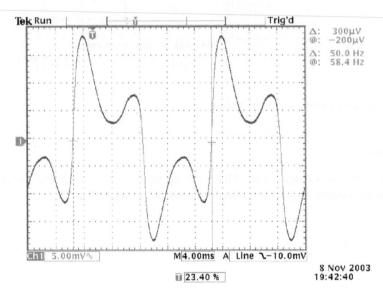

Figure 1.3
Leakage flux caused by a saturating but lightly shielded mains transformer.

A little of the second example's shielding was due to the ferrous shield diverting and containing leakage flux, but the majority was due to losses. Hysteresis and eddy current losses become more significant as frequency rises (even if the loss per cycle remains constant, there are more cycles per second to dissipate energy). Further, as frequency rises, wavelength falls, and more loops become possible in a given distance. Eddy current losses are conduction losses, so the shield's electrical resistance should be minimised, making 2 mm copper or 4 mm aluminium very effective >100 kHz.

Another transformer shielding possibility that relies on losses is the Faraday shield. See Figure 1.4.

Figure 1.4
A shorted turn wound round the outside of a transformer forms a Faraday shield.

The wide copper strap provides magnetic shielding because it is a shorted turn to the transformer's leakage flux, and this is why it wraps around the entire core rather than the core's central leg. The shield's effectiveness is determined by its electrical resistance and the proportion of intercepted leakage flux, so the foil firstly needs sufficient cross-sectional area to have low resistance and be an effective shorted turn, and secondly needs to enclose as much of the transformer as possible. The ideal Faraday shield would be a tube much longer than the transformer, but practical considerations generally limit it to the width of the windings. A Faraday shield can be retrofitted to any EI transformer provided care is taken:

- Ensure that the shield does not short-circuit or contact with existing windings, perhaps by touching tags.
- Soldering the ends of the copper strip to complete the shorted turn involves heat, so ensure that the heat can't damage insulation beneath it, perhaps by placing a thin strip of sacrificial cardboard beneath the joint and removing it after soldering.

Interestingly, the copper foil need not be very thick ($0.001''$ or $0.025\,\mu$m) to constitute a shorted turn ($\approx 5\,m\Omega$), and practical Faraday shields of the same copper thickness all have roughly the same resistance, regardless of transformer size. This occurs because although a larger transformer requires a longer shorted turn, the required foil width rises proportionately.

The 0.5-mm-thick Faraday shield shown in Figure 1.4 appears over-engineered by a factor of 20, but the increased thickness may perhaps be explained by the fact that the transformer was salvaged from a receiver that would have needed careful radio frequency shielding.

Transformers and the chassis

We have seen that all transformers leak flux and that shielding is difficult. The question is whether the leakage flux is a problem. If an output transformer leaks flux into the aluminium chassis of a power amplifier, it probably isn't a problem because aluminium doesn't conduct magnetic flux. But a mains transformer leaking flux into the steel chassis of a pre-amplifier **is** a problem because the steel chassis passes the flux into sensitive signal circuitry. Fortunately, because $\mu_r \approx 1$ for non-magnetic materials, but $\mu_r > 5000$ for steel, even a small gap is able to prevent flux leaking into a steel chassis; 1.6 mm ($\frac{1}{16}''$) plain phenolic sheet is ideal, but may be hard to find, so practical alternatives include the decorative printed phenolic sheet found in DIY stores, or 3 mm ($\frac{1}{8}''$) acrylic.

Beware that although toroidal transformers are the theoretically perfect shape, practical toroids leak flux, and that mounting a toroid directly onto a steel chassis is asking for hum problems. As before, plastic sheet provides enough of a magnetic gap to significantly reduce induction. However, the danger of accidentally creating a shorted turn is considerable. Toroids are usually secured by a conductive screw pulling a large conductive washer onto the opposite face of the core to clamp the transformer tightly to the chassis. Accidentally connecting the washer or screw to the chassis by any means other than the bottom of the central mounting screw constitutes a shorted turn that could **destroy** a power transformer. If there is a possibility of the washer contacting chassis, break the conductive path through the screw. Either use a nylon screw (not ideal because the small cross-section of nylon stretches easily), or use a pair of metal screws separated by a substantial threaded plastic boss (the boss's far larger cross-sectional area is less liable to stretch and weaken the clamping force).

Beam valves and mains transformers

Beam valves deliberately focus their current into thin sheets that pass largely unintercepted between the horizontal wires of g_2, thus improving efficiency. This means that a vertical beam deflection would affect g_2 current, and because g_2 is typically supplied from a finite source resistance, Ohm's law ensures that this would change V_{g2}, thus changing I_a. One way of deflecting electrons is with a magnetic field, such as the leakage flux from a transformer. Hum due to beam deflection can be minimised by applying Fleming's left-hand rule, and ensuring that the electron beam is never at right angles to the leakage flux from the transformer. When considering induction between two transformers, it did not matter which transformer was rotated, so long as the coils were at 90° to one another. With beam valves, only one orientation is ideal with respect to a nearby mains transformer. See Figure 1.5.

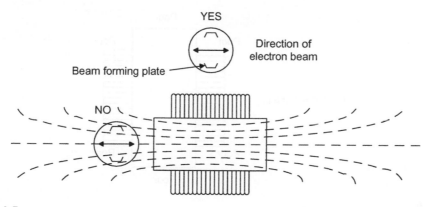

Figure 1.5
Beam valves and mains transformers.

The valve is shown in two positions, both the same distance from the centre of the mains transformer, and with correct beam orientation relative to the leakage flux from the transformer. However, leakage flux tends to be concentrated on the axis of the coil, and would also induce hum into the control grid's circuit, whereas the alternate position has much lower flux density. (Diagrams of this form portray higher flux density by having more lines in a given area.)

Input valves are very sensitive to hum fields, and should always be placed at the far end of the chassis to any mains transformer.

In theory, output transformers should leak less flux because they operate at a lower flux density (to avoid saturation and consequent distortion) and are designed for minimum leakage inductance (which translates directly into reduced leakage flux). In practice, probing output transformers and mains transformers with a search coil failed to show the expected difference. The quality of the transformer seems to be the overriding consideration, rather than its use. Thus, a Leak TL12 + push—pull output transformer leaked more flux than a good quality modern output transformer in a single-ended amplifier, despite the latter being gapped.

Although, as expected, leakage flux at 90° to the coil's axis cancels to zero, leakage at the edges of the coil can be comparable with that on axis because the coil's outermost turn is so far away from the (flux-concentrating) core. See Figure 1.6.

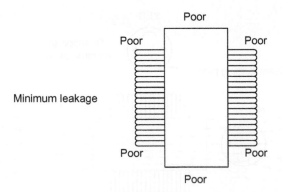

Figure 1.6
Transformers leak most flux along the axis of their coils and at coil edges furthest from the core.

Heat

Heat is the enemy of electronics. Output transformers and chokes are usually quite cool, so they can move towards the centre of the chassis if necessary (creating a mechanical problem, but we will consider this later). Mains transformers are generally warm, and mounting them towards the edge of the chassis assists cooling.

Heat shortens insulator life and causes components to drift in value. At worst, it causes fires. And we intend to use valves, which are deliberately heated...

Modes of cooling

Heat can only flow from a body having a higher temperature to one having a lower temperature, and flows until there is no temperature difference between them. There are three ways of transferring heat.

Conduction is the most efficient way of transferring heat and requires a conducting material to bond the heat source physically to its destination. An ideal conductor transfers heat with a minimal temperature drop between source and destination, and materials having free electrons (electrical conductors) such as copper and silver are particularly good. The body of a power transistor must conduct the heat generated by the (very much smaller) silicon die to an external heatsink. See Figure 1.7.

One of the reasons that the previous steel TO-3 package has been made obsolete by epoxy packages is that their copper tab interposes far lower thermal resistance between the silicon die and heatsink.

Temperature drops caused by the flow of conducted heat through a thermal resistance can be calculated using:

$$T = \frac{P}{R_{\text{thermal}}}$$

where:

T = temperature in kelvin (or °C)
P = thermal power to be transferred
R_{thermal} = thermal resistance (often given by heatsink manufacturers in °C/W).

Figure 1.7
The inside of a 2N3055 15 A/115 W power transistor. Note the relative size of the silicon device and its wires compared to the case.

If we need to conduct heat through a bar of uniform cross-section, we can calculate its thermal resistance using:

$$R_{\text{thermal}} = \frac{l}{A\lambda}$$

where:

R_{thermal} = thermal resistance (K/W or °C/W)
l = length (m)
A = cross-sectional area (m^2)
λ = thermal conductivity (W/m/K)

Table 1.1
Thermal conductivities of metals at 20°C

	Conductivity (W/m/K)
Silver	428
Copper	403
Gold	319
Aluminium	236
Iron	84

As can be seen from Table 1.1, copper is only fractionally worse than silver (the best), but much better than aluminium, and iron (or steel) is poor.

As an example, suppose that we have bonded a TO220 device dissipating 4 W to the end of an aluminium bar of ½″ by ¾″ cross-section and 50 mm long, whose far face is thermally bonded to a large heatsink. How much of a temperature drop will we suffer along the bar?

We must first convert the units into metres (1″ = 25.4 mm), so ½″ becomes 0.0127 m, ¾″ becomes 0.01905 m, and 50 mm becomes 0.05 m. We can now drop these numbers into the thermal resistance equation:

$$R_{thermal} = \frac{l}{A\lambda} = \frac{0.05}{0.0127 \times 0.01905 \times 236} = 0.88 \ K/W = 0.88°C/W$$

Knowing that we wish to transfer 4 W, a thermal resistance of 0.88°C per watt will cause a temperature drop of 3.5°C. The significance is that although the heatsink might operate at 40°C, the TO220 device will be 3.5°C hotter at 43.5°C. If an insulating kit is used, this will add its temperature drop, and if you really want to carry the calculation through, semiconductor manufacturers generally specify thermal resistance between die and mounting tab, so the die temperature could be calculated and compared to manufacturer's recommendations.

Convection relies on the heat-carrying movement of a fluid (gas or liquid) between source and destination. Fluid is heated at the source, expands, and is displaced by denser cooler fluid, forming a convection current that continuously pushes hot fluid away and draws cool fluid towards the source. Convection efficiency can be increased in various ways:

- If a greater volume of fluid moves per second, more heat can be transferred, so the forced convection of a computer fan greatly improves heat transfer over natural convection.

- We choose a fluid having a higher specific heat capacity — meaning that it absorbs more heat energy for a given temperature rise. Liquids are better than gases, and water is especially good, so pumped water cools most internal combustion engines.
- We choose a fluid that can support a higher temperature before boiling, increasing heat transfer. A few nuclear power stations pressurised water to raise its boiling point, and more extreme reactors pumped liquid sodium (883°C boiling point).
- Although some transmitter valves had water-cooled anodes, practical pipe diameters and required pump power imposed a practical limit to the maximum flow rate, and thus the heat that could be transferred. However, changing the **state** of a material requires a great deal of energy, and converting water at 100°C to steam at the **same** temperature requires **seven** times as much energy as heating that same mass of water from 20 to 100°C so, paradoxical as it may initially seem, the very largest transmitter valves were steam cooled.

Radiation (strictly, electromagnetic radiation) does not require a physical medium between heat source and destination, but it is the least efficient means of transferring heat. Radiation losses are governed by Stefan's law:

$$E \geq \sigma(T_1^4 - T_2^4)$$

where:

E = power per unit area (J/m^2)
σ = Stefan's constant $\approx 5.67 \times 10^{-8}$ W/m^2 K^4
T_1 = absolute temperature of first body = °C + 273.16
T_2 = absolute temperature of second body = °C + 273.16.

Conventionally, body 1 is set to be the hotter body, but a negative result indicates that heat is flowing from body 2 to body 1. Unless the two temperatures are quite similar, very little error is caused by neglecting the cooler body temperature. Because heat flow is proportional to the fourth power of temperature, particularly hot bodies such as the Sun (surface temperature ≈ 6000 K) can transfer heat quite effectively by radiation.

Valve cooling, positioning, and cooling factor

Output valve anodes are hot and, unless adequately cooled, will heat their micas sufficiently to cause outgassing of water vapour that poisons cathodes. Typical audio valves isolate the hot anode within an evacuated glass envelope, so the anode cannot lose heat by convection.

The supporting wires from the anode to the outside connectors are quite thin, so the anode cannot lose heat by conduction. The only remaining method of heat transfer is radiation.

Radiation obeys reciprocity, so a good emitter is also a good absorber. Thus, domestic kettles are shiny metal or white plastic because reflective surfaces that don't absorb heat are also poor emitters. Conversely, matt black absorbs heat well, so anodes are darkened to enable them to radiate infrared (heat) more efficiently. Although we think of glass as being optically transparent, some light is inevitably lost in transmission, and glass is imperfectly transparent to infrared radiation. Radiation that is not passed is absorbed and heats the glass envelope, so some of the received heat from the anode can be lost by convection if air is allowed to flow freely past the valve envelope. Very roughly, the glass envelope splits valve thermal losses equally between convection and radiation. Thus, efficient cooling requires that we consider how the anode can radiate, and how easily convection currents can flow past the envelope. Many valve data sheets specify the maximum tolerable envelope temperature — usually 180 or 200°C for consumer valves, but some industrial valves (such as the 6528) will tolerate 250°C.

Power valves should separate their centres by a spacing of at least twice their envelope diameter, otherwise they heat each other by radiation. But a useful trick can be applied to the radiant heat received by a valve. Many valves, particularly small-signal valves, have an anode with quite a narrow cross-section. Total received radiant heat is proportional to the area seen at the destination, so rotating a valve to present a narrow cross-section to the source can reduce heating from adjacent power valves. By reciprocity, if the source also has a narrow cross-section, rotating the source to present a small area to the destination also reduces transmission. This technique is particularly useful for circuits such as differential pairs or phase splitters where electrical considerations dictate that the two valves almost touch, but it can also be used to safely reduce the distance between a driver valve and a power valve. See Figure 1.8.

Placing power valves in the middle of a chassis is unlikely to be a good idea because the chassis severely restricts convection currents. Mounting a valve horizontally can improve convection efficiency because it exposes the valve to a larger cross-section of cooling air. However, this could conceivably cause a hot control grid to sag onto the nearby cathode, with disastrous results, so check the manufacturer's full data sheet to see if there are any strictures about mounting position. If it doesn't cause other problems, align the socket so that the plane of the grid wires is vertical [2], preventing them from sagging onto the cathode.

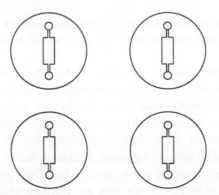

Figure 1.8
Careful anode orientation allows closer spacing along the anode's narrow axis.

Early valves had cylindrical electrodes, so viewed down their axis, electrons strike the anode equally from all points of the compass. Since the electron density is equal at all angles, the anode temperature is also equal at all angles. However, beam tetrodes do not have axial symmetry, and direct their beam of electrons along a single diameter, causing the anode to heat unequally. Bearing in mind that the glass envelope converts half of the radiant heat loss from the anode to convection loss from the envelope, a horizontally mounted beam tetrode should be rotated on its axis so that the hottest parts are at the sides, allowing them to be efficiently cooled by convection, rather than at top and bottom. As an example, see GEC's recommended alignment for the KT66 [3]. See Figure 1.9.

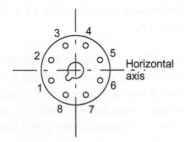

Figure 1.9
GEC recommended orientation of valve base for horizontally mounted KT66.

Because the sections of a KT66's envelope between pins 1 and 2, and 5 and 6 are the hottest, when mounting a push–pull pair of KT66 vertically, it makes sense to ensure that pins 7 and 8 of one valve face pins 7 and 8, or 3 and 4, rather than 1 and 2, or 5 and 6.

Mounting valve sockets on perforated sheet greatly assists cooling by allowing a convection current to flow past the valve, but as the hole area of sheet metal is typically only $\approx 40\%$, it is still not perfect. If better cooling is needed, the valve socket can be centrally mounted on a wire fan guard, and if even that isn't sufficient to keep the envelope temperature below the manufacturer's specified maximum, a low-noise fan can be mounted on pillars underneath the chassis using the same screws that secured the fan guard. See Figure 1.10.

Figure 1.10
With care (and large washers), a valve socket can be mounted on a fan guard, allowing almost unrestricted airflow.

The degree of care we must take over convection currents is dependent upon the ratio of envelope surface area to anode dissipation, so a very useful figure of merit based on these factors may be calculated as follows:

$$\text{Cooling factor} = \frac{\pi dh}{P}$$

where:

d = envelope diameter (mm)
h = anode height (mm)
P = anode dissipation plus heater dissipation.

A wide range of valves was deliberately selected, anode height measured, then sorted by cooling factor. See Table 1.2.

Table 1.2
Comparison of cooling factor for a range of glass envelope valves

	P_a (W)	V_h (V)	I_h (A)	P (W)	d (mm)	h (mm)	Cooling factor
6J5GT	2.5	6.3	0.3	4.39	30.1	21	452
7N7	5	6.3	0.6	8.78	30.1	21	226
KT66	25	6.3	1.3	33.2	53	30	151
6S4A	8.5	6.3	0.6	12.3	22	23	129
KT88	35	6.3	1.6	45.1	53	34	126
EL84	14	6.3	0.76	18.8	22.2	29	108
EL34	30	6.3	1.5	39.5	33.3	40	106
ECC81	5	6.3	0.3	6.89	22.2	9	91
13E1	95	26	1.3	129	65	44	70
6528	64	6.3	5	95.5	53	35	61
6C45II	7.8	6.3	0.44	10.6	22.5	8	53

Unsurprisingly, older designs have more conservative cooling factors, and putting two identical triodes in the same envelope halves the cooling factor (7N7 vs. 6J5GT). More significantly, based on the table and practical experience, provided that the valves are mounted on an open chassis (not inside a box where ambient temperature can rise), the following observations may be made relating to cooling factor:

> 150: No special precautions other than common sense necessary.

$\approx$ 150: Enforce exclusion zone* of 2 $\times$ envelope diameter.

$\approx$ 130: Enforce exclusion zone of 3 $\times$ envelope diameter.

$\approx$ 110: Allow 40% unrestricted air flow. This might be done by moving the valve to the very edge of the chassis (almost half the flow unrestricted), or mounting the valve on perforated metal (typically 40% hole area).

$\approx$ 90: Enforce >3 $\times$ exclusion zone **and** 40% air flow.

$\approx$ 70: As above, but lift the chassis >20 mm above supporting surface to allow cooling air to convect freely from underneath.

< 70: As above, but add a fan directly below the valve to force cooling.

Table 1.2 sets $P_a = P_{a(max)}$, but valves are not always operated at full anode dissipation, so calculation at a valve's actual anode dissipation might usefully improve its category.

Referring back to Table 1.2, it will be seen that whereas GEC specified 4″ between KT88 centres (corresponding to 2 $\times$ envelope diameter exclusion zones), the author suggests a slightly more conservative 3 $\times$ for these rare and valuable NOS valves. It will also be seen that **if** the 6C45II's rating is continuous, it is either wildly optimistic or assumes forced cooling.

Using the chassis as a heatsink

We have previously considered how we can transfer heat from source to destination, but if the destination has finite mass, we must eventually raise its temperature, reducing the temperature difference and consequent heat flow. Early transmitters transferred their waste heat to outdoor cooling ponds, giving them access to the (by comparison) infinite mass of the planet's atmosphere ($\approx 5.3 \times 10^{18}$ kg). Domestic amplifiers necessarily treat the surrounding air as an infinite mass, but we must ensure that access to this air is not restricted.

Some small components, such as power resistors and regulator ICs, unavoidably generate significant heat. Small resistors are commonly mounted on stand-offs to allow an unimpeded air flow, and regulators are often fitted with small-finned aluminium heatsinks. Neither of these strategies is ideal because they attempt to lose heat by convection to the very limited mass of air trapped within the chassis, quickly raising its temperature. Eventually, the trapped air loses

---

* An exclusion zone is defined as the circle diameter coaxial with the valve where nothing, including other exclusion zones, may enter.

heat by conduction to the cooler chassis, and thence the surroundings, so an equilibrium results with a high internal air temperature and hot components.

A high air temperature within the chassis is undesirable because:

- The components causing the high air temperature are unnecessarily hot, and even though they may have been designed to withstand heat, their working life is invariably reduced.
- Electrolytic capacitors are especially sensitive to heat, and a very rough rule of thumb is that their working life halves for each 10°C rise in temperature. Capacitor manufacturers' data sheets include extremely useful charts that allow lifetime predictions to be made from ambient temperature and capacitor ripple current, so it is well worth looking up the full data sheet for your particular capacitor at the manufacturer's website.
- Components having a critical value, such as in equalisation or biasing networks, will drift from their initial value as a consequence of heating.

Ultimately, we can only lose heat to the surrounding air, and a large surface area cools more efficiently. This means that the best way to cool components is to ensure that they are thermally bonded to the chassis, using a **thin** smear of heatsink compound. Heatsink compound is not a particularly good conductor of heat, but it is far better than air. The purpose of heatsink compound is to fill the tiny insulating **air** gaps that result from placing two imperfectly smooth surfaces together. Excessive compound worsens cooling. Most people (and this includes manufacturers) use far too much. A thin, even smear applied to both mating surfaces is all that is required.

Beware that mismatch between the thermal coefficient of expansion of screws and heatsink bonded to the chassis could loosen them at operational temperature, so check tightness when fully warmed, otherwise the compromised thermal bond can cause semiconductors to be hotter than expected and, worse, an erratic bond causes drift. Conversely, mismatch in thermal expansion between the semiconductor's package and its fasteners that causes tightening could crack the package; spring washers and clamps are available for maintaining optimum force at all temperatures.

A useful secondary advantage of mounting aluminium-clad or TO220 resistors directly onto the chassis is that they provide convenient mounting tags for other components. It may seem unnerving to touch a chassis with hotspots due to local heatsinking, but this technique minimises the internal air temperature, and thus minimises the heating of sensitive components.

Efficient convection requires a free flow of cool air to replace hot air. Although most designers recognise the importance of allowing adequate ventilation by providing holes in the top of a chassis near hot components, air must also be free to enter the chassis from the bottom if an efficient convection current is to flow. Thus, the ideal solution is to make the **entire** underside of the chassis from perforated steel or aluminium, and support it on feet ≈ 20 mm high (to allow air to flow unimpeded into the underside of the chassis). See Figure 1.11.

Figure 1.11
Perforated sheet and tall feet allow excellent cooling.

If necessary, individual components can be cooled even more efficiently by bonding them **directly** to a finned heatsink fitted outside the chassis, as is common with transistor amplifiers. See Figure 1.12.

Although heatsinking most obviously springs to mind when considering large power amplifiers or power supplies, precision pre-amplifiers must minimise their internal temperature rise in order to prevent equalisation networks drifting in value, so it may be worth thermally bonding anode load resistors to the casing or an external heatsink.

When finned heatsinks are used, it is essential to orient them correctly. Heatsink manufacturers specify thermal resistance (°C/W) with the fins vertical in free air because this maximises the surface area available to the natural cooling convection current flowing up the fins. Despite this, the author has lost count of the number of commercial amplifiers having horizontal heatsinks. Heatsinks cost money, so why degrade their performance? The prettiest example that the author can find showing the importance of correct fin orientation is a motorcycle. See Figure 1.13.

The engine is an air-cooled V-twin. The pistons are identical and run in barrels that are detachable from the crankcase. Despite the increased production cost, the two barrels are different, one having longitudinal fins, the other latitudinal, and this is done solely to optimise cooling. Your amplifier fins cannot achieve a forced 100 mph horizontal convection current, so they need to be vertical.

Figure 1.12
Cutting a hole in this steel chassis allowed the power transistors to contact the heatsink directly. Note also the clamp that applies pressure to the epoxy package rather than the protruding tab.

Figure 1.13
This air-cooled Ducati motorcycle has barrels with fins aligned in the direction of air flow.

Cold valve heater surge current

The resistance of a conductor such as a valve heater filament changes significantly with temperature in accordance with the following equation:

$$R_T = R_0(1 + \alpha T)$$

where:

R_0 = cold resistance (Ω)
R_T = resistance (Ω) at temperature change T
α = thermal coefficient of electrical resistance (≈ 0.005 per °C for pure tungsten)
T = temperature change (K or °C)

Normally the temperature rise of conductors in electronics is too small for the effect to be noticeable, but a thoriated tungsten filament operates at ≈ 1975 K, so at an ambient temperature of 20°C (293 K), its resistance is much lower, and theory predicts that it very briefly draws 8.6 times the operating current. In practice, surge current is limited by heater supply output resistance, perhaps to only half the predicted value.

Indirectly heated valves operate their filaments at ≈ 1650 K, resulting in a theoretical surge current of ≈ 7 times the operating current, although measurements suggest that a ratio of $\approx 5:1$ is more appropriate. More significantly, the thermal inertia of the cathode sleeve slows heating, so the surge current lasts for a few seconds, and could be sufficient to blow a poorly chosen mains fuse. Further, it would not be prudent to use heater wiring of a rating only just sufficient to cope with the steady-state current if long-term reliability were required.

Wire ratings

Wires have ratings related to heat because electrical insulation deteriorates with increased temperature and the internal conductor has resistance, so passing a current causes self-heating $(P = I^2R)$.

Arcing and insulation failure should not be a problem at the voltages found within most valve amplifiers, but it is still advisable to maintain 2—3 mm separation between high voltages, and it is particularly important to prevent wires carrying high voltages from touching hot components — leads to anode top caps must not touch the (hot) valve envelope. Less obviously, avoid routing wires near hot resistors.

If a wire becomes too hot, its impaired insulation may leak current between circuits leading to impaired performance and possibly fire, so it is important to ensure that the wiring is rated appropriately for the current to be passed. This means that wire current ratings are determined by ambient temperature, ability to cool, and the temperature rating of the insulator. See Table 1.3.

Table 1.3

Typical current ratings for PVC insulated wire

Conductor diameter (mm)	Maximum current (A)	AWG
0.6	1.5	22
1	3	18
1.7	4.5	14
2	6	12

PTFE has a higher melting point than PVC, permitting greater self-heating and thus a higher current rating than might be expected from a given conductor diameter, but the penalty of using a small-diameter conductor (having higher resistance) at high currents is that voltage drops along that wire are proportionately higher, and this can become significant within the capacitor/rectifier/transformer loop of a power supply.

Unwanted voltage drops

All wires have resistance that is proportional to their length and inversely proportional to their cross-sectional area. Although the currents in valve circuitry are typically quite low, making the voltage drops proportionately low, once we consider that we want a signal-to-noise ratio of >90 dB, the voltage drops caused by small resistances become significant.

The highest currents, and therefore highest voltage drops, occur in the loop from transformer via rectifier to reservoir capacitor and back again. A capacitor input filter draws pulses of current at twice mains frequency from the transformer that are typically four to six times greater than the DC load current. It is essential that the wires carrying these pulses are as low resistance, and therefore as short as possible, which means that the rectifier and associated reservoir capacitor should be close to its mains transformer. It's the same logic that puts the battery in a car's engine compartment. (The 1959 Mini was intended to be conventional rear wheel drive from a 500 cc longitudinal engine, but this proved to be underpowered, so the only way to fit a larger engine was to mount it transversely and adopt front-wheel drive, but that left insufficient room for the battery, forcing it into the boot, requiring a long, very thick wire to the starter motor.)

When an output stage enters Class B, it draws current pulses at twice the audio signal's frequency from the power supply. To prevent these pulses breaking into driver circuitry and increasing distortion, the loop area from audio load to reservoir capacitor should be as small as possible. This means that output transformers should be close to their supply capacitor.

The potential effects of unwanted voltage drops usually determine the 0 V signal earth scheme, often known colloquially as earthing or grounding. There are two fundamental methods of dealing with earthing:

- "Earth follows signal." The 0 V signal earth wire follows the path of the signal. In order to minimise unwanted voltage drops along this (necessarily long) wire, it has a large cross-section so this brute force strategy leads to 1.6 mm tinned copper bus-bars.

- Star earth. All connections to the 0 V signal earth are made at a single point. Because the distance between individual connections is so small, the common impedance is small, so unwanted voltage drops are also small.

The significance of these two wiring schemes is that the choice between them needs to be made at a very early stage. "Earth follows signal" tends to favour long, slim mechanical layouts, whereas an ideally implemented star earth tends to favour square or even circular layouts centred about the star earth. The significance of unwanted voltage drops is so great that it dominates the gross mechanical layout of an example later in this chapter, whilst detailed electrical implications will be covered in Chapter 3.

More powerful amplifiers are heavier, and in an effort to split the weight into a pair of manageable chassis, you might consider having a remote power supply, necessitating an umbilical connecting cable. Beware that the umbilical wires need a larger cross-section than within a single chassis in order to minimise the unwanted voltage drop over this increased distance, perhaps needing an umbilical connector having a larger diameter cable entry.

Electrostatic induction

Electrostatic coupling is capacitive coupling. Minimising the capacitance between two circuits minimises the interference. Remembering the equation for the parallel plate capacitor:

$$C = \frac{A\varepsilon_0\varepsilon_r}{d}$$

where:

C = capacitance (F)
A = area of the plates (m^2)
ε_0 = permittivity of free space $\approx 8.854 \times 10^{-12}$ F/m
ε_r = relative permittivity of dielectric between plates (typically $2-3$ for plastics)
d = distance separating the plates (m)

When capacitance must be minimised between two points, all parts of the equation should be attacked. Reducing plate area means keeping wires short and crossing them at right angles, whilst increasing plate distance means maximising separation of parallel wires and not lacing them together into a neat loom. In terms of chassis layout, this means that output valves should be reasonably close to the output transformer and that driver circuitry should be reasonably close to the output valves.

A layout having minimum electrostatic induction tends to have such short wiring that the circuit looks very simple. If your planning seems to require many long wiring runs, try another arrangement of the large components or input/output connectors. In particular, mains wiring should be kept as short and compact as possible, not just to minimise interference but also to minimise the risk of accidental contact if faultfinding becomes necessary.

In a high impedance circuit such as a constant current sink, a chassis-mounted transistor causes a conflict between thermal and electrostatic considerations because the collector is invariably connected to the transistor case. Fastening a TO-126 transistor such as an MJE340 to an earthed chassis using an insulating kit adds 6–8 pF shunt capacitance, degrading any efforts we may have made to design a wide-bandwidth constant current sink. In this instance, we are forced to compromise our thermal considerations and lose heat directly to the air within the chassis using a small finned heatsink.

Fortunately, capacitance to chassis from aluminium-clad or transistor-style resistors used as anode or cathode loads is rarely a problem at audio frequencies. Because one end of the resistor is at AC earth, but the capacitance is distributed along the resistor element, the total element to earth capacitance is effectively halved, and because the impedances are lower than in a constant current sink, the stray capacitance becomes insignificant.

Valve sockets

A wide variety of valve sockets is available for a given valve base. See Figure 1.14.

All three of the chassis-mounting International Octal sockets in the photograph have different spacings for their securing screws, so it is vital to make a firm decision about which socket type is to be used, and whether the circuit is to be hard-wired or printed circuit board. One point that may be worth considering when choosing Octal sockets is that NOS McMurdo phenolic sockets have the same hole spacings as Loctal sockets, allowing an easy change from 6SN7 to 7N7 at a later date.

Another consideration is heater wiring positioning. Taking the B9A base as an example, most valves using this base have their heaters connected between pins 4 and 5, so the socket should

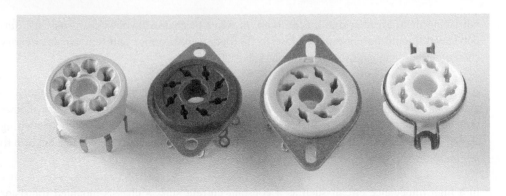

Figure 1.14
Octal sockets. Note that the chassis-mount types all have different mounting centres.

be rotated to position these pins closest to the heater run. Note, however, that operating the popular ECC83/12AX7 from 6.3 V requires a **link** between pins 4 and 5, and for minimum hum the other heater wire should be taken across the centre of the socket to pin 9 — do **not** loop it round the outside. McMurdo B9A valve sockets were designed so that if the line joining the two socket fasteners was aligned at 45° to the edge of the chassis, pins 4 and 5 were closest to the chassis edge, minimising hum from heater wiring. See Figure 1.15.

Traditional valve amplifiers always had their valve sockets aligned so that heater pins were closest to the chassis edge, thus minimising the length of exposed heater wiring.

How electrostatic screening works

Electrostatic screening places an earthed conductive barrier between the source of interference and the sensitive circuit. See Figure 1.16.

As can be seen from the diagram, the screen diverts interference currents from the source to earth. If there is an impedance between the screen and radio frequency earth, the interference current must develop a voltage across it, allowing the screen to capacitively couple interference to its enclosed (sensitive) circuitry.

The impedance from the screen to radio frequency earth can be considered to be the lower leg of a potential divider, and the capacitance from the screen to the interference source is

Figure 1.15
McMurdo B9A sockets position the heater pins (4, 5) close to the edge of the chassis provided that the mounting holes are at a 45° angle to the edge.

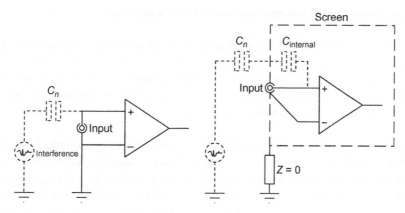

Figure 1.16
Screening breaks one capacitance into two, centre-tapped by an earth.

the upper leg. The lower leg could be a length of conductor, which has inductance, thus forming a 12 dB/octave filter in conjunction with the capacitor. To maximise screening efficiency, the screen needs a path to chassis having low resistance and low inductance. When resistances and inductances are connected in parallel, the total value falls, so the screen should ideally contact the chassis at multiple points to minimise impedance and maximise screening. This is why radio frequency designers cut the flanges of the lids containing their circuitry — it ensures that each finger firmly contacts the case, and minimises radio frequency impedance to all points of the screen. See Figure 1.17.

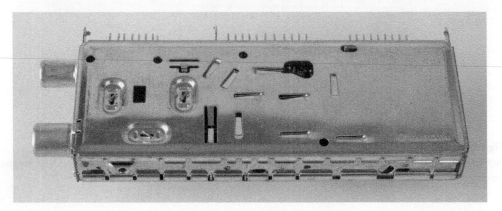

Figure 1.17
Lids with fingers allow multiple low-resistance, low-inductance contacts.

From an audio point of view, screening cans should be firmly screwed to the chassis at multiple points using serrated washers to ensure a gas-tight connection that does not deteriorate over the years. Pre-amplifiers using choke inter-stage smoothing should ideally use oil-filled chokes, not because the oil confers any advantage, but because the metal can that contains the oil provides electrostatic screening.

Valves that are sensitive to electrostatic fields can be enclosed by an earthed metal screening can, but screening cans are not created equal. Perfect screening would enclose the valve completely, but that would prevent heat from escaping, making the valve significantly hotter, and reducing its life [4]. It is easy to raise a valve's temperature, but lowering it is much harder, so a screening can must always strike a compromise between screening and cooling. See Figure 1.18.

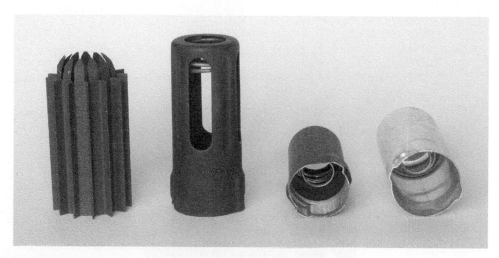

Figure 1.18
Selection of valve screening cans.

From left to right

The first "screening can" was actually manufactured as a heatsink, and slightly reduces envelope temperature. All that is needed to turn it into a screen is a short wire to bond it electrically to the chassis. The dungaree screening can has a bayonet fitting onto a skirted base and holes allow air to flow, which reduces its screening efficiency and temperature rise. Note that the can's inside is painted matt black to absorb radiant heat from the valve, which is then re-radiated by the black outside. The third can is for small-signal valves only as airflow is very restricted, but the black paint inside and outside assists cooling. The fourth can is an abomination that deserves to be crushed because not only does it restrict airflow, but it also reflects radiant heat back to the valve.

Very occasionally, screening cans are deliberately made to aid heat losses, and have fingers to contact the glass envelope. See Figure 1.19.

Input sockets

Input sockets should be kept apart from mains wiring (hum) and loudspeaker terminals (instability), but as both of these problems are caused by capacitance from the input socket

Figure 1.19
Screening can with internal fingers to aid heat dissipation.

to the offending connectors, screening the input socket easily cures the problem if space is limited.

An extremely useful facility on a power amplifier is to add a muting switch that applies a short circuit to the input of the power amplifier. See Figure 1.20.

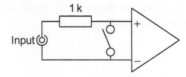

Figure 1.20
This muting switch at the input of the amplifier enables audio leads to be swapped without the necessity of switching the amplifier off.

Not only does the muting switch enable the amplifier to be plugged to different sources without having to switch it off and then on again, but it is very handy when the amplifier is being tested. The 1 kΩ resistor protects the source electronics from the short circuit.

Mechanical/safety

The purpose of the chassis is to support the components mechanically, some of which are heavy, and to **enclose all the dangerous voltages**, thus eliminating the risk of electric shock. The safest form of chassis for the home constructor is the totally enclosed earthed metal chassis. Placing a mains transformer in the centre of a chassis made of folded aluminium is asking for the chassis to sag. Heavy items should be placed towards the edges, where the nearby vertical section adds substantial bracing.

It's often worth making a chassis long and slim not just because it will be more rigid, but because you stand a better chance that the throat of your pillar drill or drill stand will allow you to reach all the proposed holes than if you make a large square chassis. If a heavy item **must** be moved towards the centre of a chassis, a bulkhead or two can be added across the chassis to brace it, giving the further advantage of breaking the inside of the chassis into electrostatically screened compartments. This allows noisy circuitry (rectifiers and smoothing) to be placed in a "noisy" compartment, and sensitive audio in a "quiet" compartment. See Figure 1.21.

When planning the positions of the major parts, it is important that one part should not obscure electrical connections or securing screws of another unless it can be easily moved aside to gain access.

Think carefully about accessibility of fasteners. It's time to reconsider if you can't easily apply a spanner or nutdriver to a nut, or a screwdriver to a screw head. If there's really no choice about nuts being inaccessible, consider securing the part with a single 3 mm plate having tapped holes in lieu of individual nuts. Conversely, consider hex head fasteners if limited axial access blocks a screwdriver but a spanner or hex key could obtain access from the side.

If the chassis is made of separate panels, how easily can each panel be removed? Wherever possible, wires should pass panels via edge slots rather than through holes to avoid those wires preventing subsequent panel removal. Failure to observe these considerations can make subsequent maintenance or modification extremely difficult. Contemporary consumer

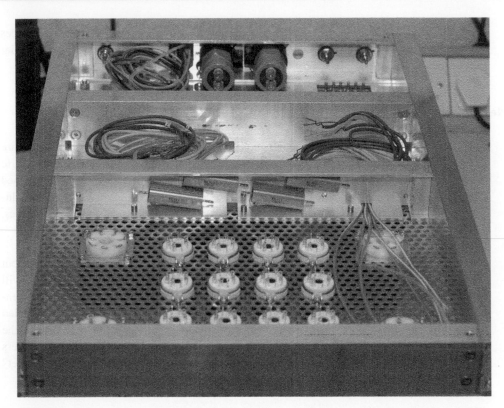

Figure 1.21
Chassis bulkheads allow segmentation into "dirty" or "clean" compartments.

electronic design places maintainability very low on its list of priorities, but not only do you want to be able to maintain your creations, you also want to be able modify them as your knowledge improves or if better parts become available.

It is often useful to take a modular approach. If a block such as a complete heater supply including mains transformer, rectification/smoothing and regulation is made as a module on a sub-chassis, it can be conveniently built and tested outside the main chassis, then installed when ready. Printed circuit boards (PCBs) are another example of modular technique.

Components directly soldered to valve sockets and tags are known as **hard-wired** or **point-to-point**, and a less obvious power amplifier technique is to hard-wire the entire driver circuitry onto a small rectangular plate that fits over a correspondingly sized hole in the main chassis. The great advantage of this technique over mounting valve sockets directly onto the main chassis is that a complete driver redesign requiring a different number of valve bases of different types simply requires replacing the small plate. Even better, the plate can be made of perforated aluminium to assist cooling.

Bear in mind that some components can be quite large. Capacitors vary greatly in size depending on the choice of dielectric, so 100 nF can range from the size of a fingernail clipping to an entire thumb. See Figure 1.22.

Figure 1.22
A selection of 100 n non-polarised capacitors. Left to right: 63 V polyester, 400 V polyester, 630 V polycarbonate, 200 V silvered mica, 1500 V polypropylene, 500 V polytetrafluoroethylene, 2000 V mixed dielectric.

Although we will look at electrical safety in detail later, large electrolytic smoothing capacitors should be considered during planning. When both electrical connections exit from an insulated base, there is no necessity for either terminal to be connected to the enclosing can, so the voltage on the can of the capacitor is indeterminate, but is generally near to the potential on the negative terminal. Although the can could be bonded to chassis if the negative terminal is at 0 V, this invariably causes a hum loop because of the internal capacitance between capacitor and can, so it is usually insulated from the chassis. The capacitor therefore has only a single layer of insulation between the high voltage supply and the outside world, but safety calls for either a double layer of insulation, or one layer plus an earthed metal shroud, so it is not good practice to have exposed cans outside the chassis (see later for explanation of Class I and II equipment).

Preventative maintenance means being able to see signs warning of impending failure, and catching it **before** it goes bang. This is one reason for having all the power valves clearly visible — if you spot a red-hot anode, you probably can get to the "on/off" switch before the fault destroys even more components. Internally, if all the components are clearly visible, you might spot a charring resistor or bulging capacitor **before** it destroys other components in addition to itself at the first application of power.

A rather more depressing aspect of preventative maintenance is to consider the consequences of failure of specific components. When traditional electrolytic capacitors fail, they spray soggy paper and foil from their bases. Although modern capacitors are somewhat tamer and vent in a more controlled fashion, it is a good idea to consider where any vented material might land. It would be particularly unfortunate if a failed heater supply capacitor vented a conductive spray over a perfectly innocent (powered) high voltage regulator PCB. Similarly, silicon rectifiers can sometimes fail explosively, and it would be particularly poor planning if the short-lived volcano was directly under a complex wiring loom. Although the author has not deliberately tested carbon resistors to destruction, industry wisdom is that they fail pyrotechnically.

When you work on the amplifier, you will inevitably turn it upside down, so it is helpful if the taller and heavier components can be arranged so that they allow the chassis to stand firmly without rocking, and without tipping onto the delicate valves.

Finally, consider the likely weight of the finished amplifier, and whether you will be able to lift it. Otherwise, be prepared to need unusual methods of moving the completed amplifier. See Figure 1.23.

Figure 1.23
The author's Crystal Palace amplifier is too heavy.

Acoustical

It is unusual to consider acoustical problems in a power amplifier, but the input valves are microphonic, and a flimsy chassis would not help. The author favours a rigid chassis with plenty of bracing, held together by lots of large screws, even if the result is somewhat stronger than strictly required by the supported weight.

Valves are microphonic because any relative movement between control grid and cathode alters the local electric field (volts per metre) and therefore the number of electrons reaching the anode. When the grid structure is placed closer to the cathode to increase g_m, the same movement becomes proportionately greater, so valves having high g_m are intrinsically more microphonic, although the more rigid frame-grid construction tames the problem. Indirectly heated valves have a rigid and firmly located cathode structure, but directly heated valves must support their heater loosely (to allow for thermal expansion), necessarily producing a cathode structure less rigid than the grid and thereby even more prone to microphony.

Pre-amplifiers may need deliberate acoustic isolation from structure-borne vibration, and anti-microphonic valve sockets with integral rubber suspension mounts used to be readily available. Nevertheless, we must take wires to that valve base, and unless the wires are flexible, they form an acoustic short circuit. A more effective way of achieving isolation borrowed from microphone mounts uses ≈ 1 mm-diameter knicker elastic to float a sub-chassis supporting an entire circuit in the same way that a trampoline is supported. Each anchoring point ideally requires a small rubber grommet to avoid chafing and to prevent the elastic slapping metal when the sub-chassis moves. See Figure 1.24.

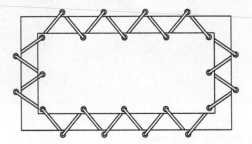

Figure 1.24
Trampoline suspension of sub-chassis using knicker elastic.

The internal mechanical resonances of an indirectly heated valve are typically above 500 Hz [5], so these are the frequencies that need to be isolated. The suspended sub-chassis is effectively a 12 dB/octave low-pass filter with $f_{-3\,dB}$ at the resonant frequency of the suspension, so if it resonated at 63 Hz (three octaves below 500 Hz), it would theoretically attenuate by 36 dB at 500 Hz. In practice, damping within the springs reduces this figure, but acoustic attenuation is generally maximised by minimising the suspension resonant frequency, given by the standard resonance equation:

$$f = \frac{1}{2\pi\sqrt{Cm}}$$

where:

f = frequency (Hz)
C = suspension compliance (m/N)
m = suspended mass (kg).

Thus, deliberately adding mass (perhaps roof flashing) to the sub-chassis can improve iso-lation. Although the resulting suspension is likely to be stiffer than an anti-microphonic valve socket, it is still important to minimise the number of wires to the sub-chassis; use flexible wires, and dress them in slight loops to avoid them acoustically short-circuiting the suspension.

Air-borne vibration is far harder to eliminate. The ideal solution would be to enclose the circuitry within a solidly built closed box, internally lined with an acoustic absorbent such as bonded acetate fibre (BAF) or fibreglass building insulation. Unfortunately, there would then be no way for the heat to escape...

Aesthetics

Although placed last on this list, aesthetics are probably close to the forefront of your mind as you will want to be proud of the results. The finished project does **not** need to look like a collision between a rat's nest and a supermarket trolley. Glowing glass is pretty, so almost all power amplifiers have their valves at the front where they can be seen. And rightly so.

Form follows function, so right-handed people generally place the most important control (volume) at the right. Symmetry is usually pleasing, so a rotary input selector switch could be placed at the left. "Retro" looks are currently very fashionable, so moving coil meters and hexagonal bakelite knobs are popular.

Sadly, blue LEDs are still popular but this aberration will doubtless pass. The author is rather fond of dual-colour LEDs (red and green) with their relative currents adjusted so as to match a valve heater's orange glow. If the green LED is powered from the heater supply, and the red from the HT, the colour adds a useful monitoring function as well as being pretty.

Parts having specific problems and solutions

Some common parts raise specific problems that have simple solutions.

IEC mains connectors

The most common (and safest) mains connector is the IEC 10A three-pin. It is far more convenient to have a fused IEC inlet than a captive mains lead and separate fuseholder, and it is often convenient to have at least one outlet. Most IEC chassis connectors are designed to be secured by M3.5 countersunk screws, but because the connectors are quite deep, fitting a nut to the back of the screw is fiddly, so a tapped hole makes sense if possible. In theory, provided the panel to be tapped is >2 mm thick, this should be fine, but IEC connectors require appreciable force to remove them, applying extra stress to the tapped threads, so 3 mm panel thickness is a more reasonable minimum.

Considering inlets, it is particularly worth planning for tapped M3.5 holes rather than clearance because:

- A tapped hole avoids having to fit a nut to the screw that is invariably in an awkward place.
- Nuts tend to foul IEC insulating boots.

See Figure 1.25.

Note also that the nearby transformer wires and associated grommets pass through **slots** cut in the channel rather than holes, thus allowing the channel to be removed with a minimum of wiring disturbance.

Considering multiple outlets, the following points make life easier later on:

- Align the outlets vertically. This uses space more efficiently on a slim chassis and allows more outlets to be fitted.
- Orient the outlets to that the live pin is closest to the chassis top plate rather than the bottom. The dangerous pin is now buried deep inside the equipment where you are far less likely to accidentally touch it during testing.
- Vertical outlets allow straight tinned copper wire bus-bars to be threaded from pin to pin — making wiring much easier and tidier.

Figure 1.25
Note that this IEC inlet is secured by tapped holes in the channel, avoiding fouling the insulating boot.

- Mark the fixing holes after cutting the hole for the body. Once the hole for the body has been cut, an outlet can be inserted and you can mark the precise position of the holes with a scriber. Provided that the securing holes are tapped, shuttered IEC outlets will just fit between the flanges of 2″ wide ⅛″ thick aluminium channel when fitted vertically. See Figure 1.26.

Modern mains transformers

Modern mains transformers are intended to be totally enclosed by a chassis, so they tend to be open flange or clamp mounting and therefore unsuitable for traditional exposed mounting. However, transformer lamination sizes and their fixing holes have changed little over the decades, so it is often possible to fit a shroud salvaged from an old transformer to a new one, allowing safe drop-through mounting. See Figure 1.27.

Figure 1.26
IEC mains outlets will just fit vertically between the flanges of 2″ channel provided that the aluminium is tapped to take the securing screws.

Figure 1.27
Adding a shroud salvaged from an old transformer permits safe external mounting of a modern transformer.

This new transformer originally had a steel clamp crimped round three sides, so this was gently prised away from the laminations (take care not to deform magnetic materials as it invariably degrades their magnetic properties). The dual chamber plastic bobbin didn't initially fit into the shroud because both sides of it were fully populated with stubs for solder tags, so the unused stubs on the unwired side were carefully sawn away, allowing the shroud to fit easily. The four mounting holes on the shroud aligned perfectly with the M5 clearance holes in the laminations.

Examples

Application of the previous rules is best demonstrated by example. Bear in mind that there are many ways of skinning cats, so if you can find a better solution, use it.

Power amplifiers

Let us suppose that we want to make a mono push–pull EL84 amplifier, perhaps a Mullard 5-10 or a rebuild of a Leak TL12+.

The major items to be positioned are:

Mains transformer
Output transformer
Rectifier valve
HT reservoir capacitor
Output valves
Driver circuitry.

Although negative feedback reduces hum, it would be best if there were no hum induced into the output transformer from the mains transformer; we draw a line between the centres of the two transformers, and rotate them so that their coils are at 90° to each other to minimise hum. See Figure 1.28.

We can also reduce induced hum by moving the two transformers apart. As soon as we separate major parts, we ought to think about whether we can use the space in between them, perhaps for the reservoir capacitor. The reservoir capacitor draws current through the rectifier from the transformer in a train of 100 Hz high current pulses. See Figure 1.29.

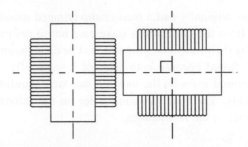

Figure 1.28

Planning a layout: A. Output transformer and mains transformer at right angles to minimise coupling.

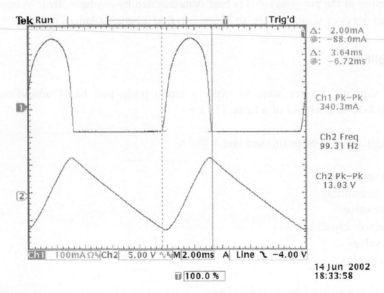

Figure 1.29

Reservoir capacitor ripple current is typically 4–6 times DC load current.

The resistance of the capacitor/rectifier/transformer wiring loop should be as low as possible, so the ideal position for the rectifier valve is adjacent to the mains transformer, with the reservoir capacitor nearby. The centre tap of the output transformer needs a low-impedance supply, so it should be close to its smoothing capacitor. If the smoothing capacitor is the traditional dual capacitor, it now makes sense for the two transformers to be moved apart just sufficiently that the rectifier and capacitor can be fitted in between with adequate room for cooling. See Figure 1.30.

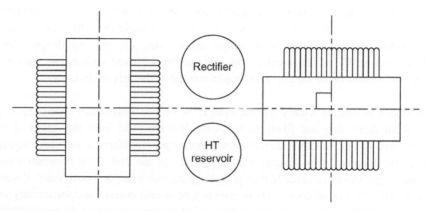

Figure 1.30
Planning a layout: B. Adding rectifier and reservoir capacitor.

The capacitor is heated by radiation from the rectifier, but provided that there is a little separation between the two, the capacitor can cool by convection and only receives radiation proportional to the angle of arc subtended by the capacitor. See Figure 1.31.

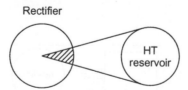

Figure 1.31
Planning a layout: C. Radiated heat from rectifier to capacitor.

Having chosen to fit a large electrolytic capacitor externally, we must ensure that its can is not exposed. Good quality capacitors tend to have quite thick insulating sleeves plus a phenolic insulator at their end, entirely shrouding the can, but older capacitors omit the phenolic and the sleeve may be thin or non-existent. Check first.

The output valves must be able to cool, so their centres should be separated by at least twice their envelope diameter – check their cooling factor. But they also need to be reasonably close to the output transformer, and it would be convenient if they were also reasonably close to the driver circuitry. The logical position for the output valves is thus on the side of the output transformer opposite to the rectifier, with the driver circuitry a little further beyond. The driver circuitry is now as far as possible from the mains transformer, and is partly shielded from it by the output transformer. The two transformers have their coils aligned at 90° to one another, but this can be achieved with either the coil of the mains transformer or the output transformer pointed at the driver circuitry.

If the output valves are widely spaced, even if the output transformer's coil axis is pointed away from them, they are likely to be in the leakage field from the edges of the coil. A separation of one envelope diameter between output transformer and valve significantly reduces the localised flux leakage from the edges of the coil. This generalisation works because larger valves produce higher power and require a larger transformer. Beam valves such as KT66 having aligned grids to reduce screen grid current are particularly sensitive to magnetic fields, so even greater spacing might be desirable to avoid modulating screen grid current and causing distortion.

The valves in the driver circuitry should be separated from the output valves in order to keep them cool, and room is needed for their associated components, so moving them to the edges of the chassis allows their heater wiring to be pushed into the corners of the chassis, and coupling capacitors (which can often be quite large) can be conveniently placed between driver and output valves. We have now arrived at the final layout. See Figure 1.32.

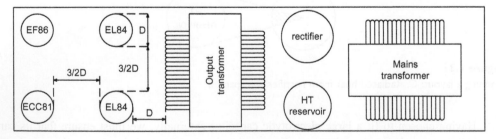

Figure 1.32
Planning a layout: D. Final, optimum layout.

Readers who are familiar with the Leak TL12 + will recognise that this layout is much longer and slimmer than the original Leak. The Leak layout requires longer wires, but the square chassis allowed easier access for wiring, so the reduced wiring time would have cut production costs. Further, for a given chassis area, a square chassis requires less material than a more rectangular chassis, making it cheaper.

Meters and monitoring points

Power valves are inevitably used by circuits capable of destroying them. To illustrate this point, consider the output valves in a power amplifier. They are connected to a substantial power supply via an output transformer having low DC resistance. If the valves are cathode biased, the cathode resistor will protect them in the event of a fault, but grid bias provides no protection. Grid bias inevitably requires some means of monitoring cathode current.

If you are in a "retro" mood, you will fit a handsome round black moving coil meter and a rotary switch having a black bakelite knob allowing each individual cathode current to be monitored, and if you are feeling really keen, you will add positions on the switch to allow key voltages to be monitored. The monitoring is important, so you will want to fit the meter and associated switch towards the front of the amplifier. The meter's permanent magnet creates a stray field, so you won't want it to be close to beam valves such as KT66, 6L6, KT88, or 6550 because it might affect DC screen grid current. The switch needs to be operated easily without obscuring the meter, and its selection and the meter must be seen simultaneously, so they have to be mounted on the same plane. If you decide to put a meter and switch on the top of the chassis, make sure there's sufficient room to operate the switch without your fingers being burned by a nearby (hot) envelope. The best solution is that used by the McIntosh MC275 that places the meter and selector on a panel sloped at 45°.

Another possibility is to fit 4 mm chassis sockets as low voltage monitoring points, and it makes sense to put these sockets as close as possible to the voltage being monitored. However, don't put them so close to hot valves that you burn your knuckles when unplugging the lead from your DVM. If the amplifier is push—pull, output currents must be balanced, and you will probably want to be able to connect two DVMs simultaneously, so remember to provide a 0 V test point for **each** DVM.

Alternatively, the traditional method was to use one meter to measure total current and another to measure imbalance current. Finding two suitable moving coil meters is tricky, but DVMs the size of a postage stamp that only need a 6 mm (¼″) mounting hole are now readily available, so they can be fitted far more easily than a moving coil meter, or you could simply provide dedicated test points for total current and imbalance.

Sympathetic power amplifier recycling

Rather than build an amplifier from scratch, you might prefer to recycle an old amplifier's chassis and transformers, but use driver circuitry of your own design, saving an awful lot of metalwork (but very little money). If you take this approach, please do it sympathetically. Randomly gouging holes and leaving others unused looks unsightly — if it's worth doing to please the ear, it's worth taking a little trouble to please the eye. See Figure 1.33.

The amplifier started life as a Leak Stereo 20, but the coupling capacitors were leaky and the electrolytic capacitors had dried out, so the author decided to recycle the chassis with a simpler driver stage, using fewer valves. This left a vacant ¾" hole, which was conveniently filled by a threaded DIN socket mounted on a small plate that was secured by the holes for the original B9A valve socket. Similarly, the holes for the original dual electrolytic capacitors were carefully enlarged for their polypropylene replacements, and a new hole was punched in front of the mains transformer for the third polypropylene capacitor. The exposed transformer links for selecting mains voltage and loudspeaker impedance were hard-wired and their connectors replaced with blank panels. The transformers were mounted on ¹⁄₁₆" phenolic board to prevent leakage flux from reaching the (steel) chassis.

Classic amplifiers invariably have appalling connectors, ranging from a lethal mains connector to awkward loudspeaker connectors, and this Leak Stereo 20 was no exception. Fortunately, the hole for the large three-pin Bulgin mains connector could be enlarged to take a flange mounting IEC inlet with integral fuse, and the hole for the mains outlets was enlarged to take an IEC outlet. Once unscrewed from the chassis, the tinned brass tags on the loudspeaker outlet boards were removed by squeezing their inside retaining lugs together using a pair of heavy-duty pliers, allowing three-way binding posts to be neatly fitted. A blanking grommet masked the hole for the original mains fuse, and the hole for the switch cable was opened out to take a mains switch. See Figure 1.34.

Output-transformerless (OTL) amplifiers

The significance of an OTL amplifier is that the absence of an output transformer forces the output valves to pass significant currents, and that these amplifiers need substantial global feedback, requiring unwanted voltage drops and stray capacitances to be minimised

Figure 1.33
Sympathetic recycling allows a new amplifier to retain its classic looks.

to ensure stability. The combination of these two factors requires true star earthing encompassing the entire amplifier and power supply, and this dominates mechanical design. As an example, a generic OTL headphone amplifier known at the Headwize forum as the Broskie–Cavalli–Jones (BCJ) will be considered. See Figure 1.35.

Figure 1.34
A little ingenuity allows the fitting of safe and convenient connectors.

Despite initial appearances, the first two valves are not a differential pair because the first valve has its anode decoupled to ground by a capacitor. The first stage has a cathode follower with a semiconductor constant current sink as its DC load. The output of the cathode follower is direct coupled to the cathode of the common grid stage, leading the stage to be known as a cathode-coupled amplifier [6]. The cathode-coupled amplifier is direct coupled to an optimised White cathode follower output stage [7]. User-adjustable global feedback to suit the particular headphone in use is taken from the output terminals of the amplifier to the grid of the common grid stage [8].

Because optimised White cathode followers develop their correction voltage across the series combination of the intentional regulating resistor at the anode of the upper valve and the output impedance of the supply, a regulator is almost mandatory, and because the 6080 is a twin triode, a separate regulator can easily be used for each channel.

All 0 V connections from the amplifier and power supply must be brought back to the single star earth and each channel's regulated high voltage supply must also be a star point. These two requirements enforce a circular layout and all other design considerations become secondary. See Figure 1.36.

The valves must lose considerable heat, yet they are close together. The only way that this can be achieved is by mounting them on perforated sheet.

Pre-amplifiers

Pre-amplifiers present different challenges from power amplifiers. It is unusual for the power supply to be on-board, and heat is far less of an issue. Conversely, signal levels are far lower.

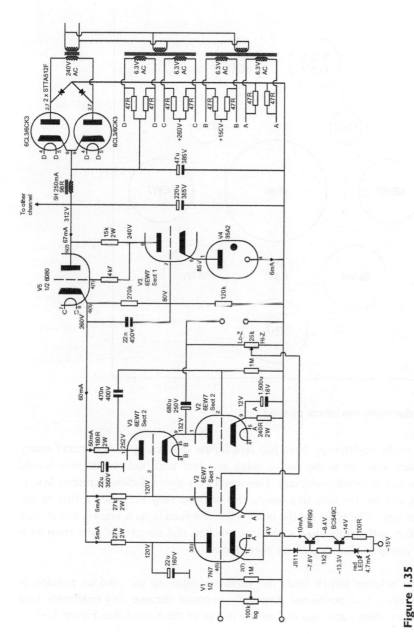

Figure 1.35
Output transformerless headphone amplifier.

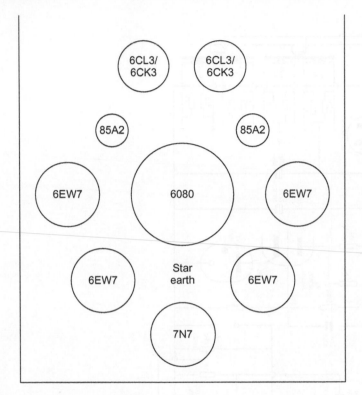

Figure 1.36
Optimum star earthing dominates mechanical planning.

A pre-amplifier might traditionally have had five inputs selected by a front panel rotary switch, taking each input up to the switch using screened lead (unscreened wire would have caused hum and crosstalk problems). This was a very poor solution. Screened lead is expensive to buy and fit, yet with five inputs, this solution only ever used a fifth of the wire at any given time. The place for the input selector switch is on a bracket at the back of the pre-amplifier, sufficiently close to the input sockets that unscreened wire from each socket to switch contacts does not pick up interference.

The output of the selector switch feeds the volume control, so the obvious position is nearby, but it is often best positioned diagonally opposite because this minimises lead length (and therefore shunt capacitance) from the output of the control. See Figure 1.37.

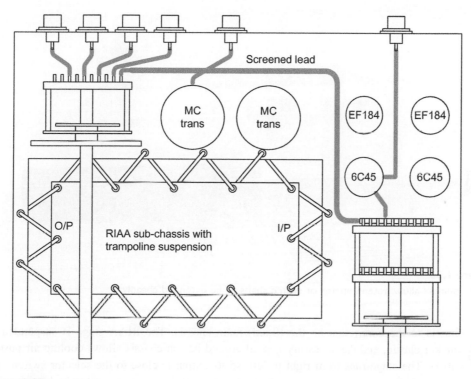

Figure 1.37
Pre-amplifier layout.

The layout assumes a right-handed operator, so you might want to mirror it if you are belligerently left-handed. The selector switch is mounted at the back left, and extended to the front with a coupling and a length of 6 mm or ¼″ shaft as appropriate (13″ lengths of stainless or silver steel rod are cheap at engineering suppliers). If necessary, a coupling could be bodged from a short length of motorcycle fuel hose and associated Jubilee clamps, but a proper coupler is better. See Figure 1.38.

This pre-amplifier has a buffer between the volume control and its output, so this circuitry is logically best positioned at the back right, forcing the volume control towards the front of the chassis. This layout also assumes an RIAA stage using input transformers, and since they generally have flying leads, it makes sense to position them between the input sockets and the first gain stage, because it allows their (flexible) leads to link the (floating) sub-

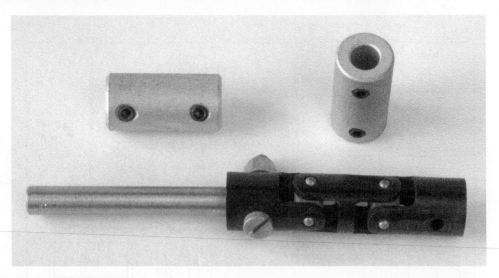

Figure 1.38
Shaft couplers allow potentiometer or switch shafts to be extended cheaply.

chassis to the connectors on the (fixed) main chassis. The RIAA stage is suspended on knicker elastic, and the necessary gap all around its sub-chassis allows cooling air past its valves. The stage runs from right to left, so its output is close to the selector switch. The output of the selector switch to the volume control is a comparatively long run, so screened lead is necessary. An electrostatic screen could be added between the RIAA stage and the volume control, but this is probably unnecessary.

Umbilical leads

Since a pre-amplifier almost invariably has a remote supply, it needs an umbilical lead to connect the two together. To allow either item to be moved, the umbilical cable needs to be unpluggable. The obvious solution is to provide a connector at each end of the umbilical lead. Once you have seen the cost of multi-pole connectors, you will quickly decide that a connector is needed at only one end of the lead. The problem is to decide which end of the lead should have the connector.

It makes better sense to make the lead part of the pre-amplifier rather than the power supply. The reason for this is that a pre-amplifier tends to be designed for a specific application, perhaps to match a particular cartridge in a particular turntable, and to drive particular power amplifiers. That being the case, its proposed physical positioning is also known, so required lead lengths are also known. The previous argument also applies to its audio leads, so why not also hard-wire the outputs, rather than adding expensive and unnecessary connectors?

Personality

A typical power supply provides a high voltage supply, one, or more heater supplies, and perhaps some means of remote power switching. All pre-amplifiers need some, or all, of these facilities, so it would be very useful if a given power supply could be used by any pre-amplifier. A pre-amplifier could then be quickly and easily replaced without having to modify the power supply. One obvious consequence of this approach is that the power supply's multi-pole connector should be chosen so as to leave pins for future expansion and that extra connectors for future pre-amplifiers must be readily available at a later date. Thus, if you spot a bargain connector, buy as many cable plugs to fit the chassis socket as you think you will need for future pre-amplifiers. (Remember that the power supply's chassis connector must be a socket to make it impossible for your fingers to contact potentially live pins.)

Pre-amplifiers often need elevated heater supplies to keep V_{hk} within acceptable limits in circuits such as cathode followers or cascode where a cathode could easily be at 200 V. The required elevating voltages are determined purely by the pre-amplifier, so any circuit for elevating the heater supplies should be within the pre-amplifier chassis, not the power supply. In this way, the power supply's personality is determined purely by the pre-amplifier plugged into it, and it remains universal. Thus, **all** individual supplies within the power supply should float from the chassis, so that their connections to chassis and to one another are determined purely by the associated pre-amplifier.

Mains/chassis earth has to be carried from the power supply to the pre-amplifier, and it is vital that this connection is reliable and low resistance. If at all possible, use more than one pin on the connector to carry this connection − this might mean that you need a larger connector with more pins, but it's safer and less likely to hum. We want a low-resistance bond between the two chassis, so that implies a good cross-section of copper conductor in the umbilical lead. The best way of achieving this is to use an umbilical cable having a braided screen, and use the screen as the earth bond connector. If you make up your own cables, use two, or perhaps even three, layers of screens because this will not only ensure that there are no gaps in the composite screen but it constitutes a robust cable having a low earth resistance. By the time your umbilical has all this screening, plus a protective nylon braid over the top, it will be quite thick, so make sure that your chosen connector has a large enough cable entry to accommodate and clamp such a cable securely.

Power supplies

Fortunately, power supply planning is much simpler than that required for amplifiers. The function of a power supply is to take AC mains, rectify it and deliver clean DC to another point, so it makes sense to lay a power supply out so that one end is deemed to be "dirty" (incoming AC mains, AC outlets, and rectification) and the far end "clean" (regulators, DC outlets).

Regulator positioning

The ideal position for a regulator is adjacent to the load because this minimises unwanted voltage drops, yet there can be powerful arguments against this positioning, particularly when the load is a pre-amplifier:

- Regulators are inevitably hot, and pre-amplifiers should stay cool.
- The raw input to a regulator is likely to be noisy.
- High voltage regulators using valves require heater wiring, usually AC, and therefore noisy. Obviously, semiconductor regulators do not suffer from this problem, and can be more easily placed adjacent to the load.
- Valve heaters draw significant current, causing a voltage drop down the umbilical cable. However, a remote heater regulator can circumvent this problem because its output voltage can be adjusted to compensate whilst measuring the voltage at the valve pins.

Once these considerations have been taken together, the outcome is that heater regulators and high voltage regulators using valves are invariably within the remote power supply, and only semiconductor high voltage regulators can be in the ideal position adjacent to their load.

Professional power supplies often permit **remote sensing**, also known as **four-wire connection**, so that each output terminal has a "sense" terminal associated with it that is connected to the (remote) load, thus automatically compensating for cable resistance. Although this technique requires extra pins on the connector and wires in the umbilical lead, the extra leads do not pass significant current. See Figure 1.39.

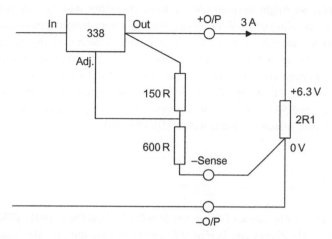

Figure 1.39
Remote sensing overcomes the voltage drop down an umbilical cable.

Note that the three-terminal LM338 is unable to accommodate true four-wire sensing, so only the bottom of the divider chain can be returned to the load. A more sophisticated supply would also return the upper end of the divider chain to the load.

Regulators and heatsinks

Power supplies may need TO-3 style regulators or series pass transistors to be fitted to the chassis. You don't need to measure and mark out all the holes precisely — simply use the mica washer from an insulating kit as a template to mark out the holes. If you are using power transistors, they will be hot, so calculate how many watts of heat must be dissipated, assume a heatsink temperature of 50°C (30°C above ambient), and calculate whether you need a dedicated heatsink, or whether the chassis will do:

$$R_\theta = \frac{\Delta T}{q}$$

where:

R_θ = required thermal resistance of heatsink (°C/W)
ΔT = temperature difference (°C)
q = power to be dissipated (W).

As an example, we might have an LM338 heater regulator passing 3 A, with a drop of 6 V across it, corresponding to 18 W of dissipation. As a very rough approximation, if a reasonably large chassis has sides made of $2'' \times 1'' \times \frac{1}{8}''$ thick channel, at any given point, the channel has a thermal resistance of $\approx 1°C/W$. Thus, mounting the regulator on the channel would keep the temperature rise to $\approx 18°C$, resulting in a local chassis temperature of $\approx 38°C$, which is acceptable. However, if two such regulators were nearby, the chassis temperature could easily exceed 50°C, so it might be worth either fitting an additional finned heatsink or separating the heat sources by a few inches (50–70 mm).

References

1. M. Gayford (Ed.), *Microphone Engineering Handbook*, Focal Press (1994), ISBN 0-7506-1199-5.
2. V.G. Welsby, *The Theory and Design of Inductance Coils*, 2nd edn. Macdonald (1960).
3. "Type KT66", Osram Valve Technical Publication TP3. GEC (February 1949).
4. "Components Group Mobile Exhibition", Brimar Valves (November 1959).
5. *Mullard Technical Communication* (November 1962). Available at: http://www.thevalvepage.com/valvetek/microph/microph.htm.
6. *Typical Oscilloscope Circuitry*, revised edition. Tektronix (1966).
7. J. Broskie, The White cathode follower. *Tube CAD Journal* (October 1999). Available at: http://www.tubecad.com/october99/page4.html.
8. A post by Alex Cavalli on page 5 of the "Building the Morgan Jones" DIY forum thread. Available at: http://headwize2.powerpill.org/.

Recommended further reading

H.W. Ott, *Noise Reduction Techniques In Electronic Systems*, 2nd edn. Wiley (1988), ISBN 0-471-85068-3.
Slightly misnamed (its main focus is interference reduction), this was one of the first books to formally cover EMC and is still one of the best.
J. Goedbloed, *Electromagnetic Compatibility* (Phillips Research Laboratories). Prentice-Hall (1990), ISBN 0-13-249293-8.
Given that Philips was a component manufacturer at the time, this excellent book contains plenty of theoretical considerations for component design and usage backed up by measurements.
E.E. Zepler, *The Technique of Radio Design*. Chapman & Hall (1943).

Although general, seventy pages over three chapters ("The Principles of Screening", "Undesired Feedback", and "Hum, Spurious Beats") tackle valve-specific problems and make this little book well worth searching out.

M.G. Scroggie, *Foundations of wireless (and Electronics)*, 7th edn. Iliffe.

Accessibly written for the amateur, any edition of this book is worth reading for useful radio frequency hints and tips that can be subverted for audio use.

Radio Society of Great Britain, *Radio Communication Handbook*, 5th edn. Radio Society of Great Britain (1976), ISBN 0 900612 58 4.

This was the last edition to cover valves in any detail and covers more material in greater depth than Scroggie but does not quite have the coherence of a single author.

American Radio Relay League, *The Radio Amateur's Handbook*, many editions.

The author has the 1947 25th edition and 1958 35th edition (however did they find the time to revise it annually?). This is the US equivalent to the UK ham radio book and is arguably slightly better, containing many informative photographs showing good layout. Any edition should be worthwhile, although fundamentals are better covered in the 1947 edition than the 1958 edition, presumably due to pressure of space caused by new developments.

CHAPTER 2

METALWORK FOR POETS

Many electronics enthusiasts hate metalwork. If they were to take a longer and more thoughtful look at their pet hate, they would realise that what they actually hate is attempting to do metalwork with **poor tools**. A skilled worker can produce good work despite poor tools, but would far rather use the best. Beginners do not have this level of expertise, and need all the help that they can get, so they **need** good tools. Buy the best that you can't quite afford. Good tools last a lifetime, and not only are they cheaper in the long run, but they're a pleasure to use.

Lighting

How can we do good work unless we can see what we're doing? Good lighting reduces the chance of expensive mistakes, so it pays to optimise it. Human beings are happiest in natural light, so optimise natural light before resorting to artificial lighting. We want to receive as much light as possible, so that means large unshaded windows, and once it is in the room we want to reflect rather than absorb it, so we want a white ceiling. Although white gloss walls would reflect light, they would also cause glare, so the ubiquitous magnolia/cream emulsion is best, whilst pale flooring helps. Conversely, bare rafters and roofing, unpainted walls and a dark floor make for a gloomy dungeon that can never be lit adequately.

Having secured at least part of a well-lit room, we should position the working area to obtain best light. Putting a bench directly in front of a window is not a good idea because

Building Valve Amplifiers. DOI: http://dx.doi.org/10.1016/B978-0-08-096638-0.00002-3

there is a tendency to look outdoors (which is bright), then back inside (which is dimmer), and the required constant adjustment of the eye's iris is tiring. Perfect bench positioning receives natural light from either or both ends.

Artificial lighting

Ideally, natural light would be sufficient, but time and weather conspire against us, so we invariably need supplementary artificial lighting. There are three distinct types of technical lighting, classified by descending area of illumination:

- General lighting illuminates the entire room, so the fitting is generally hung from the ceiling.
- Bench lighting illuminates only the bench, so the fitting might be hung from the ceiling or from a shelf directly above the bench.
- Machine tool lighting only needs to brightly illuminate the cutting point, so the fitting is always nearby and often part of the machine.

Not only do we need an adequate quantity of light, but it must be of a good and consistent quality — partly defined by colour temperature. Colour temperature is defined as the light emitted by a black body radiator heated to a specified temperature measured in kelvin. Thus, a basic incandescent lamp produces ≈ 3000 K whereas a quartz halogen incandescent lamp produces ≈ 3200 K, and a cold Northern sky is ≈ 6500 K.

It is disconcerting to have different colour temperatures in one room, so we should try to match colour temperatures. White LEDs are actually blue LEDs that excite a phosphor to produce longer wavelengths, so their emission is akin to fluorescent tubes, making it easier to match LED lighting to fluorescent than incandescent. The author favours 6500 K in the workplace because it matches the UK's typical natural light, allowing artificial light to seamlessly supplement natural light on gloomy days.

General lighting

The ideal general lighting in a laboratory or workshop is the traditional long fluorescent tube in its specialised fitting because such a large source provides diffuse lighting without shadows or glare over a large area. Many people suffer a lingering prejudice against fluorescent lights – probably because the long working life and difficulty of replacing a fluorescent tube compared to an incandescent lamp means that yellowing tubes remain flickering and buzzing in place long after they should have been retired. There are two very distinct parts to fluorescent lighting: the tube and the fitting.

Fluorescent tubes ionise mercury vapour to produce ultraviolet light, which is converted to visible light by the phosphor coating on the inside face of the tube. When buying a fluorescent light fitting, everyone expects it to come with a tube, but the fitting manufacturer is not in the business of selling tubes, so they fit the cheapest possible tube. The spectral quality of the visible light is determined by the phosphor coating, and cheap fluorescent tubes use a crude phosphor mix producing a spectrum having two or three narrow peaks and a comparatively low-amplitude broadband background. The reason this stratagem works is that the eye has red, green, and blue receptors, so if the red, green, and blue peaks of the fluorescent tube are balanced correctly, the eye perceives the light as being white. However, if instead of looking at the light directly, the eye sees light reflected from a coloured object, the narrow peak stratagem fails to give correct colour rendering and a genuine broadband light source is required. Fluorescent tube manufacturers are well aware of this problem, so they also make tubes that almost eliminate the narrow peaks in favour of broadband light, resulting in a much higher **Colour Rendering Index** (CRI), some of which are sold on (often at vastly inflated prices) as medical tubes suitable for treating seasonal affective disorder (SAD). Despite this, your local electrical wholesaler is perfectly able to supply you with these far superior tubes at only a slight premium over standard tubes. The following fluorescent tube code is common in the Northern hemisphere. See Table 2.1.

Table 2.1

Northern hemisphere fluorescent tube coding

Indicates fluorescent	Power rating	First digit of Colour Rendering Index (>90 is best)	Colour temperature (hundreds of Kelvin)
L	58 W	9	65

Thus, the example code of L58W965 denotes a 58 W tube having a CRI of between 90 and 100 and a colour temperature of 6500 K. The GEC Biolux range (UK) and Sylvania Activa 172 range (US) both have CRI of >95 whereas standard tubes typically have a CRI <60. Further, the better tubes can afford to be configured for a higher colour temperature without looking harsh, so they tend to operate at >5000 K, producing a light closer to the 6500 K of daylight (which is why they work for treating SAD). Conversely, the cheap tubes must be configured for a lower colour temperature, typically 3000 K, which ought to match incandescent lighting, but looks nasty because the poor spectral distribution results in a low CRI.

The fitting supports the tube and includes the circuitry needed for starting and stabilising tube current, traditionally achieved by a neon starter and an iron-cored choke known as a **ballast**. Fluorescent lights do not have the thermal inertia of an incandescent tungsten filament, so they produce flicker at twice their supply frequency, and because they are such large sources, some light is seen from the corner of the eye (which is more sensitive to flicker). Flicker can be eliminated by energising the tube from a high frequency (HF) supply rather than mains frequency, and because this replaces the iron-cored ballast, it is known as an **HF ballast**. HF ballasts apply >40 kHz to the fluorescent tube, eliminate flicker and are typically 5% more efficient than a choke ballast. Buying and retro-fitting an HF ballast to an existing fitting is cheap, but fiddly on small fittings, and your local electrical wholesaler can supply complete high frequency fittings, although they are presently more expensive than traditional fittings. Removing that heavy iron choke makes a high frequency fitting so much lighter than the traditional fitting that it is easily positioned and secured to a ceiling by a single person.

Summarising, appropriate fluorescent tubes driven via HF ballasts produce shadow-free lighting akin to natural daylight without flicker and having sufficiently good colour rendering that resistor colour codes are easily read. What more could you want?

Bench lighting

Work benches need a lot of light, but are often sited against a wall, so bending over the work for close inspection promptly casts a shadow from the (usually centrally mounted) ceiling light. One common solution is to add a local spot lamp, but because we may work anywhere on a large bench, it needs constant repositioning, which is a nuisance. An alternative is a fluorescent tube running the entire length of the bench in front of the user, but this raises the obvious problem of it glaring directly into the user's eyes. There are commercial solutions, but they can be cumbersome, wasteful of their light, or expensive — usually all three. Fortunately, a superior solution is easily made. See Figure 2.1.

Figure 2.1
This bench-mounted fluorescent light produces excellent shadow-free lighting.

Although not directly visible in the photograph, an L70W965 fluorescent tube powered via a high frequency fitting has been fitted to the front of the equipment shelf, and this is shaded by an inverted length of domestic guttering supported by brackets at either end. The brackets were cut from 3 mm sheet aluminium and have semicircles of 18 mm MDF that fit and support the guttering's internal radius. The rotation of the guttering upon the semicircles was set so as to just mask direct sight of the fluorescent tube at the working position.

To maximise lighting efficiency, the guttering is internally lined with aluminium kitchen foil (shiny side out), and the foil is bonded to earth to prevent interference to test leads passing over the light. Unsurprisingly, 70 W of fluorescent lighting this close to the work produces plenty of shadow-free light that makes it easy to see exactly what you are doing.

However, there is a slight downside. Don't attempt to test high-sensitivity, high-impedance circuits such as condenser microphone amplifiers with their covers off while the light is on — 40 kHz interference will be capacitively coupled into the electronics. On the bright side, it's a convenient test of effective screening.

Machine tool lighting

Intense light at the cutting point is essential for accurate work using machine tools — this is the one time we do want glare, albeit carefully controlled. Whilst a machine light to industrial safety standards may be only a small proportion of the cost of an industrial lathe or mill, it could easily double the cost of a cheap pillar drill intended for light amateur use. The typical amateur solution was to subvert an articulated desk lamp, but these are expensive and bulky.

Industrial lamps are low voltage because the flexibility of positioning given by the articulated arm means that it is not possible to guarantee that the lightbulb cannot be moved into the plane of centrifugally ejected swarf. If the lamp could be mounted in a fixed position guaranteed to be well away from the swarf plane, a low voltage bulb would no longer be necessary, and mains lighting could be used, avoiding the problem of mounting a bulky transformer. The light still needs to be switched, and the easiest way to do this is to wire it directly across the machine's motor so that it comes on only when it is needed. Not only does wiring the lamp directly across the motor windings save a switch, but it also damps the back EMF of the motor's inductance when the motor is stopped, preventing arcs and extending switch life.

Quartz halogen downlighting has been very fashionable in kitchens, not because it provides a good working light (it doesn't), but because it enables moody promotional photographs in glossy magazines. Individual downlighters are generally fitted under wall cabinets or as a central ceiling cluster of adjustable lights, but the overall lighting is so uneven that both strategies are usually needed, resulting in punitive power consumption. However, individual adjustable fittings are easily removed from a redundant cluster, giving a single fitting that tilts and swivels, and that uses a GU10 fitting quartz halogen bulb having two independent layers of glass between the user and the mains filament, making it far safer than a traditional bulb. See Figure 2.2.

Figure 2.2
An adjustable downlighter salvaged from a ceiling cluster makes a cheap machine tool light.

These lights are good in an amateur's workshop because they occupy little space in an area where space is always at a premium, yet they can provide machine lighting exactly where it is needed, and a pillar drill is the ideal candidate.

The ideal mounting position on a pillar drill is on the underside space between quill and pillar, taking care to avoid the bulb being caught by the chuck as it descends from its rest position. Mounted in this way, the bulb is well away from the swarf plane and produces a bright soft-edged circle of light about 50 mm (2″) in diameter centred upon the drilling area. Fitting requires removing the head from the pillar and drilling the casting, but as the drill's switch is nearby, subsequent wiring is easy. An unexpected bonus is that by fitting it behind the quill, the quill and chuck obscure the bulb and there is no direct glare. However, there is a single disadvantage. A 50 W quartz halogen light radiates a good deal of heat and when you put a small (and therefore short) drill in the chuck, the work comes much closer to the light. If you hold the work with your fingers they get very hot, very quickly.

The heat problem can be solved by replacing the wasteful quartz halogen bulb with an LED. LED plug-in replacements are available for the mains voltage GU10 fitting, and incorporate a switcher to convert mains voltage to the low voltage needed by the LED. However, because the whole fitting must be the same size as a quartz halogen lamp, there is limited room for smoothing prior to the switcher, potentially allowing the LED to flicker at twice mains frequency. For workshop use, LED lighting should be supplied from ripple-free DC because LEDs switch fast enough to produce strobe effects that could cause a spinning chuck to appear stationary.

Enlightened LED manufacturers quote colour temperature for their white LEDs, but it is more common for them to be described as:

Warm white: Somewhere around 3000–4000 K
Cool white: Somewhere around 6000 K.

The reason for the vagueness is that the spectrum emitted from a white LED tends to be composed of a sharp peak at blue wavelengths (the natural colour of the LED) and a somewhat broader peak at longer wavelengths due to the added phosphors, with a dip in between that causes white LEDs to struggle to achieve CRI > 65, making colour temperature a moot point. As blue LEDs become more efficient, there will be less pressure to maximise phosphor efficiency at the expense of colour rendering.

The twin problems of heat and flicker mean that the best machine tool light is a low voltage LED fitting powered from DC and containing a single-chip 3 W cool white LED, **not** a cluster of LEDs — a point source produces harsh light that highlights scribed lines and throws shadows, making it much easier to align a drill on centre punch dimples. Because the MR16 standard for low voltage electrical fittings was set before LED lights were introduced, it assumed AC and did not include indexing for polarity, so MR16 LED lights include a rectifier. For a white LED $V_{\text{forward}} \approx 3.6$ V, but LEDs require constant-current drive and this is most efficiently derived from a switching regulator, with the happy consequence of reducing sensitivity to supply voltage and anywhere between 9 and

15 V_{DC} is likely to allow correct operation. Fortunately, short-lived electronic junk such as telephone answering machines and ADSL routers generally comes with a substantial wall wart power supply, and this is often ideal. We must first check that the supply produces DC (to guarantee flicker elimination), then check whether it can provide adequate voltage and current.

A variable bench supply and DVM can be used to quickly determine an LED fitting's minimum acceptable voltage. See Figure 2.3.

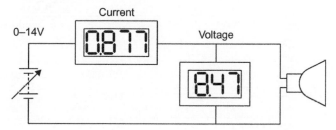

Figure 2.3
Testing a low voltage LED fitting for minimum working voltage.

Increase voltage from zero and monitor the load current, which initially increases with voltage, but starts falling once the minimum acceptable voltage has been exceeded. Assuming that the required voltage is less than the wall wart's nominal voltage, increase voltage to the wall wart's nominal voltage and check that the current drawn is less than the wall wart's current rating. Provided that the wall wart's nominal voltage exceeds the fitting's minimum voltage and its current rating exceeds the fitting's current at that voltage, it is suitable. In general, you are looking for ≥ 9 V and >400 mA for a 3 W LED fitting. MR16 is nominally a 12 V_{RMS} standard and MR16 LEDs contain a 16 V electrolytic after the bridge rectifier, so don't apply $>16\,V_{DC}$ from the supply (that 16 V limit allows the two diode drops in the bridge rectifier to provide a safety margin).

Modifying four 3 W MR16 LED fittings to fit directly onto goose necks and homemade magnetic bases took time, but the results were so good that the author's three articulated desk lamps were promptly donated to a charity shop. See Figure 2.4.

Figure 2.4
A single-chip LED in MR16 fitting fitted to a gooseneck mounted on a magnetic base is the ideal versatile machine tool light.

The goose neck and base came from a hideously kitsch but cheap clock spotted in a discount store (often a useful source of modifiable junk, and if the idea doesn't work, you haven't wasted much money). Neodymium magnets are cheapest from specialist magnet suppliers, and it is usually cheaper to buy a large number of small magnets than a few large ones, so the author bought a cheap pack of two hundred 8 mm diameter by 2 mm thick button magnets. These magnets are astonishingly powerful but difficult to glue securely, and we don't want them ripping off when we move the lamp. The solution was to glue them to the reverse side of a disc of 1.2 mm aluminium sheet. The magnets still want to clump together and the 1.2 mm disc was not very rigid, so a 2 mm disc having ten 8 mm holes to locate the magnets was sandwiched to the first to give a rigid structure. Assembly was very easy. The 1.2 mm disc was thinly coated with epoxy adhesive then placed on a sheet of steel and the 2 mm disc pressed on top. The magnets were then added, and their powerful attraction to the steel sheet anchored everything in place, allowing fine adjustment and removal of excess glue before it set. See Figure 2.5.

Figure 2.5
The magnetic base is formed of two layers of aluminium; the 2 mm layer is drilled for the ten 8 mm × 2 mm neodymium magnets.

Once glued, the composite disc was screwed to a boss screwed to the gooseneck base so that the magnets act **through** the 1.2 mm aluminium, eliminating stress on their (presumed weak) glued joint. Obviously, acting through a 1.2 mm (magnetic) gap reduces the magnets' holding power, yet the ten 8 mm $\times$ 2 mm magnets are easily strong enough to hold the complete light fitting on a vertical steel sheet without sliding, and the gap makes the inverse square law snatch as the base approaches a magnetic surface less uncomfortable.

The author made his lamps in 2010 but notes that a commercial product combining a 3 W single-chip LED with goose neck and magnetic base appeared in a 2013 tool catalogue. No doubt its price will soon fall.

Personal protective equipment (PPE)

Injuries are expensive. The rising cost of healthcare has led to a raft of workplace legislation insisting that employers provide appropriate PPE, and that workers use it. Don't expect sympathy at a hospital if you arrive with a preventable injury. Invest in:

- Really good earmuffs: You're doing this work for the pleasure of your ears, so why damage them?
- Safety glasses: All-enveloping goggles tend to steam up, but are essential if using a chop saw on aluminium. If you wear prescription or reading glasses, consider a pair having safety glass.
- Steel toe-capped shoes: Expensive, but the author recently heard of a young lady who broke a toe when "tidying up a pile of hammers". Ouch.

Marking out

This is where the mistakes are made, so don't rush; 10 minutes saved here could cost hours later.

"Think thrice, measure twice, cut once."

Measurements

Before you can mark out, you need detailed measurements of the parts you want to fit. Commercial manufacturers are supplied by their subcontractors with dimensioned drawings including tolerances, but you might use salvaged parts, which **never** come with drawings. You need to be confident of making accurate measurements. Digital callipers are far cheaper than they used to be, and make life so much easier. Even better, they have a button that changes them from inches to millimetres, or vice versa. See Figure 2.6.

Figure 2.6
Digital callipers are cheap, so there's no excuse for not using them.

Locations on metalwork are traditionally found using a clean grey steel rule in conjunction with an engineer's try square. See Figure 2.7.

Figure 2.7
Engineer's rule and try square.

Although you need both a 150 and a 300 mm rule, it is better to use the shorter rule whenever possible because it is usually thinner, which reduces the parallax error that occurs as a result of looking at graduations some distance away from the work. When buying rules, carefully check the clarity of the graduations — cheaper rules have etched rather than engraved graduations, making them less clearly defined, and more difficult to read. Likewise, a rule with a dull grey finish is far easier to read than a shiny (or even worse, rusty) rule. Although it seems obvious, check before buying that a rule is straight. The author once wasted considerable time correcting what he assumed was his shoddy woodworking, but later discovered that he had sawn lines faithfully matching a curved 600 mm rule.

Small try squares (75 mm) are easier to use on sub-assemblies, but a larger square (150 mm or greater) is needed for a chassis.

A measuring instrument that has recently appeared is the digital protractor. Using exactly the same technology as digital callipers, but adapted for angular measurement, a pair of steel rules pivot about a boss that includes a 0–360° LCD display having a resolution of 0.1°. Very useful.

Marking out manually

With the best will in the world, your marking out will never be perfect, so choose a reference edge from which to measure, and only use the try square from this edge to minimise errors.

The centre of drilled holes is marked by the intersection of two lines, and these lines are made using a sharp scriber under good light. See Figure 2.8.

Figure 2.8
Traditional scriber (lower) and modern (upper). Both are equally good.

With all the construction lines that you will need, there will be a lot of these intersections, so when you are marking the position of a hole, use the scriber to draw a circle around the correct intersection. Even better, draw the circle a little smaller than the intended hole. This will prevent you from drilling holes in the wrong place, and might stop you drilling a hole oversize. If you have a pair of dividers, use them to lightly mark the size of larger holes accurately — any mistakes in marking out will become apparent instantly.

You will often have to cut irregularly shaped holes for transformer connections. Cross-hatch the metal to be removed with a marker pen to avoid confusing construction lines with cutting lines. The reason for using a marker pen rather than a scriber is that if you make a mistake, it can be removed with methylated spirits, whereas scribed lines are permanent.

The CAD solution

Alternatively, you can rely on the precision of a printer to do your marking out for you, although it is a good idea to check the accuracy of your printer in both horizontal and vertical axes against a good quality engineer's steel rule. Modern printers print with a resolution of at least 300 dots per inch (dpi), and usually far more, so an engineering drawing package can easily produce a precise template that just needs to be spotted through with a scriber to ensure perfect marking out.

Admittedly, learning to use a CAD drawing package takes time, but it will be followed by a step improvement in the quality of your metalwork.

Checking and centre punching

At this point it is still possible to correct mistakes. Having just carefully drilled a large hole exactly on the markings, but 8 mm away from where it should have been and thus ruined a day's work, the author can't emphasise enough the importance of offering up the parts to be fitted as a check before drilling.

Given half a chance, the drill will skid randomly around the surface of metal before cutting — probably in the wrong place. Punching a dimple helps the drill cut where you want. Having checked your marking out, use a centre punch to indent the centres of all drilled holes. The modern punch is the automatic spring-loaded punch, whereas the traditional punch has to be struck with a hammer. Although the autopunch is quicker to use, you will find that it is less accurate than the traditional punch. See Figure 2.9.

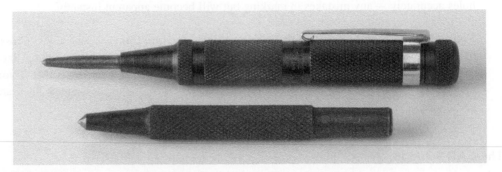

Figure 2.9
Traditional centre punch (lower) and autopunch (upper).

Whichever punch you use, it must be **sharp**, and should be ground to an included angle of 90°. A punch is easily sharpened on a bench grinder, or a grinding wheel in a drill. (Always wear goggles when using a grinder.)

Centre punching sheet metal is noisy. Noise is minimised and marking accuracy increased by supporting the area to be punched directly above a leg of the bench or table. This work is being done for the pleasure of your ears, so buy some really good earmuffs (cheaper than you would think), and wear them.

Transfer punches are tools that locate through one hole to punch axially onto the part beneath, and are typically used for marking where a screw hole should be drilled. Imperial or metric sets of individual punches are available to fit holes snugly. See Figure 2.10.

Sadly, the commercial transfer punches that the author has seen have a marking cone so long that it prevents the shaft from engaging the hole in sheet metal. See Figure 2.11.

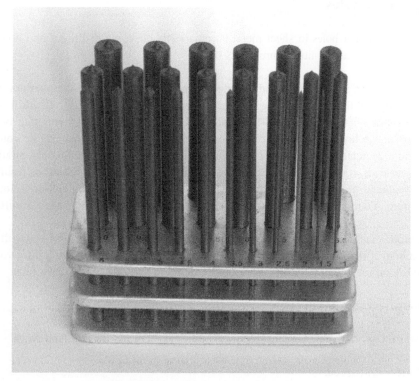

Figure 2.10
A set of transfer punches might assist marking out.

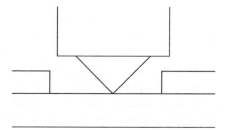

Figure 2.11
Despite being exactly the right size, this transfer punch's overlong marking cone prevents the shaft engaging with the hole.

Drilling round holes

Drilling triangular holes in sheet metal is remarkably easy, but drilling round holes in the correct position takes a little more care.

A drill mounted in a stand, bolted to a bench, invariably means that the operator bends or sits to carry out the work. Drilling produces swarf that flies radially from the drill, and may be hot (although if it is hot whilst you are drilling aluminium, then something is wrong). You are at eye level to the swarf plane, so wear safety glasses. Whether you do your metal-work in a dedicated workshop or on the kitchen table, you will get hot. If perspiration runs into your eyes, it stings and can cause you to blink — perhaps with disastrous results. Steal the sweatband that your wife or daughter wore when she was feeling self-conscious about her figure. Your ears are near your eyes, and drills are noisy, so wear those earmuffs.

The work must always be supported directly beneath the drill — otherwise it will deform or vibrate. If the work moves, the drill will snatch and you will suddenly find your work spinning on the end of the drill. This is most alarming and potentially dangerous.

Drilling round holes in the correct position without snatching boils down to:

- Preventing relative movement (especially vertical) between the drill and work
- Lubrication
- Correct drill speed.

Preventing relative movement

Don't attempt to drill into the middle of an unsupported chassis. Always support the work directly under the drill with a scrap of wood, often known as a **drilling block**, otherwise the force required to make the drill cut will make it suddenly bite deeply and snatch. If the work isn't level when drilling, the drill will wander, so favour off-cuts of manufactured wood as drilling blocks rather than natural wood that may have warped or not been planed accurately. Ensure that the work remains firmly in contact with the drilling block by deburring the underside of each hole immediately after it is drilled. Drilling blocks soon become scarred with part holes and you will find that far more burr is produced when there is no support directly under the drill, so use a fresh drilling block for particularly important work. The author uses the scrap circular cut-outs from loudspeaker projects as drilling blocks.

Always treat a handheld power drill as the last resort — a drill in a stand bolted solidly to the bench is far better. This is much easier than it used to be, because power drill manufacturers have standardised on a 43 mm collar to grip the drill in the stand, so a choice of stands is available. Alternatively, small pillar drills are now available at remarkably low prices, and although they do not withstand comparison with a genuine workshop pillar drill, they are perfectly good for amateur work. If you have not used such a tool before, you will not believe how much easier it makes your work!

The most important specification when buying either a pillar drill or stand is the depth of the **throat**. The throat is the distance between the drill axis and the outside of the pillar that supports the drill. A shallow throat prevents the drill from reaching the centre of your chassis, forcing you to use a handheld drill. In this instance, a deep throat is very desirable.

Drilling brass is especially awkward because it invariably snatches as the drill breaks through the far side, so it must **always** be clamped to the drill table to prevent vertical movement. Aluminium and steel are much more forgiving and should ideally be clamped, but if you do hold the work by hand, press firmly, and avoid having your fingers near a part of the work that would cut you if the work were to spin. If the worst comes to the worst, do not try to fight the drill; it is much stronger than you are. Let go, and switch the drill off. Ideally, you would have a foot-operated stop switch, but few amateur workshops achieve this level of sophistication.

All pillar drills and most stands have slots in their table to take the T-nuts used by engineering clamp sets, but the T-nuts (and securing screws) come in different sizes, so check they will fit your slots. A useful clamp set includes triangular step blocks akin to a Mayan pyramid onto which the clamp locks. See Figure 2.12.

At least two clamping points are needed to secure work firmly, and given that the step blocks are coarse, a clamp will rarely be perfectly level — so select a step a little higher than the work to avoid it exerting pressure on the work's edge, deforming it. A thin scrap of wood or plastic under the clamp tip stops it marking the work. Align the work with only light pressure, then tighten when correctly aligned. It's quite hard to gauge where the

Figure 2.12
Drilling a 22.5 mm hole in mild steel (let alone brass) requires the firm clamping provided by an engineering clamp set.

axis of a drill is, so the author substitutes a large needle (with its eye cut off) during alignment, and spins the chuck to check tip concentricity. If part of the work protrudes, fouling the chuck, and preventing the short needle from reaching the dimple for comparison, use the tip of a large transfer punch instead.

Lubrication

Lubrication greatly aids cutting efficiency. If you use a lubricated 3 mm **centre drill** or **spotting drill** to start each hole, you will find that you can immediately use the final size of lubricated drill, without intervening pilots. This not only speeds your work, but actually produces more accurately aligned holes. See Figure 2.13.

For aluminium, the author keeps a small glass jar containing sufficient depth of methylated spirits to wet the bristles of a ½″ paintbrush. Before drilling, wet the tip and flutes of the drill, and spot a droplet at each hole to be drilled. In time, you will knock the jar over and discover why you only wanted a shallow depth of spirits. When you have finished work for the day, screw the lid on the jar, otherwise the spirits will evaporate — leaving you with an empty jar and a lingering fragrance that decays to a distressing aroma of stale socks. (Avoid jars having a plastic lid as the vapour may dissolve it.)

Figure 2.13
Centre drills enable holes to start in the correct place.

A dispensing alternative is to recycle a perfume or cat-clawing-deterrent atomiser. Since the liquid in both is largely ethanol, the container is impervious to methylated spirits, and although the spray is somewhat haphazard, the liquid is always clean (the brush and jar method tends to transfer swarf to the jar).

The traditional **cutting fluid** for steel was a 50/50 mix of lubricating oil and water with a healthy dash of washing-up liquid. The water cools, the oil aids cutting, and the washing-up liquid allows the two to form a temporary emulsion. Unfortunately, it is extremely messy and rusts tools. Modern proprietary cutting fluids are less destructive, but still require cleaning after use.

Only a little lubrication is required — too much will spray you and your surroundings.

Otto Klemperer (1885–1973) conducted some of his finest Beethoven performances seated owing to having further paralysed himself in an attempt to extinguish his burning bed using spirits of camphor, but metalwork is difficult sitting down, so treat flammable lubricants with care.

Correct drill speed

Use the correct drill speed. A 1 mm drill must run fast to clear swarf or it will clog and break, so it should run at 2500 rpm or faster, whereas a 12 mm drill should run at the slowest possible speed, 200 rpm or slower if possible. This is not nearly as much of a problem as it used to be for the amateur, because the better quality power drills have a two-speed mechanical gearbox **and** electronic speed control, and even cheap pillar drills allow drill speed to be changed by moving the belt from one pair of pulleys to another, typically allowing five speeds.

A rather nasty cost-cutting trend seen on recent small machine tools is the fitting of a brushless DC motor plus electronic speed control rather than a synchronous AC motor plus gearbox or belt and pulleys. The practical difference is that altering gearing to achieve a low speed increases the torque available (which is exactly what is needed for a large drill), whereas the DC motor running at low speed needs to be powerful to provide the required torque, so a 1 kW motor must be fitted whereas 250 W would have sufficed via gearing. Thus, the manufacturer saves production costs but you pay for that saving ever after in excessive power consumption.

Drilling very small holes

Drilling <1 mm holes for component leads requires a very high drill speed and a chuck having low run-out (deviation from concentricity). Although it is possible to put a pin chuck in a power drill or pillar drill, the main chuck is likely to have a run-out of >0.1 mm, which is a substantial proportion of a 1 mm drill, causing oversize holes and excessive drill breakages. Further, because the machine was designed for larger drills, it is difficult to feel what is happening using <1 mm drills, making breakages even more likely. Nevertheless, if needs must, it can be done.

When you tire of breaking <1 mm drills, a dedicated small drill is much better. High-speed multi-tools with sets of dedicated cutters are very popular and their manufacturers often offer dedicated stands to convert them into a pillar drill. Don't bother. The stand has to lift and lower the entire off-centre mass of the multi-tool causing binding that means that you still can't feel what that <1 mm drill is doing. A much better solution is a miniature pillar drill. See Figure 2.14.

Figure 2.14
This miniature pillar drill is far less likely to break small drills.

This Proxxon TBM220 drill is just like a full size pillar drill in that it has three belt-selectable speeds (1800, 4700, and 8500 rpm), a set of six collets spanning 1–3.2 mm, proper bearings and, most importantly, a quill with a decent feel. A three-jaw Jacobs-pattern chuck is available as a manufacturer's accessory, but the author rarely seems to use it.

Disadvantages are that the (brushed) motor is somewhat noisy, and the collets won't grip below 0.9 mm. Fortuitously, the significantly longer collets from an Eclipse 160 pin chuck fit, enabling use of 0.3 mm drills if necessary. Although it's no longer the bargain it was when the author bought his, it allows you to drill little holes exactly where you want them and is essential if you want to make your own PCBs.

Tungsten carbide drills having a 3 mm mandrel are normally used for drilling PCBs because they remain sharp despite the abrasive glass dust produced by drilling FR4 (Fire Resistant type 4). Unfortunately, not only are tungsten carbide drills much more expensive, they're very brittle. Beware also that a tungsten carbide cutter having a 3 mm mandrel could be a slot drill intended for a PCB plotter — the significance is that the end of a slot drill is perfectly flat and there is no self-centring action, making it impossible to use in a conventional drill. Unless you are certain of your drill and skill, you will find it cheaper to blunt and break a few good quality HSS (high-speed steel) drills than break a single tungsten carbide drill.

Drill selection for metal

Use sharp, correctly ground drills — **never** attempt to re-sharpen drills (if you genuinely know how to do this correctly, you don't need to read this chapter). A good way to find decent drills is to look in a model engineering magazine. All engineering suppliers stock drills made by reputable manufacturers, but individual drills are expensive, so buy a set of 1–6 mm in 0.1 mm steps — and another from 1 to 12 mm in 0.5 mm steps is useful. These are virtually all the drills you will ever need, and they will come in a protective steel box with each size marked. See Figure 2.15.

Whatever you do, don't buy drills from anywhere but a proper engineering tool supplier — if they also sell lathes and milling machines having suds pumps and needing a three-phase supply, then their drills are probably fine. There is a world of difference between the inaccurately ground rubbish made of muckite that is adequate for putting up the occasional shelf, and a true engineering drill. The author once disregarded his own advice by buying some unbranded centre drills at a model engineering exhibition that blunted quickly, and life became so much easier when they were replaced by proper centre drills made by Dormer. Buy cheap, buy twice.

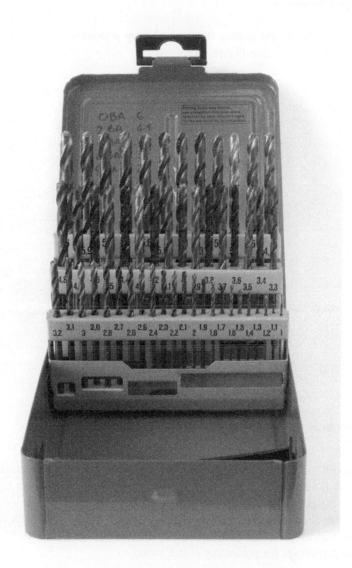

Figure 2.15
Set of engineer's twist drills in case.

Handheld taper reamers

These have eight straight cutting flutes running up a shallow cone, allowing enlargement of holes in sheet metal. See Figure 2.16.

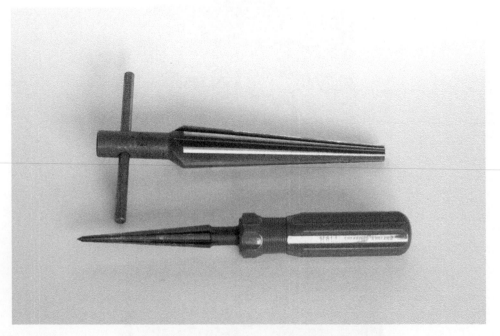

Figure 2.16
Tapered reamers allow any diameter of hole to be cut in sheet metal.

The 1″ and ½″ taper reamers are commonly available, allowing manipulation of hole size from ¼″ to 1″. As shown, they are available either with a handle or a bar – the handled version is far more controllable. Although very useful, the downside of these reamers is that they can sometimes dig in and cut an octagonal hole. Practice on scrap before your first important use, and don't attempt to use them to enlarge a hole already having a notch cut out of it (perhaps for preventing switch rotation).

A ¼″ two-flute handled variant is also made and this is extremely useful on the electronics bench when you discover at the last moment that a screw doesn't quite fit through its intended hole.

If possible, use tapered reamers from both sides of the hole to reduce their tendency to dig in and to minimise the taper on the resulting hole.

Deburring

Whenever you drill a hole in metal, there will always be a small burr on the upper surface and a larger burr on the lower surface. The best way of deburring is to minimise the burr in the first place. Provided that the work is firmly anchored and supported directly below a well-lubricated drill, very little burring will occur.

All metalworkers have their preferences for burr removal, but a rose countersink in a handle or a specialised "wiggly" deburring tool both work well. See Figure 2.17.

Figure 2.17
"Wiggly" deburring tool (lower) and rose deburring tool (upper).

You could simply use a large drill without a handle to deburr holes, but you will find that the flutes of the drill tend to cut the outer layer of your skin as you grip it, and that the drill tries to bite deep into the work. If you are lucky enough to know a toolmaker who sharpens their own drills, politely ask them to regrind a ½″ and a ¼″ drill for deburring. The leading edge is ground to a negative rake of $\approx 1°$, so that when gently rotated by hand against a drilled hole, it doesn't quite bite. Correctly ground, these tools deburr superbly. See Figure 2.18.

Figure 2.18
Half-inch drill ground to form deburring tool.

Screw heads and why to avoid countersinking

Stainless steel countersunk hex screws do look nice, and countersinking is sometimes essential to avoid the screw head fouling other parts, but countersinking introduces new problems. The included angle of screw heads is typically 90°, so we can forget any idea of countersinking using a twist drill (typically 118° or 135°) or centre drill (60°). There are various different types of countersinks. See Figure 2.19.

It is essential to lubricate countersinks before cutting. Unlubricated rose cutters clog, and when forced into the work, suddenly bite and cut an octagonal countersink. Countersinks

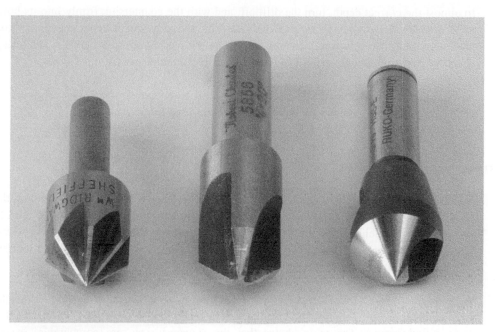

Figure 2.19
Selection of countersinks.

having only three cutting edges are less likely to clog but are not quite as controllable. Conical countersinks are the most controllable and produce the best finish, but are the most expensive.

Although countersinks are nominally self-centring, they easily wander, causing the screw head not to seat properly, so be careful to align the countersink accurately before cutting. The best solution is a dedicated countersink for each screw size having a centring spigot that guarantees perfect results, but these are rare, so snap them up if you ever see any. Spigoted counterbores produce a flat-bottomed hole for submerging cap head screws, and are readily available, but require a greater material thickness than ordinary countersunk fasteners and their spigot may require a surprisingly large clearance hole – check before use.

Countersinking one hole to the correct depth is merely tedious, but countersinking a series of holes to matching depths is astonishingly difficult. The solution (if your pillar drill has a depth stop and the work is of uniform thickness) is to carefully countersink one hole to almost the correct depth, turn the drill off, and with the countersink firmly into its hole, set the drill's depth stop. If, as expected, the first and subsequent countersinking is too shallow, rather than adjusting the depth stop, shim the work up from the drill's table using sheets of paper until countersink depth is perfect, then countersink the other holes.

Nobody produces perfect metalwork, and cap head screws are remarkably forgiving of minor inaccuracies. Countersunk holes are not in the least forgiving.

Screw threads and tapping

Sometimes, despite careful planning, we know that a fastener will be inaccessible from the far side, so a screw and nut cannot be used. Although self-tapping screws can be used to secure parts, the deliberately sharp thread protruding on the blind side easily nicks wires, while the weak thread cut in the chassis can't withstand the repeated removal and reinsertion demanded by the development of engineering prototypes.

A better alternative to a self-tapping screw is to cut a proper engineering thread into the chassis, using a **tap**. Tapping and clearance sizes are given (in mm) for ISO metric, BA, and unified screws. The table shows that the standard sizes are comparable, except for 4BA and 8BA, which do not have direct counterparts. The significance of including the obsolete BA thread is that many NOS parts were intended for BA fasteners. See Table 2.2.

Table 2.2

Clearance and tapping drills for common threads

ISO metric	Clear	Tap	British Association (BA)	Clear	Tap	Unified	Clear	Tap*
M6	6.2	5.1	0BA	6.2	5.1			
M5	5.2	4.3	2BA	4.9	4.0	10–24	4.8	3.7
M4	4.2	3.4	4BA	3.9	3.0	8–32	4	3.4
M3.5[†]	3.7	2.9				6–32	3.6	2.7
M3	3.2	2.5	6BA	2.9	2.3	4–40	2.9	2.2
M2.5	2.6	2.05	8BA	2.3	1.8	3–48	2.6	2.0
M2	2.1	1.6				2–56	2.3	1.7
M1.7	1.8	1.4	10BA	1.8	1.4	1–64	1.9	1.5

*Unified tapping sizes in this table are specified for ≈95% thread engagement – a larger tapping drill reduces thread engagement.
[†]Be warned that although M3.5 is notionally a popular thread in the electronics industry, even the electronics factors seem reluctant to stock it, so this size is best avoided if possible.

A rule of thumb for the threads in the table is that the thickness of the material to be tapped should not be less than half the thread diameter. Thus, up to M6 can be tapped into the 3 mm channel, but the largest thread that should be tapped into 1.6 mm sheet is M3.

Note that for M3 upwards, the recommended clearance size is the nominal thread size plus 0.2 mm. In practice, most plated screws will pass through a hole of nominal size minus 0.1 mm. Thus, the author drills 3.9 mm for M4, and if his drilling should be imperfectly aligned, a little work with a needle file and the ISO recommended clearance drill leaves a perfect hole. Stainless steel fasteners have superior thread engagement and genuinely require ISO clearance size.

Unlike drills, it is not usually worth buying a packaged selection of tap sizes because electronics work rarely uses the three largest sizes from the common engineering selection of M12, M10, M8, M6, M5, M4, and M3; the author most commonly uses M5, M4, M3.5. It is best to buy a set of three for each particular thread size; starting (or first) tap, second tap, and bottoming (or plug) tap, although for aluminium, you only really need the starting tap and bottoming tap. See Figure 2.20.

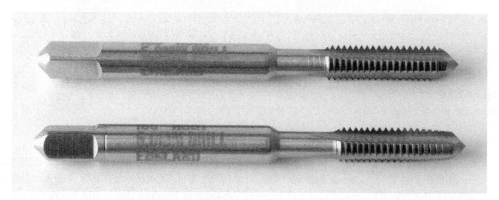

Figure 2.20
Starting (lower) and bottoming (upper) taps.

Taps are held in a **tap wrench**. The chuck wrench is best for small taps and is easier to keep aligned, whereas the larger bar wrench allows more force to be applied and is best for taps >M8. Although taps are somewhat self-aligning, it is important that the tap is first aligned with the axis of the hole — so check from side to side and from front to back.

A guided wrench takes a little longer to use, but the resulting thread is stronger and has noticeably better engagement. See Figure 2.21.

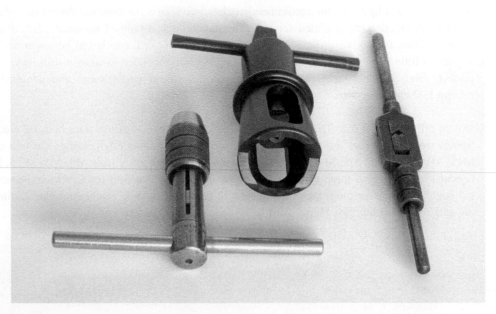

Figure 2.21
Chuck tap wrench, guided tap wrench, and bar tap wrench (left to right).

Beware that a hole drilled in soft plastic can have a smaller diameter than the drill that produced it — check the drilled size before tapping and, if necessary, use a larger drill to achieve the correct tapping size and prevent subsequent binding that could break the tap.

The way that you cut the thread depends on the hardness of the material and whether the hole to be tapped is clear (you can see through it) or blind. A clear hole allows swarf to clear easily, but a blind hole quickly clogs the tap.

To tap a clear hole, align the tap so that it is vertical, lubricate it, and press down gently as you screw it into the hole. Never force a tap. If everything has been done correctly, the tap will glide through the work and emerge on the other side covered in swarf. Use a stiff brush to remove all the swarf, **then** unscrew the tap from the work. If the metal is only a few millimetres thick, you will only need a starting tap to cut a perfect thread. See Figure 2.22.

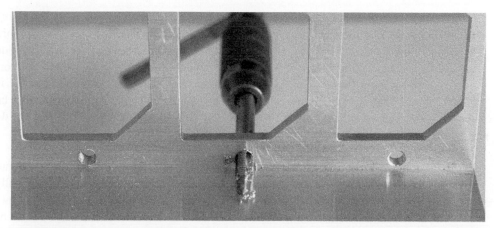

Figure 2.22
The tap emerges from tapping 3 mm aluminium covered in swarf. Remove the swarf with a brush, **then** unscrew the tap.

If the hole is blind or deep, the swarf can't clear easily when tapping, so use plenty of clean lubricant, and cut by turning the tap half a turn forwards followed by a quarter of turn backwards to clear the swarf. It may be necessary to withdraw the tap and clear the swarf from tap and hole several times before the entire hole is tapped. A vacuum cleaner is handy here. It is vital that swarf is cleared regularly, otherwise you may not notice when the tap touches the bottom of the hole, breaking the tap. Removing a broken tap from a blind hole is virtually impossible. Blind holes invariably require at least a starting tap and bottoming tap to be used, so they are time-consuming. Nevertheless, don't be tempted to rush — it is so easy to wreck the work.

If you see a tap wrench identical to one you already have (and like) going cheap, don't walk away; work proceeds significantly faster with starting and bottoming taps in their own wrenches rather than alternating taps in one wrench.

Tapping invariably leaves a very small burr that is best removed by a rose deburring tool. Be very gentle when deburring, the tool can easily dig in and remove a significant depth of thread. An alternative technique is to very lightly countersink the hole before tapping so that the burr occurs just below the surface and need not be removed, thus maximising thread depth.

A tapped thread can be destroyed by a screw encountering even the slightest remaining fragment of swarf, and aluminium is particularly vulnerable because the bulk metal is soft. Worse, aluminium swarf has sharp edges that have oxidised and because aluminium oxide is hard, it is essential to remove it, first by vacuuming, then by squirting the hole thoroughly with a cleaning solvent such as isopropyl alcohol. Spray-can isopropyl alcohol is expensive, but abrasive aluminium dust/swarf remaining after vacuuming can also be removed by washing in plenty of hot soapy water — obviously, you wouldn't want to do this with steel!

Die-cast aluminium boxes have lids secured by screws passing into threads tapped into the body of the box. The threads in the box are rarely fully formed because nominally self-tapping screws are used, and casting alloy is weak, so discard those screws and carefully run a lubricated tap of the same thread into the holes — you will be amazed by how much swarf is produced. Clear the swarf thoroughly. Fit good quality replacement fasteners and treat the box threads with extreme care — they strip easily.

If you break a tap, don't attempt to drill it out. Taps are made of tool steel that is harder than a drill so there is no possibility whatsoever of drilling the tap out, only of chipping the drill and further damaging the work. If you are very lucky you might be able to grip the broken tap with a pair of pliers having serrated jaws and undo it, but it is more likely that you have to accept that the work is ruined and start again. The most likely reason for a tap breaking is that it was blunt and required excessive force to cut, so (despite their cost) be prepared to bin and replace them as they blunt — it's cheaper than replacing a broken tap **plus** the ruined work.

Remember that a tapped hole does not allow for any subsequent manoeuvring of the screw, so you need to be confident about the accuracy of your marking out and drilling. You need astonishing confidence to combine tapped holes with countersunk screws.

Screw choice

Fastener specialists delight in amiably correcting the author's inaccurate requests for a screw. Screws are specified by:

- Thread type: Metric, unified, BA, Whitworth, etc.
- Thread diameter: Almost always the thread's outer diameter, so M8 = 8 mm outer diameter. Plumbing is full of traps — although ¼″ BSP is a thread size, the ¼″ dimension refers to the pipe's bore, and the (tapered) thread is nearer to ½″.
- Thread pitch: Coarse (UNC = UNified Coarse), fine (UNF = UNified Fine). Metric threads may explicitly specify their pitch, M8 × 1.25 = 8 mm metric 1.25 mm pitch (coarse), M8 × 1 = 8 mm metric 1 mm pitch (fine). Although frequently unspecified, coarse is usually the default pitch, so the 12 in M8 × 12 means 12 mm long, not the pitch!
- Screw length: This ought to be easy, but countersunk screws are specified by their entire length whereas all other screws are specified by the length of thread.
- Type of driver needed: Slot, Phillips, Pozidriv, Supadriv, hex socket, Torx, etc.
- Shape of head: Cheese, pan, button, countersunk, etc.
- Material: Zinc-plated steel, plated brass, stainless steel, high tensile steel, nylon, etc.

Although electronics factors stock fasteners, it's worth being humiliated by the fastener specialists because their prices are so much lower. Because there **is** such a variety, fastener specialists tend not to have catalogues, instead they ask you to specify what you want and quote you by return. They're used to dealing with proper engineering firms, so don't try their patience by asking for silly small quantities; they deal in hundreds and preferably thousands. Provided you play the game by their rules (full specification, sensible quantities), you will be pleasantly surprised by their prices. Emailing your request as a spreadsheet is particularly effective because it minimises their work on your trivial (to them) order and allows you to subsequently revise your quantities upwards or downwards and immediately see what it's going to cost you.

As an example, the author recently needed some stainless steel hex socket cap head screws M3 × 6, and a well-known UK electronics factor had them for £7 per 50, whereas the fastener specialist quoted £2.50 for 100. Admittedly, the fastener specialist charged £25 post and packing, but their prices justified buying an entire range of metric stainless steel socket head fasteners in both cap and button head.

Shortening screws

Occasionally, a screw is too long and needs to be shortened. Cutting to length is no problem but tidying the cut end so that the screw passes easily into a (possibly delicate) thread can be. Fibre vice jaws can grip even a countersunk screw head without damaging it and allow the cut face to be filed flat, but the remaining burr is most easily removed by holding the screw on the end of a screwdriver and firmly driving a flat needle file around the cut end with the file aligned to the angle of the thread. Provided that a circular (and apparently eccentric) flat results, the screw will pass easily into a thread.

Another useful clamp in this situation is a nut cut radially from one face so that when clamped in a vice, or lathe, or loose drill chuck, it closes onto the thread and grips it. Annoyingly, many metric screw heads have a larger diameter than the dimension Across Flats (A/F) of the thread's corresponding nut, but fibre vice jaws can usually take up the discrepancy without damaging the screw head.

Washers

Washers are stamped from sheet producing a rounded side and a burred side. Face the burred side towards the screw head or nut to avoid marks that will become visible if the fastener has to be undone then refastened.

Sheet metal punches

Attempting to drill holes larger than 10 mm in sheet metal is simply asking for trouble.

The solution for round holes is to use a sheet metal punch. Although rarely available from high street shops, engineering suppliers and electronics factors in particular generally stock a reasonable range, and they're much more affordable than they used to be. If the distributor doesn't have the size you need, contact the manufacturer directly (Q-Max in the UK, Greenlee in the USA). See Figure 2.23.

The punches are of a two-part construction drawn together by a bolt. Provided that they are kept well greased on all surfaces, they cut a beautifully neat hole and last for years. Useful sizes tend to be Imperial:

Figure 2.23
Selection of sheet metal punches.

⅜″: Imperial potentiometers and rotary switches, grommets for small cables.

½″: Imperial toggle switches (but measure first − modern switches are often smaller than their nominal size), 32 A loudspeaker terminals, larger grommets.

⅝″ (16 mm): B7G sockets, DIN sockets.

¾″: NOS B9A valve sockets, some cable clamps. Modern ceramic sockets are 21.8 mm, so use 22 mm or ⅞″ (22.2 mm).

1⅛″: NOS International Octal and Loctal sockets. Modern ceramic sockets are 26.1 mm, but the nearest size guaranteed to clear this is 1 1⁄16″ (27 mm).

Beware that sheet metal punches are usually specified by clearance size, so a 26 mm punch might not cut an exact 26 mm hole, but a hole that is clearance for 26 mm and a tight fit for a 26.1 mm ceramic valve socket. Precise work can render these distinctions important, so don't blindly believe the punch's specified size − measure its cutting diameter before use.

Larger punches require quite a large hole for the bolt, so there is no reason why you should not use a ⅜″ punch to cut the bolt hole for a 35 mm punch. Large punches have to shear a considerable circumference of metal and need a correspondingly large force. Surprisingly, much of the torque you expend actually goes into overcoming rotational friction between the shoulder of the bolt and the top of the punch. Distributors do not usually stock the punch manufacturer's entire range of punches, and they very rarely stock the thrust bearings that can be retrofitted to large punches – contact the manufacturer.* Thrust bearings take all the strain out of punching holes >30 mm, and are thoroughly recommended, not only for laziness, but because by forcing less torque into the work, you are less likely to distort it.

Note that although these punches produce very little burring, they do slightly deform the surface from which the cut began. It is therefore usual to punch from the chassis inside face towards the outside to avoid this deformation being visible. However, the pressure (and inevitable slight rotation) of the opposing face on the outside can mark decorative surfaces (such as anodised brushed aluminium), but this can be avoided by placing a thin cardboard washer between the opposing punch face and the decorative surface before punching.

If, as was suggested earlier, you have used a pair of dividers to draw the exact position of the finished circle, you may be able to align the cutting side of the punch precisely before tightening up the bolt and cutting the hole. Whether or not you can do this depends on whether you can tolerate punch deformation on the marked out side. If your planning is astonishingly good, you will have realised this in advance and marked out on the reverse side.

Although it exceeds manufacturers' specifications, sheet metal punches can be used on ⅛″ aluminium. The bolt should be cleaned and freshly greased, and a thrust bearing plus long arm hex key makes life easier. Further, the metal needs to be weakened by chain drilling a series of small ≈ 2 mm holes just inside the circumference of the hole to be punched. See Figure 2.24.

Sadly, sheet metal punches eventually wear out, and when they do, they require more effort and produce more burring. The author replaced his ⅜″ punch after 25 years.

When the author first bought punches they were packaged in robust cardboard boxes that could be arranged in a drawer with their size labelling easily visible, but bubble packs have taken over, making storage and identification harder. One possibility for punch storage is to drill a sheet of >18 mm wood with holes to locate the bolts and arrange the punches in order of size. See Figure 2.25.

* Q-Max (Electronics), Bilton Road, Bletchley, Milton Keynes, MK1 1HW, UK.

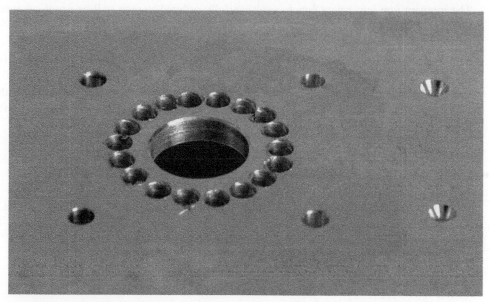

Figure 2.24
Sheet metal punches can be used on thicker metal if it is weakened beforehand by drilling a ring of
2 mm holes.

Figure 2.25
If sheet metal punches are stored in order on a drilled board the correct size can be found quickly.

The moment you buy a new punch, your carefully ordered storage will be upset, so avoid buying punches singly and buy them in batches.

You will occasionally find hole saws suggested for large holes. Hole saws are ideal for plumbers because the combined process of drilling and sawing produces a quick hole that can be concealed in a dark loft, whereas punches require a hole to be drilled before the punch can be used. However, hole saws are more expensive and produce extremely scruffy holes compared to a sheet metal punch.

Deburring holes made by sheet metal punches

Particularly as they wear, sheet metal punches sometimes throw up a curled burr that a wiggly deburring tool can't remove cleanly. A better tool, but one that requires care to avoid marking the surface, is a triangular scraper. See Figure 2.26.

Figure 2.26
A triangular scraper requires skill but deburrs beautifully.

The scraper's cutting edge is laid as flat as possible onto the work and eased round to cut the burr away flush with the surface of the metal. It may take several cuts to remove a given section of burr, but the tool removes no more material than necessary. Beware that this process leaves a very sharp edge on the hole, so you may wish to lightly soften it with 320 grade silicon carbide paper (best) or a wiggly deburring tool (quickest, but may dig in).

Using a sheet metal punch to concentrically enlarge an existing hole

Sometimes, you need to enlarge an existing hole, perhaps to replace the original phenolic B9A valve sockets (¾″) on a Leak Stereo 20 with new ceramic sockets (22 mm). If you were very patient, you could carefully enlarge all seven holes with a file or taper reamer. It would be more convenient to use a punch, but you have the problem of accurate centring. The centring problem can be overcome by fitting a thick ¾″ washer to the punch's bolt so that it locates in the ¾″ hole of the chassis. If the chassis is horizontal, the washer will fall onto the cutting edge of the punch and locate itself in the existing hole. You then start cutting, so that the tips bed into the chassis, release the punch and remove the washer. You now have perfectly aligned indentations to locate the cutting edge, so you reassemble the punch and complete the hole. Easy.

Making small holes in thin sheet

If a C-core transformer were to be mounted with windings dropped through a chassis, the lamination edges would contact the (almost certainly conductive) chassis, greatly increasing transformer losses. The solution is to fit an insulating spacer. Although the insulating material can be cut neatly to size with a scalpel or sharp scissors, making clean small holes for the screws to pass through is a problem. This quandary can often be solved by a paper punch. Mark the position of the hole, take the bottom off the punch, tip out the chads, and use the punch upside down. Gripped gently, the cutter holds the material in place, and the marking out can be clearly seen through the exit hole of the punch. The material can be moved precisely into position, whereupon a beautifully clean hole in the correct position can be punched.

Sawing metal

It might be thought that to cut a piece of metal, it is only necessary to take a few wild swings at the work with a hacksaw whilst the room rings to the shriek and shudder of the tortured saw. This is an excellent way to ruin a perfectly good saw blade, deafen yourself, and produce work of an appallingly low standard. Before using a hacksaw, check:

- Is the blade inserted the correct way round? (It should cut on the forward stroke.) See Figure 2.27.
- Does it have a complete set of teeth? If **any** are missing, discard the blade; blades are cheap — your time is not.
- Does it have the correct number of teeth per inch (TPI)? A saw should have three teeth in contact with the work at all times. Alternatively, use 18TPI for aluminium, 24TPI for

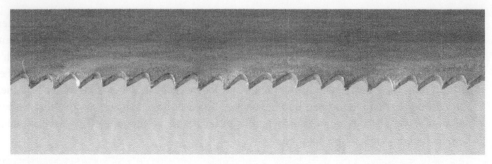

Figure 2.27
Detail of standard hacksaw blade showing cutting direction (blade cuts when moving from right to left).

mild steel, 32TPI for harder materials such as stainless steel, and adjust your sawing angle as necessary to engage three teeth.

- Is the blade properly tensioned? (Generally as tight as possible.)

Junior hacksaws are very useful for smaller, more precise work, and the sprung wire frame type is cheap, common, and nasty. **Proper** junior hacksaws have a rigid frame and a screw to tension the blade. Hacksaws are commonly available with pistol grip handles because this makes it easier to apply maximum force, which is important for a full-size hacksaw, but you will find that the old-fashioned axial handle permits more accurate cutting, making it the best choice for a junior hacksaw. See Figure 2.28.

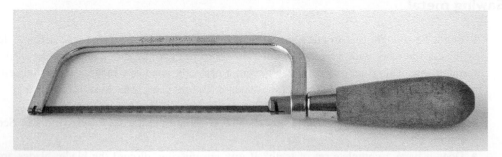

Figure 2.28
A junior hacksaw is ideal for more precise work.

A superior design of pistol grip full-size hacksaw has recently appeared having a number of advantages over the classic design:

- The frame is a one-piece casting, and therefore extremely rigid.
- The combination of screw adjustment plus lever permits far greater blade tension, reducing blade flex.
- There are no tensioning protrusions at the front of the saw, enabling more of the blade's length to be used when sawing in a confined space.
- The shaped rubber grips allow better control.

Against these considerable advantages there are two minor disadvantages:

- The blade cannot be rotated by 90° (this can be useful when sawing holes in a chassis).
- Only one length of blade can be fitted (the author has only ever fitted standard length blades, so doesn't see this as much of a disadvantage).

The author finds that he cuts far more accurately with this new hacksaw, but the traditional frame will be retained for those occasions when the blade needs to be rotated by 90°. See Figure 2.29.

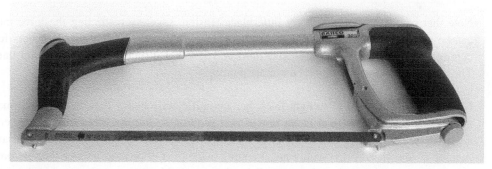

Figure 2.29
This hacksaw's cast frame is very rigid, allowing it to cut accurately.

Sheet metal and saws

Cutting sheet metal is a problem because a hacksaw blade's teeth are insufficiently fine to cut at right angles to the work, so the only way to cut sheet metal is to cut at an extreme angle, and if this means crouching on the floor whilst the work is held vertically in a vice, so be it. Another method is to clamp the work horizontally to the bench, and use a **panel** saw, which looks like a wood saw, but takes a hacksaw blade. See Figure 2.30.

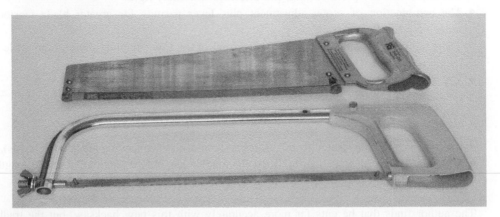

Figure 2.30
Traditional hacksaw (lower) and panel saw (upper).

Note that panel saws provide very little blade tension, severely restricting the force that can be applied while cutting. A drop or two of lubricant whilst sawing does wonders for cutting efficiency but creates a terrible mess and makes it difficult to see the cutting line. The author only saws with lubricant under duress.

Guillotines cut beautifully straight lines, but are sufficiently heavy that they need permanent bench space rather than being moved into position as needed. They are also expensive. The most versatile and economical tool for cutting sheet metal is a power bandsaw running at low speed fitted with a metal-cutting blade (32TPI or more). Once the sole preserve of industrial machine shops, small bandsaws suitable for mounting on benches are now readily available for a surprisingly reasonable price. The very small types (that can be lifted with one hand) need three small wheels to carry the blade and allow a reasonable throat depth, but bigger machines use a pair of large wheels. Although the author used a cheap three-wheel bandsaw for 20 years, a little more expense secured a far superior machine, and an old sewing machine table proved to be the ideal height for supporting this small bandsaw. See Figure 2.31.

Figure 2.31
A small bandsaw is ideal for the metalwork required by electronics.

Bandsaws are notorious for cutting wavy lines, but this is invariably because the machine's blade guides have been aligned incorrectly or the blade is blunt, probably both. None of the guides should deflect the blade — they are simply there to support it. Unplug power, leaving the plug in plain sight, open the covers, then remove the table and all guards so that the blade is fully exposed. Loosen all the guides and withdraw them from the blade, then adjust tilt and tension of the upper wheel so that the blade runs centrally and freely on the wheels when the motor wheel is turned with one finger.

Starting with the guides below the table, bring the rear guide in so that it just kisses the back of the blade but never deflects it — spin the motor wheel to check. Now bring in the two side guides just behind the blade's teeth and ensure that they kiss and may rotate but must not deflect the blade or snag on its butt welded joint or teeth. Spin the motor wheel to check.

Withdraw the upper guide post fully into the machine and repeat the previous procedure with the upper guides, then check that alignment does not change when the post moves between fully withdrawn and fully extended. If it does (and it probably will), you need to adjust the alignment and movement of the post.

Fully withdraw the upper guide post and replace the table. Adjust table tilt so that the table is at right angles to the blade using a square, then rotate the eccentric washer beneath the table to touch it, allowing the setting to be quickly replicated. Replace all the guards and fully extend the upper guide post.

As can be seen, aligning a bandsaw's guides takes time, but the result is a machine that cuts predictably and precisely. The great advantage of a bandsaw over any reciprocating saw is that because the bandsaw cuts continuously (no reverse stroke), it is far less likely to snatch. Unfortunately, a bandsaw can't cut blind holes.

Chain drilling and why to avoid it

Chain drilling refers to marking out the hole to be cut, then drilling a chain of small holes that almost touch inside the marked line, somehow removing the central waste metal, then filing the resulting (very rough) hole to precise size. To minimise subsequent filing, the holes need to be drilled as close to the marked line as possible, but they produce significant burr on the reverse side that may not be removed when the hole is filed to the correct size. Drilling all those holes is tedious, and drilling that number without one overlapping the marked line is tricky. All in all, chain drilling is slow and likely to produce work of a very poor standard. Don't do it.

Sawing blind holes by hand

The best way to cut blind holes is by hand, using a tension file or rod saw in a standard hacksaw frame or coping saw frame. See Figure 2.32.

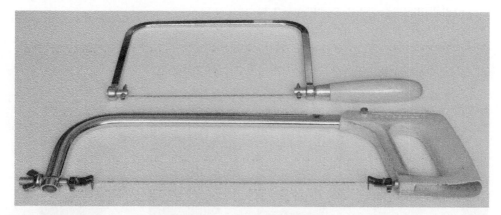

Figure 2.32
Tension file fitted to hacksaw frame (lower) and coping saw frame (upper).

The process starts by drilling a pilot hole in the material that is to be removed to allow the file to be passed through.

The file cuts on the forward stroke, so run the file gently through your fingers to determine the cutting direction. See Figure 2.33.

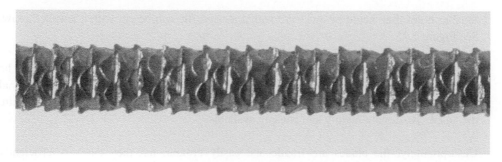

Figure 2.33
Detail of tension file blade showing cutting direction (right to left).

Now fit the file to the handle end of the frame, but pass it through the hole before fitting it to the other end of the frame and tensioning it. Beware that as nobody presently makes tension files, you need to be careful with yours – the most common cause of breakage is kinking caused while applying frame tension with the blade passing through a pilot hole that is too small ($<$6 mm). With care, the hole can be cut so accurately that very little remedial filing is needed. See Figure 2.34.

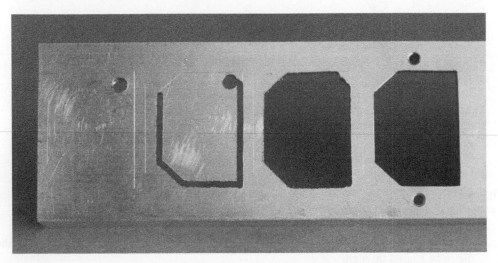

Figure 2.34
The sequence of events for cutting an irregular hole. Left to right: mark out and drill a pilot hole, cut irregular hole with tension file, tidy hole with file, drill and tap fixing holes.

Smaller holes that would be awkward with a tension file can be cut with a jeweller's saw. See Figure 2.35.

The plain blade is clamped between square washers at each end and tension can only be applied by compressing the frame against your tensed chest whilst tightening the final washer. This is fiddly and appears to require at least three hands, but you will soon acquire the knack. Blades are readily and cheaply available.

Unsurprisingly, the blades are fragile and easily broken, but they cut very efficiently because they have virtually no back to cause friction – so don't apply any weight to the

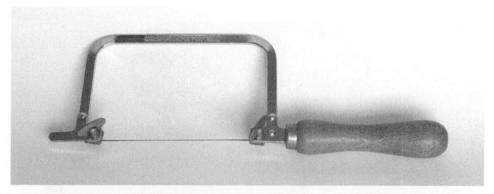

Figure 2.35
A jeweller's saw is ideal for small irregular holes.

saw, just glide it smoothly back and forth and let the cutting edge do the work. However, be warned that the negligible stabilising back makes cutting straight lines harder, so you need to keep a close eye on the cut, and if you are right-handed, you will want to cut holes in a counterclockwise direction in order to see the cutting line. Irritatingly, some connectors require a "D"-shaped hole, and there's nothing for it but to cut the hole by hand, so it's just as well that a jeweller's saw does such a good job. See Figure 2.36.

Sometimes the frame prevents the blade from reaching the proposed hole from any direction.

Single-ended hole-cutting files are also available fitted with a handle. To stop them bending, they must have a larger diameter than tension files, requiring you to remove a great deal more material to cut a given distance, making the work harder. Nevertheless, because a frame is not needed, they can reach any part of the work and produce a reasonably neat result.

Sawing with a power saw

A faster alternative is to use a power jigsaw at low speed with a **metal** cutting blade having the finest possible teeth. This is not as easy to control as a hand saw, and is potentially dangerous. It is also extremely noisy, and ear protection is essential. The drumming of the saw on chippings leaves marks on the work unless a piece of thin cardboard is carefully

Figure 2.36
"D"-shaped holes are a nuisance, but are quite easily cut using a jeweller's saw.

fitted to cover the saw's steel sole, although the author notes that more recent saws come supplied with a replaceable plastic outer sole. See Figure 2.37.

As before, a hole is drilled in the material to be removed and the saw blade passed through. Ensure that the teeth of the blade are **not** touching the edge of the hole, and whilst pressing down firmly, start the saw. When you reach the end of a cut, back the saw off a little before switching off, otherwise the teeth will snatch and the saw will try to jump up from the work, deforming it. A jigsaw is the quickest way to cut a large square hole. Drill four holes large enough to easily clear the blade (probably ≈ 10 mm), then cut in a gentle curve to one end of the marked line, back the saw off, switch off, withdraw the saw, and starting from that end go back up the line to complete the cut. See Figure 2.38.

A less alarming powered possibility is a scroll saw, and these have such a deep throat (typically 16″) that they can reach the centre of any reasonable chassis. Although intended for decorative cutting of thin wood, blades suitable for cutting metal can be fitted. Scroll saws are typically one-tenth as powerful as a jigsaw, so accidents are less likely, but beware

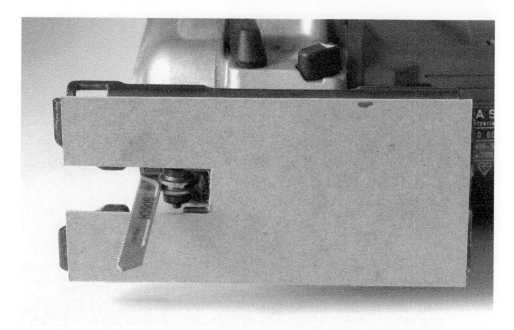

Figure 2.37
Fitting a cardboard sole to a jigsaw reduces surface damage when cutting metal.

that they tolerate only very light pressure on the blade for an accurate cut. Fitting plain blades is astonishingly fiddly and cutting metal is slow.

A technique that can be useful when a rectangular hole with radiused corners is needed is to use a chassis punch at each corner, then use a saw to cut straight lines between the punched holes, leaving a rectangle with neatly radiused corners. Subsequent filing with a flat file easily ruins the radiusing, so use flat files to ≈ 2 mm of the radii, then finish filing with round files.

Making large round holes without a chassis punch

Mark out the hole using a scriber, saw it out as carefully as possible, and remove obvious errors with a half-round file. Take a piece of 160 grade emery cloth and wrap it tightly round a short length of broom handle, and use it to scour a few turns inside the hole as if

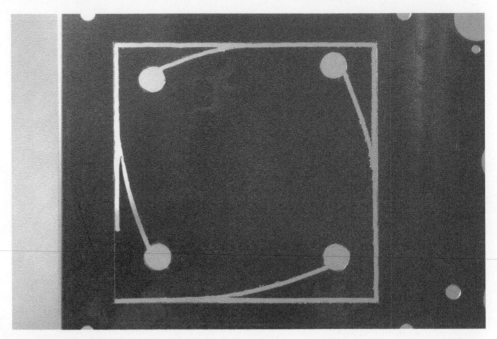

Figure 2.38
Cutting a square hole using a power jigsaw requires a sequence of eight cuts.

you were turning a handle. Rotate the work by 90° and repeat, then twice more, so that the work has been scoured in all four orientations. (The reason for rotating the work is that if all the cutting is done in one orientation, a slightly irregular hole results because you are able to apply slightly more force in one direction than another.) Inspection will reveal significant burring on both sides, requiring careful removal using a needle file or scraper. Change to 240 grade silicon carbide paper, and repeat the scouring process, then deburr for the final time. With only a little effort, it is perfectly possible to produce a hole by hand that appears to have been made by a precision boring machine.

Although the author uses this technique mainly for very large holes, it can be used for smaller holes by substituting a round file. Round files will cut using the scouring action, but they clog quickly, so it is best to move them gently back and forth as you scour.

Making holes in perforated sheet

Perforated sheet is wonderful for cooling, and makes subsequent wiring easy, but it is very difficult to make small holes that are not aligned with the existing holes. For this reason, you might choose to allow the geometry of the sheet to determine mounting positions of valves etc., rather than attempting to enforce your own geometry. Forget about using a twist drill. The only way to make small holes (≈ 4 mm or so) is to mark them out with a scriber and carefully file them using a needle file. If this sounds tedious, it's because it is, although a single-ended Abrafile in a handle speeds work. See Figure 2.39.

Figure 2.39
Single-ended Abrafile.

A cheat that you can use (that does your tools no good at all, but saves time) is to spin a round needle file in a battery drill, and by plunging it gently back and forth, allow this to cut the hole. Used this way, needle files don't last very long, but you may feel that the cost of a new needle file is an acceptable trade against your time.

Fortunately, larger holes can be marked out and sawn using a tension file, jeweller's saw, or scroll saw. Large holes are much easier.

Sheet metal punches still work well on perforated sheet, but it's essential to align the cutting edge so that it is supported equally at all its cutting tips before tightening the bolt and beginning cutting, otherwise it will pull itself out of alignment. Deburring can **only** be done with a needle file, and is tedious.

Sawing perforated sheet to size

Hand saws aren't really suitable because they try to follow the path of least resistance — which is usually straight down the middle of the holes, and not necessarily where you want the cut. A bandsaw is best, and a jigsaw will do, but the essential is a sharp fine blade running quickly and minimal cutting force.

It's also very difficult to file the cut edge because the file tends to be trapped by the holes. However, 160 grade emery cloth supported on a block of wood 3″ by 5″ does a superb job very quickly. Top and bottom burrs can be removed using 240 grade silicon carbide paper (again supported by the block), but the vertical burrs at each of the (many) holes still have to be removed individually with a needle file.

Despite all these caveats about perforated sheet and its associated metalwork, the end results are well worthwhile.

Sawing, filing, and deburring

There are two philosophies to sawing and subsequent filing. The lazy way (which requires more effort in total) saws quickly but makes little effort to guide the saw cut properly, necessitating a saw cut well away from the final line, requiring considerable remedial filing. The more efficient way is to cut slower and more carefully as close to the final line as possible so that minimal remedial filing is needed.

Files come in different shapes (triangular, square, flat, half round, round), sizes (150 mm needle files to 300 mm engineer's files), and cuts (coarse to fine). Files are precision cutting tools and should live in a proper stand akin to a kitchen knife block that prevents them knocking and chipping teeth. One day, the author will make one. Until that happy day, his files live in a dedicated drawer. On the materials used by electronic projects, files last a lifetime — the author still uses files he bought on the way to his first job interview.

The essential file is a large flat coarse file for roughing, supplemented by a smaller flat fine file for more precise work. Half-round files are also useful, as are round, but square and especially triangular files are far less useful. Needle files often come in sets of six and with a very wide range of quality. Two or three well-chosen premium quality files are far more useful than copious poor files — buy files at the same time as drills from the same engineering supplier.

Traditionally, files came naked, so the tang was heated to very dull red heat using a blow-lamp, then hammered into a beech handle to the accompaniment of smoke and noise.

Contemporary files come fitted with ergonomically designed handles that include rubber inserts and allow a better grip than wooden handles. Needle files are still supplied naked, and their shafts are knurled so that they may be used in this way, but beech handles having brass collets specifically made for needle files are available and permit much more accurate filing.

Deburring is often accomplished using a fine file (perhaps a flat needle file), but when the burr is large and uneven, a file can sometimes dig in, spoiling the final edge. 240 grade silicon carbide paper supported on a hard wooden block is less likely to spoil an edge. For really precise work, lay a fresh (and uncreased) full sheet of abrasive paper on a flat surface, angle the work, and gently draw the burr against the tensioned stationary paper, so that the force on the burr tends to pull and snap it away from the supporting edge. The author keeps sheets of 240 and 600 grade silicon carbide paper on the lower table of his pillar drill specifically for this purpose.

As it is used, abrasive paper inevitably collects dust, so periodically shaking it loosens surface dust and maintains efficient cutting, whilst simultaneously creating a dust cloud that coats the surroundings and that you inhale. A much better solution is to periodically vacuum it. Tension the paper on a flat surface and draw the end of the hose over the paper as if you were trying to carefully smooth the end. A trail of cleaned paper will result.

The traditional folded chassis

The traditional construction was the folded aluminium chassis, and classic valve designs included beautiful engineering drawings complete with exact dimensions, folding lines, and all holes clearly marked and dimensioned. The author can only assume that there were many more folding machines available in the early 1960s, and that all constructors had access to a full sheet metal workshop.

With care, it is possible to fold aluminium sheet ≤ 1.6 mm by hand. We need to apply the bending force precisely along a line, so the metal must be firmly gripped in hard jaws, but we don't want to leave vice marks, so we use a pair of clean sections of aluminium angle ($\frac{1}{4}''$ or thicker) and longer than the metal to be folded as jaws. Firmly grip the jaws in a bench vice or Workmate™ all along the proposed fold so that the folding line is just visible along the line of the jaws. Simply applying force at the far end of the sheet would produce a very poor fold — the force must be applied directly at the folding point. The way to do this is to use the heel of the hand via a 19 mm thick block of flat wood, and to progressively apply force all the way along the fold, gradually folding it. See Figure 2.40.

Figure 2.40
Thin aluminium can be folded by hand.

When a right-angled fold is made using this method, its fold tends not to be as sharp as when made by a true folding machine, but the fold can be sharpened by moving a piece of MDF along the fold whilst firmly tapping it with a lump hammer (the MDF prevents the hammer from bruising the metal). It is essential that the metal is gripped as tightly as possible before folding and that it is not disturbed before the fold is sharpened by tapping. It is also essential that the MDF is held down firmly and tapping force is not applied over any edges, otherwise unwanted fold lines will be added. Be warned that duralumin does not fold well — you want a more conventional aluminium alloy. Short folds are much easier than long ones. Practice on some scrap before committing to important work.

Although it is sometimes possible to buy an undrilled folded aluminium chassis, even a small chassis needs to be 1.6 mm thick in order to be able to support the weight of transformers without bending. Although a pre-folded chassis might seem convenient to use, it is always awkward to drill holes in the sides of the chassis because it is difficult to support the metal whilst drilling.

Making a chassis from extruded aluminium channel and sheet

A better alternative to the one-piece folded chassis is to assemble it out of separate pieces. See Figure 2.41.

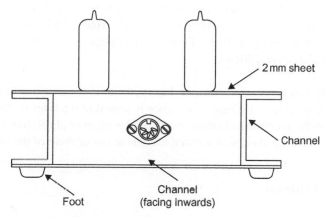

2 mm sheet

Channel

Channel
(facing inwards)

Foot

Figure 2.41
A very rigid chassis can be made from aluminium channel.

The top plate, to which most of the components will be fitted, is made of 1.6 or 2 mm aluminium, which is readily available either as off-cuts from an aluminium stockholder or (more expensively) from one of the electronics factors. Whilst it is tempting to use even thicker metal, many of the holes will be cut using chassis punches that may be damaged by thicker metal. Additionally, most valve sockets were designed for a 1.6 mm chassis, and whilst they can tolerate 2 mm, clearances become problematic if the plate is thicker.

The front, back, and sides are made from extruded aluminium channel cut to length. The sides have the channel facing out, thus providing convenient handles with which to lift the chassis, with the front and back fitting into the remaining space between the sides. The whole construction is fastened with engineering screws and nuts. This form of construction has many advantages over the folded chassis:

- The chassis can be cut to any convenient size using hand tools.
- Cutting holes in the chassis is now easy, because each surface can be properly supported whilst it is being worked upon.

- If modifications are required later (not uncommon), individual parts can be replaced if necessary.
- Aluminium channel tends to be quite thick (⅛″ or ¼″), making it a good heatsink, and threaded holes may be tapped into it, which is often convenient.
- If access is needed at one edge, that piece of channel can be temporarily removed.
- Looking at the bottom of the chassis, the channels are rigid load-bearing members to which feet and the safety cover plate can be easily fixed (which should ideally be perforated, to allow a cooling air flow). Fitting a cover plate to the bottom of a folded chassis is usually rather more difficult.

There are only two minor disadvantages. Firstly, the total top area is a little larger than the folded aluminium chassis, because some space is wasted at the sides by the outward facing channel. Secondly, for electrical safety, each separate piece of aluminium should be reliably earth-bonded to the top plate with serrated washers at one or more of the fixing points.

Cutting extruded channel

It is perfectly possible to cut extruded stock using a hacksaw in a vice. Vice pressure easily deforms open channel, so cut and plane a block of wood to be a tight fit within the channel to prevent deformation when clamped. See Figure 2.42.

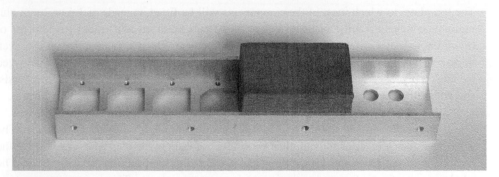

Figure 2.42
A tight-fitting block of wood prevents deformation when clamping.

With practice, you will learn to cut quite accurately whilst expending little effort and making minimum noise. Nevertheless, you will always have to tidy the cut face. Chop saws are primarily used by woodworkers, but blades designed to cut aluminium (not steel) can be fitted, and the advantage is that they leave a perfectly clean cut edge that needs only minimal deburring. They are expensive and unspeakably noisy but save a great deal of time, and are the only easy way to produce perfect mitred joints. Wear goggles that seal fully to the face rather than simple safety glasses because a chop saw cutting aluminium produces a spray of chips that flies everywhere. Consider using foam earplugs as well as earmuffs because the ringing shriek from that blade seems a perfect way of provoking tinnitus.

Corner pillars

A larger chassis is less rigid, yet is invariably required to support more weight. Although a chassis can be constructed by simply screwing channels to the top plate, a huge increase in rigidity can be gained by fitting 16 mm square section corner pillars so that the channels become a self-supporting rigid picture frame onto which the top plate (or plates) is screwed.

The author faces each end of the pillars in his lathe — which allows the pillars to be made a snug fit inside the channel. Assuming you **can** gain access to a lathe, the sequence of events for machining a pillar to length with maximum accuracy and minimum tears is as follows:

- Carefully measure the internal width of the channel at the corners — not the open ends. The reason for this is that channel is rarely square (perhaps as a result of being clamped unsupported in a vice whilst being cut to length). See Figure 2.43.

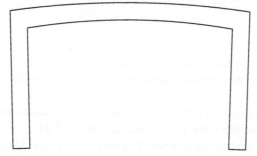

Figure 2.43
Clamping channel for sawing without support deforms it.

- Cut four lengths of 16 mm square bar to be 1–2 mm longer than your measured internal dimension.
- Take a pillar and face **both** ends in the lathe with minimum wastage (centring is not important).
- Remove the pillar from the lathe, and measure its length with callipers.
- Subtract the required length from the measured length to find out by how much it is oversize. If you use digital callipers (easiest), you will probably obtain fractionally different answers depending on where you measure — indicating that the faces are not perfectly square. This doesn't matter, it's still far better than can be achieved by hand.
- Add 0.1 mm to the previous answer. In theory, this means that the finished pillar will fit into the channel with a 0.1 mm gap. In practice, it means that the pillar is guaranteed to fit in the channel, and because the channel is usually slightly distorted, it will probably grip the pillar snugly.
- Put the pillar back in the lathe, engage power, and gently ease the lathe tool up to the face until it **just** begins to cut.
- If you now set the dial on the lathe slide to zero, its graduations can guide the removal of the required amount of material quickly and precisely.

A metalworking lathe was traditionally an expensive piece of precision engineering, restricting their purchase to keen steam enthusiasts, but light (<40 kg) tabletop lathes are now surprisingly cheap, and whilst they're not really good enough for clocks or steam engines, they're adequate for electronics hardware (and unipivot pick-up arms). Beech butcher's tables are frequently available second-hand as people discover that there's more to cooking than looking glamorous and having shiny tools, and these solid tables make excellent bases for small lathes and pillar drills.

Fitting the pillars

The neatest way of securing the pillars to the channels is with a pair of M5 screws from each channel into tapped holes in each pillar. This is done as follows:

- Two of your channels will have pillars inside them. Decide which these channels are, and drill M5 clearance holes at each end. See Figure 2.44.
- Gently ease a pillar into the channel. You may need to spread the channel slightly using finger and thumb to ease it in. Provided it is aligned squarely, the pillar can be gently tapped with a soft hammer until it is perfectly flush with the end of the channel, or the

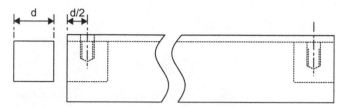

Figure 2.44
Positioning drill holes for pillars inside channel.

same result can be achieved by pressing the channel onto a flat surface. Running the tip of a finger over the join checks correct alignment far better than peering at it.

- Without disturbing alignment, use a small G-clamp to hold the pillar securely in place **without** obscuring the channel's 5 mm holes. G-clamps easily mark the work, but this can be prevented by scraps of cardboard or plastic.
- You can now mark through the 5 mm holes to determine where the pillar should be drilled and tapped. The ideal way of marking through is to make a dedicated transfer punch out of a piece of scrap steel having a 5 mm diameter stub faced by a poorly aligned lathe tool so that a slight dimple is left at the centre. The punch can then be dropped through each hole in the channel and tapped lightly with a hammer to leave a precisely positioned dimple ready for drilling. See Figure 2.45.
- It is **vital** to **scribe** an identifying number onto each pillar and its corresponding position so that you know its location and orientation (methylated spirits and/or finger grease removes pen or pencil).
- Remove the pillar, use a small centre drill to start each hole, then drill a 4.3 mm (M5 tapping size) hole to a depth of 10−12 mm, and tap it M5.
- Slide the pillars into their correct positions in the channels, and secure them with M5 screws. Their alignment will be near-perfect, and a little nudging will achieve perfection.

You now have a pair of channels with pillars secured at each end, and want to fit these to your other two channels to form the frame. It is very easy to make a mistake at this point, so be careful:

- Looking down onto the finished frame, the M5 clearance holes must be offset from the end of the channel by the half the pillar cross-section, plus the channel thickness. See Figure 2.46.

Figure 2.45
A custom-made 5 mm transfer punch allows perfect marking out and drilling.

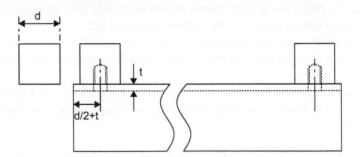

Figure 2.46
The pillars are offset by the thickness of the adjacent channel so the 5 mm holes must also be offset.

- Having drilled the M5 clearance holes, lay out the entire frame on a clean flat surface, and use small G-clamps to secure it at the corners without obscuring the M5 holes. Check alignment carefully and adjust as necessary.
- Mark through the M5 holes to the pillars, ideally using the dedicated M5 transfer punch.
- Remove the clamps and unscrew the pillars from the first pair of channels.
- Drill and tap the pillars.

You can now proudly assemble a perfectly aligned picture frame.

Making the picture frame with mitred joints

You will have realised that if all channel sections faced inwards, there would be no wastage of valuable chassis space and that this could be achieved by mitred joints. See Figure 2.47.

Although judicious addition of sawdust and wood glue disguises an imperfect joint in wood, nothing disguises poor joints in metal. Worse, mitres invariably try to pull themselves out of alignment as they are tightened. Save humanity from profanity and don't attempt mitred joints in metal. If you insist on ignoring this advice, buy a top quality chop saw, fit a blade specified for aluminium, spend time aligning it perfectly, and plane a block of wood to fit snugly inside the channel so that when the saw's clamp is tightened, the channel isn't distorted. Having cut the mitres, slip the corner block or angle inside, and align and clamp the joint so that holes can be marked using a transfer punch. See Figure 2.48.

If you are a real glutton for punishment, you will countersink your fixing screws, probably fractionally displacing the joint. See Figure 2.49.

A fiddly butt joint (but easier than a mitre)

Having turned the air blue making a mitred picture frame, you will be amenable to any suggestion that involves right angles. One channel to be joined is simply cut off square, but the other piece has two squares removed, requiring a little careful filing. See Figure 2.50.

The joint can be secured using ½″ square stock, but 1¼″ × 1¼″ by ¼″ thick right angle extrusion is better because it assists channel alignment and, if necessary, one edge can be cut down to allow a little more room for subsequent parts to be fitted within one channel. See Figure 2.51.

Figure 2.47
Mitred corner joints avoid wasting internal chassis space.

Figure 2.48
Align the sections on their corner block or angle and use a transfer punch to mark the tapped holes.

Figure 2.49
The completed joint.

Figure 2.50
This alternative corner joint avoids mitres but retains internal chassis space.

Figure 2.51
The corner angle can be foreshortened on one edge if necessary to allow more room for connectors etc. on that panel.

Finally, the joint can be assembled and you will find that it **can** be manoeuvred to perfection. See Figure 2.52.

Figure 2.52
The completed joint.

Fitting the top plate

The top plate inevitably needs lots of holes for all the valves and subsidiary components. It also needs to be secured to the picture frame. The easiest way to do this is to drill securing holes in the top plate, then clamp it to the picture frame with a couple of G-clamps and punch through to mark where holes need to be drilled in the channel. Rather than use nuts and screws, the author prefers to tap the channel and use 6 mm M5 screws because once they have passed through a washer and the top plate, they barely protrude from the far face, but you might have a different opinion.

Fitting carpet-piercing spikes

Corner pillars provide an ideal depth of material to be axially tapped M6 and fitted with loudspeaker carpet-piercing spikes. The spikes might not confer any sonic advantage, but they certainly allow a heavy amplifier to sit on carpet without leaving a mark. The obvious way of drilling the axial hole in the pillars is in the lathe during facing, but it is fiddly to centre square stock perfectly (even in a four-jaw chuck), and subsequently drilling the channel so that it aligns correctly with the (blind) hole becomes even more difficult. The author has found it easier to face the pillars without worrying about centring, then offer the assembled frame up to the pillar drill and drill the M6 tapping holes (5.1 mm) straight through the channels and into the pillars. Because two channels at right angles rest flat on the drill table, this ensures that the drilled hole is square, if not perfectly central.

You could tap M6 straight through the channel and into the pillars, but you might prefer to remove the pillars for tapping, and drill the channel M6 clearance. This reduces the amount of thread supporting the finished amplifier, but because the clearance hole guides the spike into the thread, it makes it much easier when fitting a spike blindly from underneath.

If you don't have access to a lathe

If you can't gain access to a lathe, corner brackets made from 1¼″ × 1¼″ extruded aluminium right angle ¼″ thick work just as well, although they reduce the room available for panel hardware such as switches and sockets, but they can't take spikes. Instead, the channel can be drilled and tapped to take carpet-piercing spikes, but be aware that ⅛″ channel has a much shorter depth of thread than a pillar, so it is not as strong.

If you have very limited tools/patience

Mechanical prototyping systems employing extruded aluminium and custom corner blocks are common for small-scale industrial projects. Further, there is usually an extensive range of compatible accessories enabling you to build everything from a walking robot to a tool bench. Such systems aren't cheap, but they save time.

Finishing

Aluminium can be spray painted, but paint tends not to stick very well to aluminium, and tends to chip off unless etching primer is used. Buying aerosol cans of car primer and top coat is quite expensive, and the fumes are most unpleasant. Nevertheless, this is one way of finishing the chassis.

A better method, but one that requires rather more planning, is to have the chassis anodised by a professional anodiser. Note that only aluminium can be handed to an anodiser; no foreign substances whatsoever are allowed. Surprisingly, this is actually quite cheap, because the pieces that you will hand over will be very small compared to the main batch that is being anodised. It may mean that you need to wait until a batch of your chosen colour is to be anodised, but the finished result will be far superior, provided that you have prepared the work properly.

An awkward problem with anodising is that it slightly increases the size of your work, and reduces the size of holes. If you have done a really nice piece of metalwork that snugs together beautifully, it is somewhat disconcerting to discover that it no longer fits. Commercial manufacturers use their prototypes to discover the amount by which their work should be undersize before anodising.

Both painting and anodising show up every imperfection of the underlying surface, so the surface cannot be too well prepared. A "brushed" finish can be obtained using reducing grades of silicon carbide (colloquially known as "wet and dry") rubbed along one direction only. If the final stage is lubricated with soap and water, a very smooth finish can be obtained. Alternatively, soap-filled wire wool scouring pads soaked in hot water can be very effective.

It is far better to begin with a good surface, so most stockholders keep aluminium that has one face protected with a plastic film. Keep the film on for as long as possible, and do not allow objects to fall on the sheet; aluminium is soft and easily marked. If using a scriber, keep your construction lines to a minimum, and score lightly with a **sharp** scriber – the gouges made by a blunt scriber are much more difficult to remove.

Whether you painted, anodised, or simply brushed your chassis, make sure that you really have drilled **all** the holes you need. Drilling holes afterwards is invariably messy, and easily spoils your finish.

It's common for transformers to have scruffy laminations. Varnish globules from haphazard impregnation can be removed mechanically or with chemical paint stripper. Matt black blackboard paint is ideal for brushing onto laminations but be warned that older transformers were often impregnated with beeswax rather than varnish and seeping wax eventually lifts the paint. Satin black car spray paint is best for transformer cans.

Recommended further reading

B.E. Jones (Ed.), *The New Practical Metalworker*. Waverley (1938).
This comprehensive pre-WWII book comprises three volumes, and covers everything from hand tools to the use of walnut shells for polishing, but make sure it is complete.
A.G. Robson, *Engineering Workshop Principles and Practice*, 6th edn (1942).
A slimmer book primarily concerned with machine tools, but still useful.
Unsurprisingly, the previous books are pre-CNC, so they explain hand techniques in copious detail, supported by beautiful line drawings and quaint photographs. The value of old mechanical engineering books is that model engineering fairs are a superb source of second-hand quality metalworking tools at knock-down prices.

CHAPTER 3

WIRING

The purpose of wiring is to translate a theoretical circuit diagram into a practical, working circuit. Unfortunately, there are many pitfalls, and careless implementation can ruin a good design. **Fortunately**, most of the pitfalls can be predicted easily, and therefore avoided.

Tools

The cheapest tools are the most expensive ones. Cheap tools make it difficult to do a good job, so they waste time, may damage components, and they need periodic replacement. Conversely, good tools mean that the quality of the job is limited purely by your own skill, they're nicer to use, and they last a lifetime. Your aim when working on anything should be to leave no trace whatsoever of your intervention — your repair/modification/construction should look as though it left a quality factory in that condition.

Soldering irons

Obviously you need a soldering iron, but what is the difference between them, and why are some so cheap while others are so expensive?

The job of the iron is to heat the parts to be soldered to a temperature such that once the solder is applied, it melts quickly and flows to form a perfect joint. Almost anything will

Building Valve Amplifiers. DOI: http://dx.doi.org/10.1016/B978-0-08-096638-0.00003-5

do this, but the component may not work afterwards. Two thermal properties characterise the iron: thermal mass and temperature. Thermal mass is simply the mass of the hot part of the iron, and the greater this is, the more difficult it is for the proposed joint to cool it.

A cheap iron determines its temperature by only generating sufficient heat to match its losses to the environment, whilst keeping the tip hot enough to melt solder. For heat to flow into the joint, the iron must be hotter, but the moment it touches the joint, heat flows until the two temperatures equalise. The joint's temperature has to be raised by $\approx 180°$ to melt solder, so the iron must be significantly hotter. The excess temperature can be minimised by maximising the iron's thermal mass, but that makes the tip clumsy to use. The upshot of the two arguments is that such an iron initially runs **too** hot, burns the flux in the solder, and may well damage the components. It will almost certainly cause printed circuit board tracks to lift if used for desoldering.

A better iron is temperature controlled, and has an oversized heat source (≥ 45 W as opposed to $12-25$ W). The iron is at the correct temperature all the time, but when the joint cools the tip, the control trips and the oversized heat source quickly restores correct temperature, making the iron appear to have a much higher thermal mass. Additionally, temperature-controlled irons rarely operate directly from the mains, avoiding the thin wire and consequent fragility of a mains voltage heating element. Because the tip doesn't over-heat, it is less likely to damage fragile components like polystyrene capacitors, and it lasts longer whilst being easier to keep clean.

The most common heat source is a direct electrical heating element, although radio frequency eddy current heating directly into the tip is also used. Irons inevitably leak unwanted current from the element to the work via their tip, and reputable manufacturers specify this leakage current. The electrical insulation between element and tip must be thin in order to transfer the heat efficiently, but because it is thin and hot, this insulator cannot be perfect (insulators become more leaky as temperature rises). Ohm's law dictates that the leakage current is determined by the electrical resistance of the insulation and the voltage across it, so low voltage irons intrinsically have lower leakage than mains irons. Leakage can kill digital semiconductors and degrade analogue noise performance, so it's just not worth the risk — choose a low leakage iron and assume that leakage is high if unspecified.

The additional complexity of temperature control and a dedicated power supply means that a low voltage iron designed for electronics work is invariably more expensive than the cheap mains iron made for soldering tin plate. However, the better iron pays for itself in time, because elements and bits last far longer, and it is less likely to damage a printed circuit board or semiconductor. And in all that time, it is nicer to use.

Earth bus-bars need a large cross-sectional area to minimise their electrical resistance, but this inevitably makes them difficult to solder because they conduct heat away so efficiently that it can take time for the iron to raise a part of the bar to soldering temperature, by which time the nearby polystyrene capacitor has already been destroyed. The solution is to use a larger iron that provides faster local heating – the author uses a 200 W temperature-controlled mains iron. See Figure 3.1.

Figure 3.1
A 200 W iron makes short work of heavy soldering.

Tips

The part of the iron that contacts the work is the **tip**, or in more old-fashioned parlance, the **bit**. Old-fashioned irons had solid copper bits whose working surface would gradually be dissolved by the solder to become concave, and would then need to be filed flat. Filing was the accepted way of cleaning such irons.

Modern irons use iron-coated tips to protect the copper, and should **never** be filed.

The normal method of keeping the tip clean is to wipe it on a moistened sponge or brass wool (both supplied for the purpose by most soldering iron manufacturers). It is most important to keep a sponge moist, so most engineers keep a water bottle nearby. If the tip becomes sufficiently contaminated by old, dusty solder that a quick wipe on the moistened sponge cannot clean it, then a wipe on one of the proprietary tip cleaners should do the job. If that fails, then careful scraping with a knife or brass wool will cure the problem. Do not be tempted to try silicon carbide or glass paper as the heat melts the glue to make the tip even dirtier.

Tips come in many different shapes, sizes and temperatures, and when a brand new iron-coated tip is fitted to an iron, it must be tinned the instant that it can melt solder, otherwise it never wets properly afterwards. The best tip is conical, with an oblique cut across the end to produce an elliptical soldering surface. These tips are usually specified by the width across the minor axis of the ellipse, and a good tip width for valve electronics is 2.4 mm. A wider tip allows you to get more heat into the work, and is better for heavier jobs but is clumsy, whereas a fine 1.2 mm tip is excellent for pick-up arm wires or ≥ 1206 package surface mount components, but is unable to heat larger jobs. Ideally, you need a wide range of tips, and should be prepared to change tips with each soldered joint if necessary. See Figure 3.2.

Figure 3.2
Selection of useful tips.

Irons that use a magnetic thermostat, such as the Weller Magnastat, need different tips for different temperatures. Their tips rely on the Curie effect whereby a magnet temporarily loses its magnetism at its Curie temperature, and this releases a spring-loaded ferrous shaft coupled to a micro-switch in the handle. Originally American, Weller tip temperatures are numbered in hundreds of degrees Fahrenheit, and for most work, a No. 6 (600°F/315°C) tip is ideal, but when working inside old amplifiers, a No. 7 (700°F/370°C) is better at burning away the dirt that otherwise insulates ancient solder. Very occasionally, these irons overheat, and the cause is a dud tip. In the event of this happening to you, try changing the tip before diving in and replacing the switch (much harder). Tips are crucial and the Weller Magnastat was the industry standard for decades, partly because of its excellent range of tip shapes.

Be aware that the newer Weller Magnastat TCP-S iron uses LT series tips plus a Magnastat tip adapter to provide temperature control, making the composite tip dimensionally and functionally the same as a PT series tip except for the LT series tip's smaller cone diameter. This German design change renders the Weller TCP-S slightly inferior compared to the original TCP because:

- Fewer LT series tip shapes are available than PT series.
- The junction between the tip and adapter inevitably adds thermal resistance compared to the single-piece PT series tip, making the composite tip fractionally less thermally efficient.
- Your stock of PT series tips can't be fitted to a TCP-S iron unless the removable barrel that allows tip changes is either replaced with Weller part number 0051031199 (barrel for the old 50 W TCP), or drilled out from 4.9 to 6.1 mm. Obviously, drilling invalidates any guarantee and prevents subsequent use of LT series tips, so buying and fitting the older barrel design is best.

Alternatively, irons can sense tip temperature using a thermocouple and control their heater electronically, allowing quick adjustment during use, but this facility is rarely needed and such irons are surprisingly expensive considering their component cost. The most important question if you spot a "bargain" iron is the price and future availability of tips — if this seems at all questionable, buy a lifetime stock of tips at the time.

Portable irons

Portable gas irons use a catalytic converter to flamelessly burn butane gas at their tip. See Figure 3.3.

Figure 3.3
Although crude, portable gas irons can be useful.

Temperature is crudely set by adjusting the gas flow. Because of their haphazard temperature control, gas irons have no place in quality work, yet there are times when they are invaluable. Very occasionally, a single joint needs to be made (or broken) but a mains socket isn't nearby, so rather than finding an extension lead and electric iron, a gas iron can be quite handy. The other use is when repairing the electric iron.

Very occasionally, you may see portable electric irons powered by an internal NiMH rechargeable battery, but these batteries self-discharge quite quickly, increasing the chance that the iron won't work when you need it. Conversely, the author remembers sitting in a car and noticing that the leather tool wallet resting against his thigh was becoming warm — the iron's switch had been knocked when too many tools were squeezed in.

Summarising, portable irons are horrible, but gas irons are far and away the least horrible.

Solder and flux

The traditional electronic solder was 60/40 flux-cored solder. The 60/40 refers to the ratio of tin to lead, and the flux is a chemical, that when heated by the iron, chemically cleans (corrodes) the surfaces to be soldered to allow a good joint. The ratio of 60/40 is chosen because it is the eutectic point at which all of the tin is dissolved in the lead and vice versa. Further happy qualities are that it has the lowest melting temperature of any combination of tin and lead, and that the lead prevents the outgrowth of tin whiskers that might ultimately contact other parts of the circuit, causing failure.

A great deal of contemporary consumer electronics has a design life of 18 months, after which the consumer is goaded into buying a replacement, thus keeping the manufacturer in business. This business model ensures a steady flow of waste electronics, potentially to landfill sites, where rain would leach lead from the solder into the water table, and thence into drinking water. The brain is vulnerable to damage by lead during early development, so the European Union banned the sale of electronic equipment containing leaded solder and others have followed their lead.

Unfortunately, lead-free solder has a higher melting point, and the increased soldering temperature oxidises the work more, requiring a more vigorous flux to enable good joints, which raises two further issues:

- If the flux is more vigorous and cleans the work better, it must be more corrosive to the iron's tip. Irons designed for lead-free soldering recover acceptable tip life by having thicker iron plating and/or by backing off tip temperature when the iron is replaced in its rest. The thicker iron plating increases thermal resistance, so these irons tend to be 80 W rather than 50 W, and the need to back off temperature makes control electronics a necessity, increasing cost.
- If the flux is more vigorous and cleans the work better, its fumes must be even less healthy to inhale, so fume extraction irons include a pipe positioned close to the tip for connection to a fume extraction and filtering system. The fume extraction pipe and associated plumbing makes the iron somewhat clumsy, so they're not popular, and one alternative is a conventional iron plus extraction hood over the working position.

Industry wisdom is that making a good joint using lead-free solder requires complete elimination of lead, so components have to be guaranteed lead-free and the soldering iron's tip must never have been contaminated by contact with leaded solder.

Whilst lead elimination is possible during manufacture of new equipment, it cannot be guaranteed during maintenance of old equipment, so 60/40 solder is currently available for maintenance, but many service engineers have hoarded lifetime stocks against the day it disappears from sale. New old stock (NOS) components such as valve bases or Soviet-era polytetrafluoroethylene (PTFE) capacitors can be guaranteed to have their pins tinned with leaded solder, so they must be soldered using 60/40.

Other solders are available, some of which contain sufficiently powerful fluxes to cut through surface aluminium oxide and enable soldering to aluminium. **Never** attempt to use aluminium solder for electronics work, and just like lead-free, once a soldering iron tip has been contaminated by aluminium solder, it should never be used for anything else.

Manufacturers consume solder on a production line at a sufficient rate that their solder is always clean and fresh, but a 500 g or 1 lb reel could easily last a decade elsewhere, during which time it can become dirty. The soldering difference between dirty and clean solder has to be seen to be believed, so keep your solder clean by storing it in an airtight jar when you're not using it. Drawing solder through a clean cloth moistened with isopropyl alcohol before use removes surface contamination, and significantly improves the quality of subsequent soldered joints.

Solder is expensive in small quantities, so buy it in 500 g or 1 lb reels. In this quantity, there are various different sorts of solder in various diameters; 0.7 mm flux-cored solder having $\geq 2\%$ silver content is excellent for most uses. Surface mount components require silver-loaded solder to prevent the silver in their contact plating from leaching out – but the solder needs to be <0.5 mm diameter to allow the iron's fine tip to melt a controllable quantity quickly enough to produce a good joint. Similarly, the ceramic tag strips found in old Tektronix oscilloscopes must be soldered using silver-loaded solder to prevent the plated foundation pulling away from the ceramic. Silver-loaded solder produces far better soldered joints than conventional solder and you will be seduced by its superior properties.

Your lead choice hinges on your manufacturing status. If you are a manufacturer you have no choice: your components must be lead-free and you must use lead-free solder. On the other hand, if you are a hobbyist or service engineer then unless you can guarantee the lead-free status of the equipment's original components **and** your replacements you need 60/40.

Soldering and wetting

Assuming that you have an iron at the correct temperature, with a correctly sized clean tip, and clean solder of an appropriate type, how do you make a perfect soldered joint?

So far as soldering is concerned, cleanliness really is next to godliness. The surfaces to be soldered must be clean, and the iron's tip must be clean and shiny without dross. Soldering requires intimate combination at the surface of the metal, and dirt hinders this process. Don't dab at the solder, and certainly don't use the iron to carry solder to the joint. The solder should be applied to the point of contact between the work and the iron such that it melts and flows immediately, and the iron's tip should be wiped clean before and after each joint. See Figure 3.4.

Figure 3.4
The tip should be wiped on a moistened sponge before and after every joint.

Clean surfaces will **wet** perfectly, and surface tension causes the solder to flow instantly across the heated work to form a perfect joint. Dirt causes the solder to form visible globules on the surface that do not wet the joint, and defective joints are therefore known as **dry** joints. There are many variations between these two extremes, but it can generally be said that good joints are made quickly, whereas dry joints are more likely if the iron needed to be in contact with the joint for more than a second or two.

After cleanliness, the other essential for perfect soldering is good heat transfer. Solder does not conduct heat particularly well, so firm contact with the brass or copper to be soldered is far more effective than vaguely dabbing and hoping that a dirty dollop of molten solder will magically make the joint if you hold it there for long enough. Start making the joint by firmly scraping the flat of the iron's tip onto the part with the larger thermal mass and better conductivity (PCB track, solder tag, etc.), maintaining firm contact, tilt the iron up a little, push a little solder firmly into the gap at the back of the tip − it should flow instantly across the joint, then lightly scrape the iron's tip to the component or wire and withdraw. The whole process should take less than two seconds, and the scraping technique is particularly effective with surface mount devices. Practice makes perfect.

The best joints are **mechanical** joints. If the parts to be soldered are already unable to move relative to one another, then they cannot move as the solder solidifies, and a perfect joint should result. Movement whilst the solder is cooling/solidifying causes dull dry joints − although be warned that **all** lead-free joints look like this. Don't blow on the joint to cool it − the occasional burn is the price we pay for good joints.

The best joint is the **first** joint. Joints are degraded by reheating because it encourages further oxidisation of the heated materials. If you are forced to resolder an old joint, remove the old solder and replace it with new. The fresh flux ensures clean surfaces, and the fresh solder will not be contaminated with dross.

Shrink-back

When you solder insulated wires, you are likely to encounter a phenomenon known as **shrink-back**. When you strip the wire, you make a circumferential cut into the insulation, but to avoid nicking the conductor, the cut is not made sufficiently deep to cut all of the insulation. The uncut bridge of insulation breaks as the unwanted insulation is pulled away, but inevitably stretches the remaining insulation along the wire. When the conductor is heated for tinning, the stresses in the insulation relieve, and the insulation shrinks back, leaving more conductor exposed than expected. There are various ways of managing shrink-back:

- Minimise the stress that causes mechanical shrink-back. Cut the insulation using a sharp scalpel. To avoid nicking the conductor, the blade has to be rolled around the wire without any sliding motion. The wire can then be bent very slightly at the cut to break any remaining insulation, rotated half a turn and bent slightly until the insulation is completely broken. The act of bending to break the insulation stretches one side of the insulation but compresses the other, averaging the stretch to zero, and minimising mechanical shrink-back.

- Minimise mechanical shrink-back before soldering. Because mechanical shrink-back is caused by cutting, stripping and stretching insulation, if the conductor is subsequently gripped with pliers, the insulation can be pushed back with fingers, perhaps compressing it and pre-compensating for thermal shrink-back, thereby eliminating shrink-back.
- We accept that there will be shrink-back, and pre-tin the conductor so that shrink-back occurs, **then** solder the wire in place. This method precludes mechanical joints using stranded wire, but it is satisfactory with solid core wire, although not as effective as the previous method.
- The degree of thermal shrink-back is proportional to the time the iron is in contact with the work and its temperature, and this is why it is so important to make fast joints using the correct size tip at the correct temperature. A tightly wrapped mechanical joint helps surface tension to make the solder flow quickly, reducing time and thermal shrink-back. Some older insulation is notorious for thermal shrink-back because the plastic melts and surface tension tries to minimise its area, resulting in a congealed burnt sphere at the end of the insulation. Test solder the wire first, and if unacceptable thermal shrink-back occurs, junk the wire and find something better — it will be cheaper in the long run. If you're forced to work with existing wiring suffering from excessive thermal shrink-back, plan carefully before each joint and use all the previous techniques to manage shrink-back.
- Deliberately choose wire that doesn't suffer from shrink-back. Silver threaded down PTFE tubing doesn't suffer because the coefficient of friction of PTFE on silver is so low that the moment the insulation is broken, it springs back to its original length, and PTFE is impervious to soldering temperatures. Again, use a sharp scalpel to make the cut, but instead of rolling the scalpel around, simply press it firmly through the PTFE onto the silver, then twist the surplus insulation to make the cut.

Solderability

Be warned that as parts age, they corrode and become more difficult to solder; printed circuit board manufacturers refer to this as **solderability** and warn that boards more than six months old may become difficult to solder unless stored under airtight conditions.

Wire deteriorates as it ages, particularly stranded wire in PVC insulation (stranding increases surface area and PVC releases plasticisers). If, when you strip it, the wire isn't bright and shiny, don't even bother trying to solder it — the solder won't wet properly and instead of a good soldered joint with the minimum of solder, you will produce a poor joint, perhaps with hanging solder globules, almost certainly with shrink-back, and possibly with molten insulation drawn between the exposed strands by capillary action. It's really not worth trying to persevere; bin the stuff and buy wire known to be recently manufactured. The problem is most likely to occur where a good variety of wire is needed yet consumption is low, so long-established research laboratories and transformer manufacturers are most at risk.

Solder tags

Solder tags are tin-plated brass tags or lugs used to make electrical connections to conducting parts that either can't be soldered (aluminium or plated plastic), or have too much thermal mass to be soldered directly (copper or steel chassis). Solder tags inevitably live in open drawers for decades before being used, so you may occasionally find a solder tag that flatly refuses to solder. A great deal of time can be wasted abrading the surface back to the brass in a futile attempt to persuade the little tag to solder. If you are unlucky enough to find a tag that won't immediately solder, discard it, and pick another, ideally from a different batch. If you know all the tags came from one batch, test a few more, and if they won't solder, ditch and replace the lot — it will save tears.

A solder tag inevitably adds a mechanical contact in series with the soldered joint, so it is always the second-best choice, and should be avoided whenever possible — the preferred industrial method of making a chassis earth bond is to spot-weld a spade connector. (Admittedly, connection to the corresponding spade connector comprises two mechanical joints, one to the spade and one to its wire, but both joints are engineered and between fully specified materials.)

Binding posts

Unlike traditional binding posts, the modern posts often used on amplifiers for the loud-speaker connection usually come with solder tags, but could and should be soldered directly. Part of the thread needs to be removed to form an exposed brass pillar suitable for soldering, which can be quickly done by a file, and although a neater job can be done in a lathe, setting up is time-consuming. One complete turn of the wire should be tightly wrapped around the exposed pillar, and the iron applied firmly to the pillar. Once the pillar is hot enough to easily melt solder, the joint can be finished by scraping the iron until it touches the wire directly, and a perfect joint will result. The thermal mass of these binding posts can be greatly reduced whilst soldering by unscrewing the knurled part that grips the loudspeaker cable so that it wobbles freely on its thread, and thus does not have to be heated by the iron. The small binding posts incorporating a 4 mm socket typically have a 4BA or M3.5 thread and can be soldered by a 50 W iron fitted with a 5 mm wide tip, but the larger 32 A posts having a 0BA or M6 thread (and therefore almost four times the cross-sectional area) need a 200 W iron.

Flux, defluxing and surface leakage currents

Now that commercial soldering must be lead-free, making a good soldered joint is much harder than it used to be, so flux is available in dispensers akin to a large felt-tip pen for cleaning questionable pads prior to hand-soldering. Like the powerful flux needed for lead-free plumbing, some of these fluxes are perfectly capable of cleaning parts even without added heat, making them useful when an NOS part looks as if it might be difficult to solder.

Unlike flux-cored solder, these loose fluxes continue to clean/corrode the terminal and its soldered joint, and must be washed away post-soldering to avoid subsequent joint failure. A very few loose fluxes specifically state that post-soldering cleaning is not necessary, but it is safest to assume that they all require washing/removal. Washing a loose PCB on a production line isn't a problem, but washing transformer terminals buried deep in a chassis is much harder. However, if the example transformer's terminals are scrubbed with the vigorous flux outside the chassis, and the residue carefully washed away, problems are less likely to occur.

Solder for electronics use is flux-cored and as the solder is heated, the flux cores heat and would boil were they not constrained by the enveloping solder. When the solder melts, the constraint on the super-heated flux suddenly releases, causing a small explosion that sprays flux containing microscopic solder droplets over the surrounding area. Thick film resistors are composed of an insulating binder containing conductive particles, so the end result is the same; flux residue from flux-cored solder is slightly conductive.

Flux residue causes surface leakage currents, so most valve bases have raised ridges between their pins to prevent liquid flux from the joint being soldered meeting an adjacent pin's flux residue and forming a continuous leakage path. Despite this design precaution, flux is best removed after soldering to minimise leakage paths, and it is essential in high-impedance circuitry such as condenser microphones.

Defluxer generally comes in a spray can having a short nozzle terminating in a stiff brush. Having wetted the part to be cleaned, the brush allows the flux to be firmly scrubbed away. Flux is tough, so any chemical that can dissolve it has to be powerful — defluxer cheerfully dissolves paint. On the plus side, it's the chemical cleaner of last resort and can restore the most fearful messes to as-new appearance.

The flat, clean surface of a new PCB encourages wetting and the flow of liquid flux, making PCBs particularly vulnerable to surface leakage currents, so they must be defluxed thoroughly after soldering. This means using plenty of defluxer and scrubbing vigorously (particularly in the narrow gaps between the pins of active devices), then instantly diluting and dispersing the dissolved flux with isopropyl alcohol before immediately drying with a pre-heated hairdryer. Process speed is essential to eliminate all traces of flux. The entire process will probably need to be repeated before the board is clean. Inevitably, with so much liquid sloshing around, some will find its way onto the component side of the board, and this will also need to be washed away with isopropyl alcohol then dried with the pre-heated hairdryer. Protect the work surface with kitchen roll or elephant's loo roll to soak up all the loose liquid. Beware that axial-leaded polystyrene capacitors can be damaged by isopropyl alcohol, let alone defluxer — fit them after the main defluxing.

The reason for immediately evaporating the isopropyl alcohol with a hairdryer is that it otherwise absorbs water from the air to leave tide marks on the board. Don't be tempted to use a hot-air gun instead of a hairdryer — the hairdryer's high flow rate but low temperature is specifically designed to evaporate liquids efficiently without causing heat damage.

When the board is truly clean and uncontaminated there will be no tide marks on either side and your soldering will suddenly look much better. Sadly, because isopropyl alcohol absorbs water so readily, it is necessary to use aerosol rather than bottled isopropyl alcohol to ensure a clean board because this more expensive sealed packaging prevents absorption of water from the air.

Once again, cleanliness really is next to godliness, so once a board **has** been cleaned, it should only be handled by the edges. And then only with freshly washed dry hands.

Desoldering

Sometimes you will need to desolder a joint. If the joint is mechanical, it is best to cut as much of the wire away as possible, otherwise the prolonged heat whilst jiggling and pulling wires damages the wire and tag or board. The remaining joint has fragments of wire in solder and these must be removed.

Solder can be removed by one of two methods:

- Solder can be drawn away by a vacuum. In industry, the expense of pump-driven vacuum desoldering stations is justifiable, but they need regular maintenance and careful use. A far cheaper alternative is the handheld, spring-loaded, solder sucker. See Figure 3.5.

Figure 3.5
Handheld solder sucker.

- Desolder **wick** uses surface tension to wick the solder into copper braid that is discarded once loaded with solder. Solder wick must be kept clean if it is to work, so store it in an airtight container. This method causes the least damage to the work, but it is wasteful and expensive.

Solder suckers look like an oversized pen, and have a PTFE tip that is placed directly in contact with the molten solder, whereupon the trigger is depressed and the sucker removes the solder — hopefully. Even handheld solder suckers need to be looked after if they are to work. The sucker's internals need to be cleaned periodically or they jam, and even a temperature-controlled iron eventually damages the PTFE tip, necessitating replacement because it can no longer seal well enough against the solder to draw it into the sucker.

Solder suckers need to be held steady. Support the entire back of the solder sucker with the tips of all four fingers, and use your little finger to guide the nozzle. A controlled squeeze of the thumb on the sucker's trigger will fire the plunger without jerking the tip.

However, the real problem with solder suckers is their recoil. Pressing the trigger releases a spring-loaded piston that accelerates away from the work to create a vacuum, and atmospheric pressure pushes air (and molten solder) into the void. Piston recoil drives the PTFE tip into the work with sufficient force to break ceramic stand-offs or kick tracks off old printed circuit boards. For this reason, solder suckers should be used cautiously, and you will want to revert to braid on delicate work. Remember, it is the **board** (whether printed circuit or tag strip) that is expensive, not its components. Individual components can always be replaced, but if the board is wrecked, everything has to be replaced.

Once the solder has been removed from a joint, any remaining fragments of wire can be easily curled away with fine nose pliers even when the work is cold. But don't ever do this on a printed circuit board because applying force to a printed circuit board track will lift it. Wire fragments should be delicately removed whilst the remaining solder is still molten. Fine tweezers can be handy here as they usefully limit the force that can be applied — but they need to be the good quality type that cannot be wetted by solder (yes, reputable manufacturers specify this quality).

Never attempt to desolder and unwrap a mechanical joint on a tag strip to remove a component — the necessary heat will damage the tag strip. Use flush microcutters to snip the exposed wire where it bends over from one side of the tag to the other, leaving two easily desoldered wires. Pressed brass tags on tag strips become brittle over the years and the recoil of a solder sucker can be sufficient to break them at their right-angled bend, although positioning the sucker so that the recoil propels it **along** the face rather than **into** the tag helps. Snip the component out, then remove the remaining wire. Even using desolder wick, one in twenty tags is likely to break when you replace components wholesale.

Desolder wick needs to be hot to absorb solder, but simply placing it between the iron's tip and the solder to be removed probably won't melt the solder because of the wick's thermal resistance. It is better to first directly melt the solder, then slip the wick in beneath the tip and reheat the solder via the wick. Capillary action can only draw the solder so far, so once the solder starts flowing into the wick, fresh wick must be drawn through the joint. The resultant gentle scraping action cleans PCB pads so well that (when defluxed) they appear never to have been soldered. Align your scraping direction so that you pull along the track towards its end rather than rucking the pad up. In this way, should the iron's heat temporarily melt the glue securing the pad to the board, it will probably not lift and break.

You will find that hot wick works far better than cold, so try to arrange matters to allow you to desolder a number of joints in quick succession and thereby maintain wick temperature. Quickly desoldering a number of joints requires plenty of wick, so make sure sufficient wick is exposed before you begin, otherwise you will burn your fingers or have to pause to expose more wick, losing wick temperature and efficacy.

Desolder wick needs to be clean to be effective, yet copper oxidises quickly, so self-sealing spools are supplied in ring-pull vacuum-sealed cans having a plastic resealable top. If you only use small quantities, keep the spool in current use in a sealed container, and consider a separate screw-top jam jar for the other spools once the can's vacuum seal has been broken.

Hand tools

In addition to a soldering iron, solder, and some means of desoldering, you need hand tools to dress leads and fit components. It is easy to be seduced by all the wonderful pictures of tools in a catalogue, but you will find that for day-to-day use, you need only a few tools, provided that they are of excellent quality. Good hand tools cost more, but they last far longer and are cheaper in the long run.

Static electricity and ESD

Electrostatic damage (ESD) is a bogeyman much feared by the electronics industry because even a small percentage of failures can poison a company's previously unsullied reputation, provoking financial collapse. The problem is that people (and particularly their clothing) can become electrically charged as a consequence of brushing against insulating textiles such as carpets, then transfer that charge to earth via a semiconductor's fine internal structure, either destroying it (digital) or degrading its noise performance (analogue). All ESD solutions boil down to making any insulator that could accumulate or transfer charge electrically leaky and bonding it to earth via a current-limiting resistor (typically $1\,M\Omega$). The uncertainty as to the exact cause of a genuine ESD problem and the general climate of fear permits premium prices for ESD-safe products. Frankly, in a country as damp as the UK, you have to try quite hard to provoke ESD (synthetic carpets and clothing, excessive heating and poor ventilation, inappropriate component storage and handling), but the hot dry conditions in New Mexico (where many semiconductors are made) are an entirely different matter. Paradoxically, ESD is more of a problem during winter because the cold air is unable to support much humidity (which would discharge charged objects) and because thick winter clothing is better at accumulating charge.

Valves aren't ESD sensitive. You only need pay the premium for ESD-safe hand tools if handling low-noise or complex expensive semiconductors, and even then a simple wrist-strap and work mat tied to an ESD earth (1 MΩ in series with mains earth) significantly reduces the potential problem.

Cutters

You only **really** need two sizes, one for cutting cable from wires to heavy mains cable, and one for cutting component leads precisely. The author has many different cutters — most of which are used only rarely.

Most electronics factors stock superb cable cutters that look terrible at first sight, but are well made from decent materials and are a delight to use. Although these cutters slice cleanly through any multicore copper cable that will fit in their jaws, whatever you do, don't use them for cutting steel-reinforced cables, or even bicycle brake cable; one attempt destroys them. Curiously, vets also sell these cutters (at twice the price) for trimming dogs' toe-nails. If you have these, and use them for most work, then a pair of good quality semi-flush micro-cutters will suffice for the remaining precision work. See Figure 3.6.

Lindström's semi-flush micro-cutters seem to last the author about 10 years before blunting such that the required increased cutting force snaps a jaw at the joint, but slightly larger cutters with tungsten carbide cutting edges are now available, and they seem more durable, albeit less precise. If you have tungsten carbide-tipped cutters, use them when salvaging equipment — you will be making many cuts, but don't need precision, so why waste the lifetime of more fragile cutters?

Flush cutters are also available, and they are particularly useful when removing components. Unfortunately, the necessarily smaller included angle of their cutting edge makes it sharper but weaker. See Figure 3.7.

Once the cutting edge has been decided, choose an ergonomically designed handle and springing system if possible — you will use these cutters a lot. If you also buy flush cutters, only use them when nothing else will do, or you will wear them out very quickly.

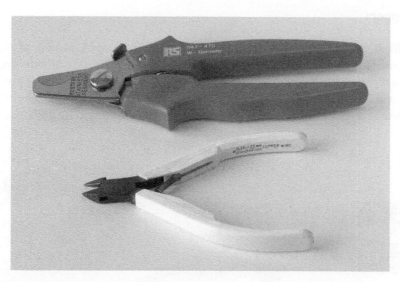

Figure 3.6
Cable cutters and micro-cutters.

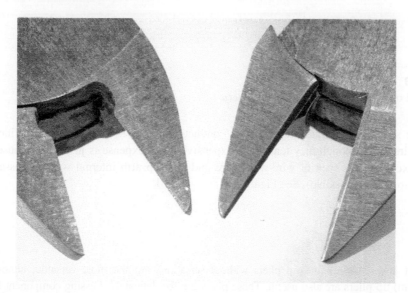

Figure 3.7
Note the difference between flush and semi-flush cutting edges.

Wire strippers

The ideal wire stripper would strip cleanly without breaking or nicking any of the fine wire strands beneath the insulation. There are many different sorts of wire strippers, and personal preference is important because none are perfect. Self-adjusting strippers are popular, and the author often uses them, but they don't always work correctly, forcing recourse to a scalpel or traditional wire strippers. See Figure 3.8.

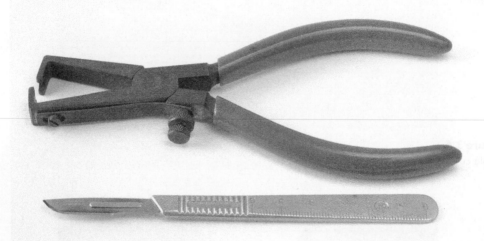

Figure 3.8
Scalpel and traditional (UK Post Office) strippers.

Wire strippers calibrated for different conductor diameters are available for finer gauges of wire and although they tend to be expensive, their expense is justified because they do an excellent job even of wire-wrap wire and pick-up arm internal wire (necessarily very fine to make it flexible). See Figure 3.9.

Pliers

Short jaw (21 mm) fine tip pliers without serrations are the most versatile, although wide (4 mm) tip pliers are also useful. These pliers are for delicately dressing component leads, not for removing your car exhaust! A pair of short-jawed heavy-duty combined pliers/cutters is handy too, but you probably already have a pair for dealing with your car or bike. See Figure 3.10.

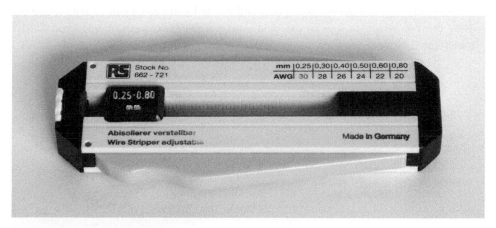

Figure 3.9
Calibrated stripper for fine wires.

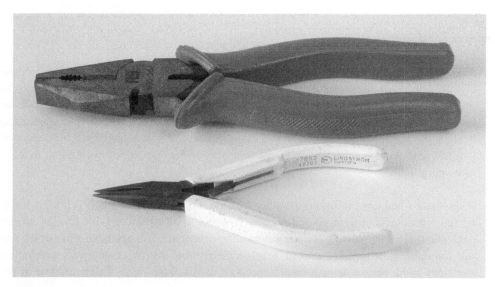

Figure 3.10
Heavy-duty pliers and micro-pliers.

Occasionally, none of the previous pliers will do, and you need a pair of long-nosed pliers. They are not nearly as nice to use, and you generally use them because you are being forced into bodging. See Figure 3.11.

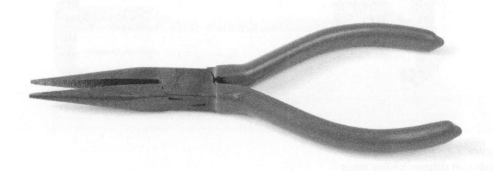

Figure 3.11
Long-nosed pliers.

Tweezers and fine work

Most tweezers are rubbish. The only tweezers worth buying are intended for genuine precision work or for manipulating surface mount components, and they are sold by the professional electronics factors. Your jaw will drop when you see their price, but lesser tools will not enable you to confidently pick up and place a low output capacitance SOT23 surface mount transistor, or delicately position a coil's fine wire. Not only will you not lend these precision instruments to anybody, but you're not even going to admit to owning them.

If the author had any, he would probably find the thin, wide tip of Ideal-Tek's 2A SA by far the most useful of their general-purpose five-tool set. Tweezers having tips specifically engineered for manipulating surface mount components are also available. You won't dare ask for these as a birthday present because their price will inevitably raise unwelcome questions about your other electronics spending. If you have steady hands capable of fine work, start saving.

Screwdrivers

You really can't have too many good quality accurately ground screwdrivers, and treated properly, they last forever, so buy the right ones once only. Do not use them for levering open paint tins. Avoid stubby screwdrivers because you can't see when their shaft is misaligned with the screw's axis and thus damaging the screw head. Poor quality screwdrivers are inaccurately ground and made of inferior metal. The result is that they slip and damage themselves and the screw head. That doesn't sound too bad until you realise that it happens when you must apply most force — when the screw is seized. Once you have damaged its head, removing that screw becomes far, far harder — it's so much easier to avoid damaging the screw head in the first place.

Flat-bladed screwdrivers

You need a 3 mm flat-blade screwdriver, often known as an "electrician's" screwdriver. An extra-long screwdriver (250 mm shaft length) having a 5 mm flat blade is extremely useful, and the bigger screwdrivers can be useful too, particularly 7 mm. You also need a very small screwdriver; it traditionally has a yellow handle and is about 60 mm total length, with a blade width of about 1.5 mm. Buy two — they disappear. Quite apart from its intended use, the flat of a screwdriver blade is very useful for mechanically dressing the wires of a joint prior to soldering. Check the grinding of the tip — it should be square to the shaft, of constant thickness and have well-defined edges. Poor quality screwdrivers have plated rounded edges with a tip that is not square to the shaft. They damage screws.

Phillips, Pozidriv, and Supadriv screwdrivers

There is a world of difference between Phillips and Pozidriv "crosshead" screws, so you will want to choose your purchases very carefully to avoid damaging screws and drivers. The Phillips screw (common in North America) requires a screwdriver having a sharp, pointed tip, whereas the Pozidriv screw (common in Europe) requires a screwdriver having a larger included angle and stubbier end face, yet both designs originated from the same North American company. See Figure 3.12.

Figure 3.12
Phillips tip (upper) vs. Pozidriv tip (lower). Note that the Phillips tip has a sharper angle and is longer than the equivalent Pozidriv tip.

Supadriv screwdrivers are a refinement of Pozidriv, having a secondary set of splines between the primary splines. See Figure 3.13.

Figure 3.13
Phillips tip (upper) vs. Supadriv tip (lower). Note the damage to the Supadriv tip's additional splines — Supadriv screwdrivers are **not** reverse compatible with Pozidriv screws.

A Supadriv screwdriver's extra splines enable it to transfer greater torque to a Supadriv screw. Although a Pozidriv screwdriver can be used without damage on a Supadriv screw, a Supadriv screwdriver would be damaged by a Pozidriv screw, so check carefully before use. Supadriv screws should be uniquely identified by radial lines on the face of the screw head between the splines of their screw slots. See Figure 3.14.

Figure 3.14
Pozidriv (left) vs. Supadriv (right) screw head. Note the identifying radial lines on the face of the Supadriv head between the splines.

If possible, choose an extra-long screwdriver having an ergonomic handle as this aids precise screwing. Sizes 0, 1, 2 are useful, 3 and 4 are a luxury.

The author had to tip out his large box of non-BA and non-Metric screws and search diligently before finally unearthing a small proportion of Phillips head screws (ex-Tektronix salvage). Since Phillips screws are rare in the UK, why do we have so many Phillips screwdrivers ready and waiting to mangle Pozidriv screws? Perhaps worse, the author has recently found screws bearing radial lines better fitted by a Phillips screwdriver. Further, it seems near impossible to buy a correctly ground Pozidriv screwdriver in the UK — they mostly appear to use the Phillips included angle but Pozidriv end face. Since there no longer appears hard and fast identification between Phillips, Pozidriv, and Supadriv screws or drivers, consider where the equipment came from and try that region's most likely screwdriver first. Slack quality control and ill-defined standards are making a mockery of screw head design, and when flying, the author has seen chewed (Phillips or Pozidriv?) screw heads securing wing panels.

Hexagonal keys and drivers

Buy a good quality set of long-arm hex keys in Metric **and** Imperial sizes as it is essential to have the correct size. Be careful when using long-arm hex keys because although the extra leverage reduces effort when using sheet metal punches, they are capable of splitting screw heads made of inferior metal if over-torqued. See Figure 3.15.

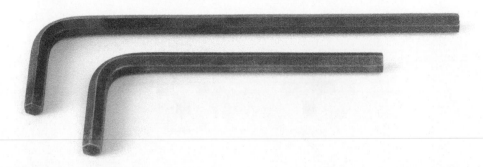

Figure 3.15
Six-millimetre long arm versus normal six-millimetre Allen key.

Be wary of ball-nosed keys — they allow easy access, but unless they are of excellent quality, the reduced contact area at the head wears quickly and is easily damaged.

If you use a lot of hex fasteners, you might want to consider buying a set of drivers — they are so much faster to use than keys.

Nuts and associated tools

Nuts are undone with spanners or nutrunners, **not** with pliers! A set of open-ended BA (British Association) was essential for traditional British valve amplifiers, but metric fasteners are the European standard. Electronic equipment uses very small nuts, so if the spanner is inaccurately ground, it slips. Nobody likes chewed nuts, making it essential to use top quality spanners. See Figure 3.16.

The external size of an open-ended spanner head for a given nut size is purely down to the manufacturer, and space is often restricted, so even if you already have a set of spanners

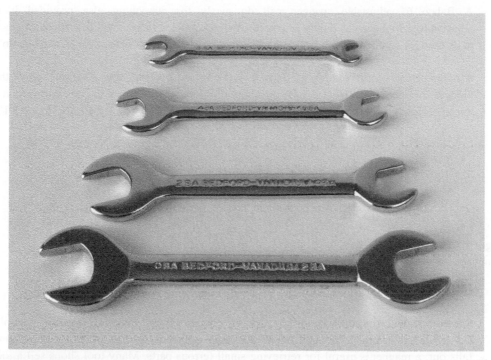

Figure 3.16
Selection of small spanners.

it's worth keeping an eye out for good quality spanners having small heads. Note that ring spanners have much smaller heads than open-ended spanners.

Nutrunners are available in all sizes, but one for potentiometer nuts is particularly useful, and prevents visible gouging of front panels by spanners. See Figure 3.17.

Figure 3.17
A nutrunner is particularly useful for potentiometers.

A Bahco 6″ adjustable spanner will do nicely for everything else provided it is adjusted to grip the nut as tightly as possible before applying torque, and is sufficiently good to be used for careful roadside repairs of bikes with parts labelled "Ducati" or "Campagnolo".

A nutlauncher is an incredibly useful tool. Looking and operated like a syringe, when the plunger is pressed, three wire hooks spread out of the far end then grip the nut, allowing you to spin it lightly onto the end of an otherwise inaccessible screw. Wonderful! See Figure 3.18.

Figure 3.18
A nutlauncher makes short work of starting nuts buried deep in wiring.

Once the nut is engaged, a nutrunner or spanner can finish the job.

Sometimes, with the best will in the world, a fastener will fall deep inside a heavy amplifier. Rather than lifting the amplifier upside down and shaking it until the offending item falls out, a magnet is useful for retrieving small ferrous parts. Many tool shops sell a tool that is effectively a powerful magnet fitted to the end of a telescopic aerial and packaged to look like a pen. See Figure 3.19.

Figure 3.19
A magnet on telescopic mount makes retrieval of small steel parts easy.

Beware that the tool's neodymium magnet has to be plated, and when this isn't done properly, it subsequently flakes off. Those thin flakes are sharp and brittle, so if the magnet starts flaking, replace it before you have to dig out painful (non-magnetic) splinters.

Scalpel

A scalpel is extremely useful, and the No. 3 small handle version is best suited for electronics, with either No. 10 (curved) or No. 10A (straight) blades, but remember that scalpel blades were designed to slice human flesh. You will find that blades lose their edge very quickly, so buy them in boxes of 100 (far lower unit price), and be prepared to fit a fresh blade the moment you notice a lack of keenness. If you have the option, non-sterile blades are marginally cheaper, and you don't intend surgery. A guard made from garden hose or flattened layers of heatshrink sleeving should cover the blade when not in use. See Figure 3.20.

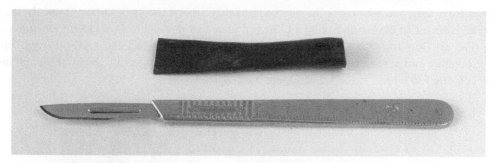

Figure 3.20
Scalpel with guard made from heatshrink sleeving.

Alternatively, surgical manufacturer Swann Morton makes a Retractaway handle suitable for these blades, but the necessary movement to allow the blade to be withdrawn means that the locked position is not as firm as the traditional handle and it's not as nice to use.

Mirror

Another useful medically derived tool is a small angled circular mirror, similar to those used by dentists, but (crucially) insulated. See Figure 3.21.

Figure 3.21
This insulated inspection mirror is invaluable.

These mirrors are particularly useful for checking that the underside of a mechanical joint is correct (and can't move) **before** soldering. Don't try to adjust the angle of one of these mirrors — you'll just crack the glass.

Hot-air gun

Tool catalogues stock all sorts of expensive thermostatically controlled hot-air guns for shrinking heatshrink sleeving, but a hot-air gun intended for stripping paint works almost as well for occasional use, and is far cheaper. Never attempt to shrink sleeving by heating it from only one side; move the gun continuously around the work to heat it evenly from all sides, or fit the nozzle having a U-shaped plate that directs hot air all around the tubing. The air flow from hot-air guns is somewhat indiscriminate, and can easily damage surrounding components such as electrolytic capacitors, but a protective shield of aluminium cooking foil works wonders. See Figure 3.22.

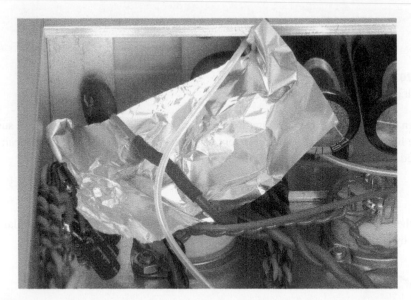

Figure 3.22
A temporary aluminium foil heat shield protects other components when using hot-air gun on heatshrink.

Not only does the shield protect other components from the hot-air blast, but the turbulence caused by blowing air at the shield means that the heatshrink sleeving is heated evenly from all sides, and shrinks in half the time, which further reduces the risk to surrounding components.

Marker pens and cameras

It may sound obvious, but if you label things **before** you take them apart, life becomes so much simpler. Alternatively, a camera is a very useful tool because if you take shots of the work as it is disassembled, you have the perfect guide to reassembly.

Lighting

Lighting was discussed in detail in Chapter 2, but it should be reiterated that you can't have too much light to work by. If you are able to have a dedicated bench, try to position it near a window, and fit dedicated lighting over the bench. The author lights his electronics bench with a 70 W 6500K high frequency fluorescent tube shaded by an inverted length of guttering lined with kitchen foil. See Figure 3.23.

In addition, the harsh light provided by the point source of a **single-chip** LED torch is invaluable for finding dropped minutiae, especially when angled almost to the plane of the search surface and swept so that even small parts cast a long shadow. Avoid the multi-LED horrors — their light is dim and diffuse.

Magnifying lamps

These devices comprise a large magnifying glass surrounded by a ring light all mounted on the end of a spring-balanced articulated arm. Their attraction increases as you grow older, but they aren't perfect:

- The reason you want one is that your eyesight isn't what it was, so you aim for maximum magnification by positioning the glass close to the work and your eyes close to the glass. Above, your nose collides and smears the glass; below, the soldering iron won't fit (although some recent irons have angled tips, presumably to avoid this very problem).

Figure 3.23
Fluorescent lighting concealed beneath inverted guttering provides superb lighting on an electronics workbench.

- Colour rendering is generally poor. Only a few fluorescent designs are fitted with daylight tubes, and their CRI is generally unspecified (so assumed poor) – and the same is true for LEDs.
- Flicker might be noticeable with power-line fluorescent ring lighting, although as the field of view is not peripheral, few people should be bothered.
- High frequency fluorescent fittings eliminate flicker but they and dimmable LEDs using pulse width modulation induce interference into high-gain amplifiers – and you've deliberately placed that interference nearby. If you need to investigate high-gain electronics whilst it is powered, the best solution is to retrofit LEDs to an old magnifying lamp and if you need them to be dimmable (why would you want anything other than full brightness?), power them in <50 V series strings from DC constant current sources and vary the programming resistance or reference voltage.

- That large articulated arm gets in the way and is not as manoeuvrable as at first appears.
- The lamps usually come with a weak G-clamp intended for the edge of the bench — make sure that a proper permanent fitting is available as an accessory and order it at the time.
- That large lens is supplied with a cover, not so much to keep it clean, but to prevent sunlight being focused and causing a fire, so remember to replace it after use.

On balance, the author's daylight fluorescent magnifying lamp on its permanent mounting has not been an unqualified success, but you might have a different opinion. Ultimately, the problem is that with age, the eye's internal muscles weaken, making them less able to distort the eyeball's shape and pull the retina into image focus. If the pupil were a pinhole, a correctly focused image would be obtained at any retinal distance, irrespective of eyeball shape. Summarising, it's easier to provide plenty of light that allows the pupil to close tighter and obtain clear focus than attempt to directly manipulate the image with a magnifying glass, so that means positioning your bench for maximum natural light from a large window and treating artificial light as second-best.

Binocular magnifiers

Exemplified by Optivisor, these consist of an adjustable head band carrying a pivoted peak fitted with a pair of glass lenses (interchangeable for different magnifications) that can be worn over prescription glasses, and the author has found them better (and cheaper) than the magnifying lamp. Beware that, as with microscopes, higher magnifications require a closer object distance, so you will probably find the typical 250 mm object distance of lens plate number 4 ($\times 2$ magnification) more usable than the typical 100 mm object distance of lens plate number 10 ($\times 3.5$ magnification). A minor Optivisor disadvantage is that although they can be combined with glasses, they can't be combined with earmuffs, so if you need them whilst drilling, foam earplugs become necessary.

Storage of precision tools

Precision tools should not be thrown in an old toolbox together with used spark plugs and oil filters. Keep them in a clean partitioned box of their own and don't lend them to **anybody**. The best storage for precision tools is intended for office use and is a small steel filing cabinet having shallow drawers that segregate tool types, allowing them to be found quickly yet be protected. See Figure 3.24.

Figure 3.24
This cabinet has ten shallow drawers that allow organised storage of precision tools.

Electromagnetic compatibility (EMC)

EMC is the new name for an old problem but splits it into two parts:

- Emission: Equipment should be designed and constructed to minimise unwanted emissions as far as reasonably practicable.
- Susceptibility: Equipment should be designed and constructed to minimise its susceptibility to unwanted fields as far as reasonably practicable.

Thus, transmitters are allowed to emit, but only on their chosen regulated frequency, whilst receivers must be able to reject all but their chosen receiving frequency.

The "reasonably practicable" legal term means that if you choose to use unscreened leads or electronics, the responsibility is yours when a neighbour's legitimate computer network spoils your listening pleasure. Conversely, if you have adhered to accepted practice but suffer interference from a poorly adjusted transmitter, responsibility lies with the transmitter operator and they are not entitled to require you to live in a Faraday cage.

EMC ranges from power-line hum at low frequencies (50/60 Hz) to transmitters operating at gigahertz frequencies. In the early days of valve audio, the most frequent problem was motorcycle ignition noise (cars had more spark plugs, but they were fortuitously screened). From the 1950s onwards, people living near television transmitters and marine radars suffered more of a problem and 1970s calculators with strobed LED displays upset nearby radios. However, the preceding problems were as nothing compared to the electronic hash generated by computers and their associated data transmission. Data transmission over power lines has become popular, so it is now common to find that the wires bringing power into audio equipment emit copious data interference as well as the traditional power-line hum.

In addition to deliberate mains interference from data, the notionally sinusoidal power-line waveform suffers distortion from imperfect loading. The ideal mains load is a pre-electronics era heater (cooker, fan heater, or incandescent lightbulb) or power factor corrected motor because the current drawn is very nearly sinusoidal, but almost everything nowadays is electronically controlled and forces current waveforms containing harmonics of mains frequency, which then propagate down mains wiring to enter audio equipment.

The previous issues mean that the majority of construction and wiring technique has to be aimed at minimising emissions and susceptibility, and the primary culprit for EMC problems is the wire because at radio frequencies it becomes an aerial.

Types of wire

The simplest form of wire is an uninsulated single copper conductor. Copper quickly tarnishes, so tinned copper wire (TCW) is commonly available in various gauges (diameters) and is often used for earth bus-bars.

The next step is to insulate the wire, perhaps with polyurethane enamel. Enamelled copper wire (ECW) is commonly used for winding transformers and electromagnets, hence its other name, magnet wire. The enamel must be stripped before soldering, either by abrasion or by burning it away with a much hotter iron (450°C, 850°F) — this is where an electronically controlled soldering iron becomes useful. Although suitable for wiring that cannot move, that thin enamel is not robust enough for general use, and a more suitable insulator for general wiring is a plastic sheath, usually PVC.

Solid copper wire is not very flexible, which is an advantage when twisting heater wiring because it retains the applied twist, but not so good for the lead to a hand tool such as a soldering iron. Breaking the conductor into a number of fine strands increases flexibility — the more strands the better, so manufacturers often describe stranded wire by the number of strands and their individual gauge, perhaps 10/0.1 mm to denote ten strands each 0.1 mm in diameter, or 7/32 AWG to denote seven strands each of 32 American Wire Gauge (AWG). Usually, the most important parameter is the current-carrying capacity of the wire and this is determined primarily by its total cross-sectional area, so the previous 10/0.1 mm wire could also be specified as 0.079 mm^2 and suitable for currents up to 500 mA. A secondary parameter is the voltage rating of the enclosing insulation, and this should be checked for very fine wires or high voltages.

Although a single conductor is useful as hook-up wire for internal wiring, we often need more conductors, and a collection of insulated conductors within a common sheath is known as a multicore **cable** (quite distinct from a multistrand **wire**).

A wire that used to be commonly seen in radio frequency coils is Litzendraht (usually abbreviated to Litz). The wire is made up of a number of insulated strands, all of which are connected together at each end, making a single conductor (which is why it is deemed to be wire, rather than cable). The significance of insulating (or serving) individual strands is that skin effect forces signal currents to the outer surface at high frequencies, so the increased surface area of Litz wire reduces high frequency (> 100 kHz) resistance, and thereby losses. The idea is periodically resurrected for audio, but the only audio signal remotely sensitive to cable resistance is the one between the loudspeaker and its amplifier, yet most tweeters are inductive unless corrected and have an impedance of $>10\,\Omega$ at 20 kHz, so skin effect would need to cause high frequency cable resistance to rise by $>1\,\Omega$ to cause a 1 dB change in level, and that simply doesn't happen at audio frequencies. The best way to improve a loudspeaker cable is to shorten it.

Wires carrying low-level signals must be protected from external interfering signals. Tightly twisting a signal's send and return legs together provides protection against magnetic fields, whereas adding an earthed coaxial conductive screen protects the inner conductor from electrostatic fields. There is nothing to stop us combining the two techniques, so twisted pair in overall screen is common for microphone cables.

A coaxial cable's conductive screen may be formed simply by wrapping uninsulated strands of wire around an insulated inner, but bending such a cable causes outer strands to move apart, permitting interference to enter, so a better solution is to braid the outer strands. Cheap domestic terrestrial TV cable has a very open braid, making the cable almost as effective (but badly tuned) as the intentional dipole array at its end. Broadcast quality coaxial video cable has two layers of tight braid to minimise interference ingress, but this is expensive, so a cheaper solution uses a single braided screen over a lapped screen of metal foil or aluminised polyester.

Coaxial cable is almost invariably intended for radio frequency use and the key parameter tends to be characteristic impedance rather than current-carrying capacity. Characteristic impedance is the impedance seen between the two conductors looking into either end of an infinite length of cable. Imagine that you have an infinite length of 50 Ω characteristic impedance coaxial cable and you cut a metre off one end. You now have an infinite length of cable and a one metre length of cable. By definition, the infinite length must still look like 50 Ω, but the one metre length also looked like 50 Ω when terminated by the infinite length, and would look no different if we terminated it with a 50 Ω resistance between the two conductors. By symmetry, the infinite length of cable looks like a 50 Ω resistance from either end, so the one metre length of cable needs to be terminated with a 50 Ω resistance at each end to maintain its characteristic impedance.

Once a cable is long enough for multiple signal wavelengths to occur along the cable, it behaves as a transmission line, and provided that it is terminated at each end by a resistance equal to its characteristic impedance, a signal propagated from one end is **totally absorbed** at the far end with no reflections. Mis-termination at the far end causes a single reflection to return back down the cable to the source, where it is totally absorbed by the matched impedance of the source. However, if the source resistance is also not matched to the cable's characteristic impedance, the reflection reflects back from the source and bounces backwards and forwards down the cable until absorbed by cable losses. The effect on analogue television was to cause a ghost image slightly to the right of the original image.

Reflections add or subtract to the intended signal but are unnoticeable provided that the cable is short compared to the signal's transitions, and this is why transmission-line definitions are usually couched in terms of wavelength and cable length. However, signals travel slower down a cable than free space, so manufacturers usually specify the velocity factor, which is the proportion of the speed of light (*c*). Typical coaxial cables have a velocity factor of $\approx \frac{2}{3}c$.

The significance of this discussion of coaxial cables and transmission lines is not that controlling cable characteristic impedance and velocity factor is important for analogue audio (it isn't), but that it leads to material choices having useful audio qualities. A key parameter for analogue audio signal cable is the capacitance per unit length, which may be calculated for any coaxial cable using:

$$C_{\text{(per metre)}} = \frac{2\pi\varepsilon_0\varepsilon_{\text{r}}}{\ln\left(\frac{D}{d}\right)}$$

where:

ε_0 = permittivity of free space $\approx 8.854 \times 10^{-12}$ F/m
ε_{r} = relative permittivity of the insulator $\approx 2-3$ for most solid plastics
D = insulator diameter
d = core conductor diameter.

Remembering that all capacitors suffer increasing dielectric loss with frequency, radio frequency coaxial cables either require a good quality solid insulator such as PTFE, or careful use of a lesser insulator. PTFE has to be extruded at a sufficiently high temperature that it would oxidise copper and melt solder, so the inner conductor is silver plated rather than tinned (nothing to do with skin effect). After a vacuum, air is the very best dielectric, so some radio frequency cables minimise the effect of a poorer quality dielectric between core and outer conductor by foaming it or arranging it into thin radial supporting spokes that reduce the average value of ε_{r}. Typical solid insulator 50 Ω coaxial cable has a capacitance of ≈ 100 pF/m, or ≈ 30 pF per foot, and this becomes significant at audio frequencies if the source resistance is significant (≥ 1 kΩ) or the cable is long (≥ 2 m).

Because an oscilloscope probe passes negligible current into an oscilloscope's 1 MΩ// ≈ 12 pF input impedance, series resistance is not a problem and the probe's coaxial cable can have a much smaller central conductor diameter, leading to significantly reduced capacitance per unit length.

A useful byproduct of the need for a thick dielectric (compared to an explicit capacitor) is that radio frequency coaxial cables tend to have >2 kV DC voltage ratings between core and screen. Thus, when you have stolen the braid screen in order to make a custom umbilical or audio cable, don't discard the (insulated) inner as it is a useful high voltage wire.

All cable has a minimum bend radius, and bending a coaxial cable so tightly that the internal insulator begins to collapse changes the characteristic impedance, resulting in a reflection from that point — which **is** a problem for digital audio. More significantly for valves, insulator deformation concentrates charge and reduces the local voltage rating, so treat cables with care and don't bend them tightly. Mains cable is supplied by its manufacturer gently coiled on a drum, yet so much equipment arrives with a tightly figure-of-eighted IEC mains lead whose kinks are almost impossible to remove. Why?

There is nothing to stop us from bundling a number of coaxial cables or twisted pairs into a single sheath. Once twisted pairs are bundled together, they might interfere with each other, so they can be individually screened, or a single overall screen added beneath the outer sheath, and the component catalogues are full of such cables and associated connectors. As an extreme example, the EMI 2001/1 early colour television camera needed ten coaxial cables for analogue video signals between camera head and camera control unit, plus more wires for control signals and power, leading to the G101 (101 conductors) camera cable.

Customised multicore cables are expensive both to make and terminate, so the later television camera solution was to modulate all the signals onto radio frequency carriers and move the power supply to the camera head. The signal cable needed to be a coaxial transmission line that could carry signals plus mains power between its core and screen, so (for safety) another screen connected to earth was placed around (but insulated from) the neutral conductor, resulting in a triaxial cable. Although the additional electronics needed for the signal multiplexing was expensive, it was outweighed by cable cost savings when miles of cable were needed, such as at outside broadcasts.

Triaxial cables and connectors are also used at the input of electrometers (ammeters whose **highest** range is only 20 mA) because bootstrapping the inner screen via a voltage follower from the signal reduces cable leakage currents whilst leaving the outer screen to perform its traditional screening function. In theory, bootstrapping a triaxial cable's inner screen could reduce cable capacitance sufficiently to connect a condenser microphone capsule to its input amplifier, but it is invariably better to solve the capacitance problem by moving the input amplifier adjacent to the source. The author has yet to find a genuine audio application for triaxial cable.

Although commercially manufactured umbilical cable quickly becomes expensive, short custom cables are easily made by bundling individual wires or cables together inside a common sheath, and nylon braid for precisely this purpose is readily available. If we wanted, we could add a braided screen taken from a video cable, enabling construction of an umbilical cable composed of twisted heavy gauge wires for heater supplies in their own screen, fine control wires, screened signal wires, plus an outer screen, and finally a retaining and insulating nylon braid. The insulating braid is needed because if a conducting screen is allowed to scrape across earthed metalwork, it creates audio crackles as substantial earth currents are made and broken.

Wiring techniques

All AC power wiring generates an external field that potentially induces audible hum into signal wiring. AC heater wiring is the obvious problem, because it is unavoidably close to sensitive signal wiring, but the more distant AC mains and high voltage AC to rectifiers can also cause problems.

To save time on assembly and subsequent maintenance, any fixing screw should be just long enough to use all of the thread in the nut and no longer, and this dictum is particularly important when securing valve sockets because overlong screws make subsequent heater wiring particularly difficult. If you choose to decouple heaters to chassis at the valve socket with capacitors, the necessary solder tags and star washers inevitably require a longer screw. You just have to grin and bear it.

Electromagnetic fields and heater wiring

The electromagnetic field is due to the **current** flowing in the power wires, which induces currents in any nearby signal wiring. This means that not only are valves with 12.6 V heaters cheaper (contrast the price and availability of NOS 6SN7 with 12SN7), but they're better; halving heater current halves electromagnetically induced hum.

Heater wiring is traditionally taken from a winding on the mains transformer to the nearest valve, and looped through from one valve to the next, until each valve has heater power. Electromagnetic fields decay with the square of distance, so heater wiring runs should be as far away as possible from signal circuitry as possible, and only come up to the valve at the last possible moment and in the most direct manner possible. The input valve is the most sensitive stage, so this should be the last in the heater chain, in order that the wiring leading to this valve carries the minimum current.

To minimise the external electromagnetic field, the heater wire should be tightly twisted. This means that although any given twist produces a field of one polarity, the twists either side of it induce opposite polarity, and so the fields tend to cancel. This twist should be maintained as close up to the tags of the valve socket as possible, and when one phase of the heater wire must reach the opposite side of the socket and return, as is the case when wiring past an ECC83/12AX7 to the next valve, the wire should go diametrically **across** the base and be twisted as it passes across. Admittedly, there is an imbalance equal to the crossed valve's heater current, but imperfect cancellation is better than none.

Valve sockets should be oriented so that the tags receiving heater wiring are as close to the chassis wall as possible, and heater wire must **never** loop round a valve (except for rectifier valves, where hum is not an issue).

The worst way to wire heaters would take the incoming pair connecting to the two heater tags from one side, then mirror on the opposite side to form a heater wiring loop having significant area enclosing the valve socket (minimising electromagnetic problems usually boils down to minimising loop area). See Figure 3.25.

Heater wiring leading to valves using B9A sockets such as EL84/6BQ5 etc. is best twisted from 0.6 mm (conductor diameter) insulated solid core wire, which is rated at 1.5 A. Octal valves generally require more heater current, so the larger tags on their sockets are designed to accommodate thicker wire that could not connect to B9A tags. When wiring to valves other than rectifiers, it is useful to use a different colour for each phase, and the author has traditionally used black and blue. When wiring to a push—pull output stage, if the same colour connects to the same pin on each valve, then the hum induced within each valve will be the same phase, and will be cancelled in the output transformer. (This argument assumes that both valves were made by the same manufacturer to the same pattern.)

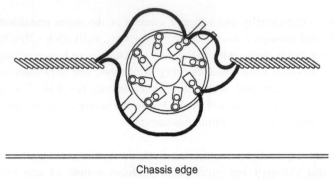

Chassis edge

Figure 3.25
Poor heater wiring at valve socket: Hum∝enclosed loop area. Note also that the wiring should have been butted up against the chassis edge.

The wire must be rated for the current it carries, but as the current changes along the chain we cannot have the optimum gauge of wire at all points, resulting in thicker than ideal wire going to the input valve. The significance of the wire's gauge is that the heavier gauges don't twist so well, yet we need a tight twist, especially near the input valve. Sadly, it is becoming harder to obtain 3 A insulated single-strand wire suitable for twisted pair heater wiring, making heater chain wiring even more difficult.

Alternatively, if we use one twisted pair per valve, each going back to the transformer, the reduced current in each wire allows it to be thinner and have a tighter twist, and fewer gauges of wire need be stocked (although more wire is used), yet the method has a further advantage. We now have multiple pairs of wires going back to the transformer, so we harness them neatly together. Each pair of wires (despite its tight twist) generates a small AC field. The twist of each pair is randomly aligned to the others so the AC fields add imperfectly. Sometimes they add constructively, sometimes destructively – they're not truly correlated. To sum sources that are not correlated, we must sum powers.

Example: We have four pairs, each carrying 0.3 A, $P = I^2R$, so we sum 0.09 + 0.09 + 0.09 + 0.09 = 0.36. We then take the square root to find the individual current that would have given that power, and find it is 0.6 A. This tells us that four twisted pairs each carrying 0.3 A will generate the same hum field as a single pair carrying 0.6 A. But we are carrying 1.2 A over our four wires, so we've halved our hum field. The quicker way to determine the expected improvement from using multiple pairs carrying identical currents is $\sqrt{n}$, where n is the number of pairs. In practice, the sources probably aren't perfectly uncorrelated or matched, but they certainly aren't perfectly correlated, so the hum field **will** be reduced compared to a daisy chain, and fine solid-core wire is more readily available.

To summarise, multiple pair heater wiring reduces hum field and wire inventory compared to daisy-chaining, but requires a greater length of more easily available wire that might take longer to implement because multiple lengths of longer wire have to be handled and laced together neatly in a loom.

Valve rectifiers not only have a dedicated heater supply, but they also have the incoming high voltage AC, so the two should be clearly distinguished; the author uses a blue twisted pair for the heater. Unlike a bridge rectifier, where two wires carry flow and return currents from the transformer, enabling loop area to be minimised by twisting those two wires together, a centre-tapped rectifier alternates current flow between the two phase wires and always returns it through the 0 V centre tap, so twisting the two phase wires together is magnetically useless, although it reduces the electrostatic field. To minimise both emissions, we must twist the phase wires together with the 0 V centre tap. Since all three of the transformer's wires are twisted together, yet need to be distinguishable, it makes sense to use red for the two phase wires and black for the centre tap.

Twisting two wires together is easy. Cut equal lengths of wire to be twisted, pair the wires together at one end and clamp them in a vice. Gently tension both wires equally at the far end, and grip them in the chuck of a cordless drill. Hold the wires reasonably taut by pulling on the drill, and start twisting. When the wire begins to accelerate you towards the vice it will have about ten twists per inch. Switch off, and **whilst maintaining tension** by holding the wire with your fingers, loosen the chuck. The wire will now try to untwist, and if allowed to do so suddenly, it will tie itself in knots. Gently release the tension in the wire, and release it from the vice. You now have perfectly twisted wire.

You will find that it is easier to achieve a perfect twist on longer lengths of wire than short ones because it is easier to equalise individual wire tension before twisting. Equal tension is important because a slack wire tends to wrap itself around a tighter wire (which remains straight). For this reason, it is worth twisting 4 m or even 6 m at a time, especially if twisting more than just a pair, because it becomes progressively harder to equalise individual wire tensions.

Electrostatic fields and heater wiring

The electrostatic field is due to the **voltage** on the wiring. Once again, twisting assists cancellation. Heater wiring should be pushed firmly into the corners of the (conductive) chassis since the extra electrostatic mirror at the corner improves electrostatic field cancellation and distance further reduces interference. Thus, heater wiring should not run exposed from one valve to the next, but ideally return to the corner of the chassis to re-emerge at the next valve. These strictures mean that good heater wiring requires considerable time/cost, so modern commercial power amplifiers sometimes skimp on the quality of their heater wiring, mandating $100 \, k\Omega$ amplifier input resistance rather than $1 \, M\Omega$, because the same interference current develops one-tenth the hum voltage ($-20 \, dB$) across the reduced input resistance.

AC heater wiring must be connected to the transformer in a balanced fashion to permit electrostatic cancellation. Unfortunately, heater wiring **must** have a DC path to HT 0 V in order to define the heater to cathode voltage, and this can be achieved in various ways. See Figure 3.26.

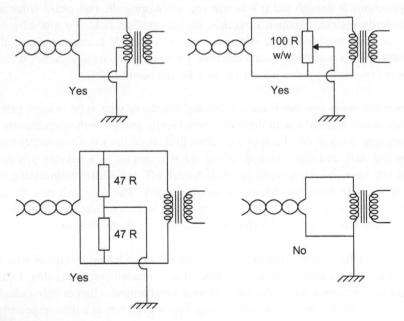

Figure 3.26
Defining DC heater potential.

The worst way to define the DC path is simply to connect one side of a transformer winding to 0 V. This ensures that one phase of the wire induces no electrostatic hum, whilst the other phase induces maximum. (Nevertheless, this technique is commonplace in high-sensitivity radio frequency equipment because although the intermediate frequency amplifier might have >100 dB gain within its chosen bandwidth, it has no gain outside, so power-line hum is not an issue.)

The ideal way of defining the DC path is to use a transformer with a centre tap on the heater winding, but if this is not available, fixed or variable resistors can be used to derive a midpoint. Accurately matched resistors used to be rare, so a variable resistor known as a **humdinger** control was traditionally fitted, and adjusted for minimum hum. Once a mid-point has been derived, and connected to HT 0 V, each wire has equal voltage (but inverted polarity) hum, and the electrostatic fields tend to cancel.

Doubling heater voltage from 6.3 to 12.6 V to enable the use of cheaper 12SN7 trades electromagnetic hum for electrostatic hum, but near-perfect electrostatic screening at all frequencies is easily achieved, whereas electromagnetic shielding is near impossible at 50/60 Hz. For ultimate reduction of electrostatic hum from heater wiring, we can screen heater wiring using copper tape, ideally having conductive adhesive (expensive), or robust aluminium foil plus double-sided adhesive tape (cheap), laid as a strip and stuck to the conductive chassis either side of the heater wiring — it is not necessary to lap the wiring.

Typical aluminium kitchen foil is ≈ 0.013 mm thick, although the author found one roll 0.018 mm and another 0.007 mm, but when the author found a roll of 0.058 mm (non-kitchen) aluminium foil, this proved to be ideal (much less fragile). Although aluminium is more awkward in terms of making a connection, unlike copper, there is little danger of electrochemical corrosion between it and the (probably aluminium) chassis. If the adhesive is an insulator (and most are), a supplementary low-resistance connection is needed to the chassis, and this is most easily provided by passing a screw through the screen and chassis and securing it with serrated washers.

Alternatively, we can use DC heater supplies, but even then it is worth treating the heater wiring as if it were carrying AC, as this will ensure that the finished project has **no** hum from the heaters.

Heaters and mains-borne common-mode interference

Up until now, we have considered interference due to the difference in voltage between the two heater conductors (differential-mode interference), although this had not been explicitly stated. Conversely, the two heater conductors might have zero interference voltage between them but be bouncing up and down in unison with respect to chassis, and this is known as common-mode interference.

Heaters are thermally coupled to cathodes, so there is capacitance between them ($C_{hk} \approx 5-10$ pF for a small-signal valve), and the electrically insulating aluminium oxide in parallel is hot and therefore leaky (R_{hk} is perhaps only 25 MΩ). Thus, C_{hk} forms a high-pass filter in combination with the resistance seen looking into the cathode and this filter's attenuation is degraded at low frequencies by low values of R_{hk}, which is why valve testers include a R_{hk} test. The heaters are connected to a heater winding that typically has primary to secondary capacitance ($C_{ps} \approx 500$ pF) to the incoming mains. See Figure 3.27.

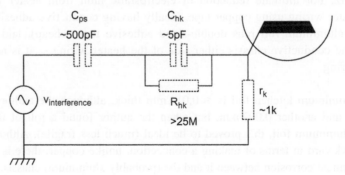

Figure 3.27
Parasitic components couple mains-borne common-mode interference into the audio path.

Once valves having inadequate R_{hk} have been rejected, there are various means by which mains interference travelling via the transformer/heater path can be deliberately attenuated, plotted and ranked in descending order of effectiveness. See Figure 3.28.

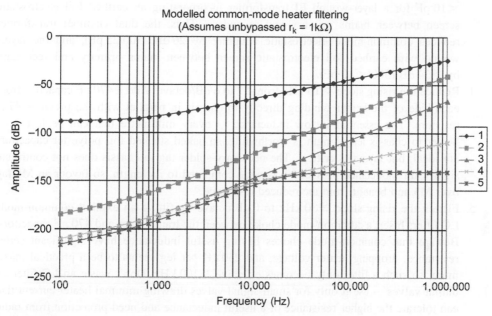

Figure 3.28
With care, common-mode interference can be significantly attenuated.

1. Reduce interference amplitude at source by using a separate heater transformer for signal valves: Directly reduces coupling of rectifier switching transients.

2. Maximise attenuation at low frequencies by adding shunt resistance: Connect a resistance (commonly 47 Ω for 6.3 V heaters) from each side of the heater winding directly to chassis. If the winding has a centre tap, also connect it directly to chassis. Common-mode: By splitting the existing capacitances (C_{ps} and C_{hk}) in two, this new resistance adds another high-pass filter, hence the change from 6 dB/octave slope to 12 dB/octave < 10 MHz. Differential-mode: This forms an LR filter in conjunction with transformer leakage inductance. Film rather than wirewound variable resistors are better for humdingers (controls that match to actual strays) because the typical centre to end 2.5 μH inductance of a 50 Ω 1 W wirewound potentiometer would degrade attenuation ≥1.6 MHz.

3. Maximise attenuation at all frequencies by minimising the mains primary to secondary capacitance, C_{ps}: For a 50 VA transformer, we can expect $C_{ps} \approx 500$ pF for a toroidal or layer-wound EI transformer, $C_{ps} \approx 25$ pF for a dual chamber EI transformer, and C_{ps}

$<$10 pF for a layer-wound EI transformer incorporating an earthed foil electrostatic screen between mains primary and secondary. Thus, the dual chamber transformer reduces common-mode electrostatic coupling by 26 dB (500/25 pF), and the layer-wound with earthed foil electrostatic screen between mains primary and secondary by $>$34 dB (500/$<$10 pF).

4. Reduce the rising slope $>$10 kHz from 12 to 6 dB/octave: Add a 470 nF capacitor from each side of the heater winding directly to chassis in parallel with the previous 47 Ω resistors. The essential is low inductance − every millimetre that can be lost from heater to chassis lead length is worthwhile, so naked stacked foil polyester capacitors are ideal, but layout is critical. The necessary solder tag to chassis does not contribute significant inductance provided it is wide compared to its length, so favour an M5 tag over M3 and benefit from the reduced resistance.

5. Flatten the rising slope $>$30 kHz to 0 dB/octave by fitting an explicit common-mode LC filter: Insert a common-mode choke between the transformer and 470 nF capacitors. Beware that common-mode chokes having useful inductance have significant series resistance, dropping heater voltage, and 250 μH per leg seems to be a practical maximum, so such a filter only becomes effective $>$100 kHz. Don't waste an LC filter on output valves − use it only for small-signal valves drawing minimal heater current that can tolerate the higher resistance of a useful inductance and need protection from radio frequency overload (that would result in demodulation and audible interference).

Note that as we progressed through the previous palliatives, the cost per dB of improvement rose sharply, and the most effective measures by far were 1, 2, and 3.

The choice of unbypassed $r_k = 1 \Omega$ in this investigation came directly from assuming a 6SN7 differential pair with $R_L = 47$ kΩ. Looking towards each anode, we see $(R_L + r_a)/(\mu + 1)$, and at the cathode the two are in parallel, so we see ≈ 1 kΩ. Practical valve differential pairs just don't achieve common-mode rejection ratios (CMRRs) of $>$50 dB, and even this degrades $>$100 kHz, justifying the use of an explicit common-mode LC filter for their heaters. Beware that lumped component models become dubious $>$1 MHz, so predictions above this frequency should be greeted with scepticism.

Once wound, a common-mode choke needs to be secured to the chassis. Not all the flux is confined within the toroidal core, some escapes, and when mounted on a conductive or magnetic plate this causes a $\approx 1\%$ fall in high frequency inductance (alternatively seen as 1% leakage inductance if the choke is treated as a transformer and one winding short-circuited). One per cent loss is not worth worrying about, so we can cheerfully secure the choke to any surface that seems convenient.

Be aware that the listed filtering strategies made two important assumptions:

- The interference source was capacitive coupling from the mains primary.
- Impedances from each secondary leg were balanced to chassis.

Very few transformers have a single secondary, and a pair of identical secondaries is the most common. Unfortunately, bifilar winding is the cheapest way to make a pair of identical secondaries, and measurement showed $\approx 2.5\,\mathrm{nF}$ capacitance between the two 12 V/2 A bifilar secondaries on a 50 VA EI transformer. The significance of such high capacitance is that any interference such as rectifier switching transients is very effectively coupled between secondaries. Whether AC or DC, a quiet heater supply requires an unshared transformer unless low interwinding capacitance can be guaranteed.

If you have the freedom to specify your mains transformer, the best way to reduce interwinding capacitance is to interpose a full width foil electrostatic screen having a low-impedance connection to chassis between offending windings. Thus, there might be a foil screen following the mains primary, then another between two secondaries if one was known to be an interference source. A single turn (not shorting) foil electrostatic screen is reasonably easily added to an EI transformer, but it leaves the ends of the winding exposed, making it ineffective in a dual chamber transformer; effective electrostatic screens require layer-wound construction on an EI core. An earthed foil electrostatic screen between two adjacent layer-wound low voltage secondaries on an EI 50 VA transformer is likely to reduce capacitance between them to $<10\,\mathrm{pF}$. The foil E/S screen is invariably slightly narrower than the windings in a transformer wound on a bobbin, leaving winding ends exposed and increasing interwinding capacitance, but stick winding allows the E/S screen to overhang the windings, reducing interwinding capacitance from <10 to $<1\,\mathrm{pF}$. Thus, the ideal heater transformer is stick wound with an overhanging foil E/S screen between primary and secondary.

Unfortunately, rectifiers and regulators destroy any assumption of balanced impedances, and if DC heaters and regulators have been chosen, interference is likely to be minimised by tying the 0 V at the reservoir capacitor directly to chassis, although this converts any remaining common-mode interference into differential mode.

Heater wiring is the first piece of wiring to go into a project, and it is obscured thereafter by signal wiring. Because it is near impossible to replace heater wiring, it **must** be installed correctly. See Figure 3.29.

Figure 3.29
Good heater wiring is pushed deep into the corners of the chassis and maintains its tight twist all the way to the valve socket.

It is a very good idea to test heater wiring the moment it is in place by plugging the valves in, applying power to the heater transformer, making sure that all the valves glow, and measuring heater voltage at each valve using a true-RMS DVM. If necessary, heater voltage can be fine-tuned at this stage:

• Choose a different mains tapping on the heater transformer.
• Reduce heater voltage by adding a resistor in series with the heater transformer's primary. Although this necessitates a resistor having insulation rated for mains voltage, the required resistance is more likely to be available and the voltage drop slightly reduces magnetising current in the transformer.
• Increase heater voltage by adding an overwinding in series with the transformer secondary. This is most easily done with toroids where a few turns of plastic insulated wire can easily be added, but those few turns should spread over the entire core. Depending on

connection polarity, this technique can also be used to reduce secondary voltage without wasting power (unlike the previous series resistor).

Quite apart from saving later tears and tantrums, a set of glowing heaters gives an important psychological boost before tackling the more complex signal wiring.

If DC heaters have been chosen, don't just check with a DVM, get the oscilloscope out and check for switching spikes — they're easier to cure now than later.

Note that the aluminium foil electrostatic screen over heater wiring mentioned earlier is an effective barrier to capacitive coupling of both common and differential interference from heater wiring into signal wiring. If ultimate interference rejection is needed it's much easier to fit this screen just after heater wiring has been installed than later on when heater wiring becomes obscured. Remember, with the rising number of chattering digital gadgets, interference can only get worse.

Twisting non-heater wiring

Although commercial sleeved copper wire can be twisted perfectly well using a drill, it breaks 99.99% pure (fine) silver wire posted down PTFE sleeving, so this has to be carefully wrapped by hand. Instead of a drill starting a light twist evenly along the entire length of the wire then tightening it, the required wrap must be locally applied by fingers either side of each wrap, starting at one end of the wire and progressing towards the other. It's difficult to impose a tight twist on PTFE sleeved wire because it slips in your fingers, but donning a pair of rubber gloves reduces the problem.

It is extremely difficult to twist even slightly different gauges of wire together, and you might wonder why this should ever be attempted. The most likely scenario is that a flying lead from a mains transformer primary needs to be taken to a single pole mains switch at the far side of the chassis, return, and then go to the mains inlet. In this instance, current flows from the mains inlet to the switch and returns along the adjacent wire, so it makes a great deal of sense to twist those two wires together to reduce hum induction. Unfortunately, unless the mains transformer has **very** long leads, it is unlikely to be able to reach to the mains switch and back, so you are forced to twist it with wire from your stock that is almost certainly slightly different. When you do this, one wire almost invariably remains almost straight, and the other wraps around it. The solution is to gently tension the wire that wanted to wrap, then wrap around it the wire that wanted to be straight. This method is a little fiddly, but enables two different wire types to be twisted evenly, thus ensuring cancellation of hum.

Although solid-core wire perfectly retains a tight twist, multicore wire tends to separate, reducing cancellation. One way to avoid this problem is to pre-tension each wire equally by individually twisting it in the **opposite** direction of the final twisted pair. Without releasing each wire's pre-tension, grip the wires in the vice and chuck, and twist them together. The effect of this is that as the wires are twisted together, the pre-tension is relieved, so there is no latent force trying to separate the final twist. Although this method works very well, solid-core wire retains the tightest twist, and can be positioned more precisely, so it is superior for heater wiring.

Mains wiring

Minimising chassis area containing mains wiring not only minimises mains shock risk, but also minimises magnetic leakage fields because it limits the length of wire having necessarily thick insulation that prevents a tight twist. Mains wiring inevitably generates considerable electrostatic interference fields, so the mains switch is ideally positioned near or on the back panel, with a mechanical linkage to the front panel if required.

Modern semiconductor equipment **sleeves** all mains wiring with insulating sleeving such that it is moderately safe to rummage inside a piece of equipment plugged into the mains, Valve amplifiers operate at such high voltages that it is **never** safe to rummage in powered equipment, and even unpowered equipment should be approached with caution. Safety is not greatly improved by sleeving mains wiring in a valve amplifier, but it remains good practice to sleeve mains wiring with heatshrink sleeving or purpose-made insulating boots that fit over IEC sockets or fuseholders. See Figure 3.30.

Mains switching

To switch a piece of equipment off, we simply break the circuit from the source of power. A mains switch would work equally well in the live or neutral conductor, and this is known as single pole switching. However, a neutral switch leaves all internal mains wiring

Figure 3.30
IEC input connector sleeved with insulating boot.

live, constituting an internal shock hazard, so single pole switching **must** switch the live conductor to eliminate this hazard.

Double-pole switching breaks both the live and neutral conductors, and ensures safety even if the live and neutral conductors are transposed. Commercial electronics often uses non-polarised figure-of-eight connectors, allowing live and neutral to be transposed, requiring double-pole mains switching.

Where there is **no possibility** of live/neutral transposal, single-pole switching of the live conductor is safer and more reliable, because the (more likely) open-circuit mains switch failure ensures a break in the live connection to circuitry. A double-pole mains switch has twice as many contacts to fail, and should the neutral contact fail open-circuit, the equipment could appear to be safe (passing no current), despite much wiring remaining connected to the live conductor and constituting a shock hazard.

Fuses

A fuse is a piece of fine wire having resistance, connected in series with the circuit to be protected. If excessive current passes through the wire, it heats in accordance with I^2R, and heats sufficiently that it melts, or ruptures. A fuse is a single-pole switch, and should therefore be connected in the live wire, **before** any other circuitry, including the mains switch.

The fusible link in a fuse has mass, so it requires a defined amount of energy ($E = Pt = I^2Rt$) to reach melting temperature. However, heat flows to the surroundings, so a mild overload could allow much of the heat that should have melted the fuse to escape, whereas a short-duration gross overload cannot lose so much heat, and ruptures the fuse quickly. As an example, a 13 A fuse to BS1362 (as fitted to a UK 13 A domestic plug), ruptures in 0.4 s at 100 A, but needs 10 s at 50 A.

To calculate the mains fuse rating, the total power consumption of each individual load on the mains transformer should be summed to find the total load taken from the mains. The current drawn from the mains may now be found, and the fuse rating should be the next rating above this. This calculation contains many sweeping assumptions and approximations, but fuses are not accurate either, so the method will be found to be satisfactory.

Some equipment, particularly if it includes a toroidal mains transformer and/or large reservoir capacitors, draws a large inrush current, but its working current is much lower. To cope with these requirements, **anti-surge** or **timed** fuses are available, which can withstand short overloads. These fuses usually have their ratings preceded by a "T", so T3.15A refers to a timed 3.15 A fuse. Conversely, **fast** fuses precede their current rating with an "F".

Protecting each output of a multiple winding mains transformer in a valve amplifier is awkward for the following reasons:

- Fuses are very rarely fitted to HT supplies because they offer only very limited protection to the output valves. In a Class A amplifier, the output valves are usually run at exactly their maximum anode rating, so a doubling of anode current quickly causes damage, and a fuse may not blow with an overload as small as this, so little protection is offered. Fuses can be fitted to Class AB amplifiers, and are advisable for OTL designs, but their non-constant resistance can cause distortion. Further, should an arc be struck as the fuse blows, AC would extinguish the arc as the cycle sweeps through 0 V, but the unchanging DC of an HT supply would maintain the arc, rendering the fuse useless.

- Fuses are never fitted to heater supplies because heater circuitry is normally too simple to warrant a fuse. Further, failure of a heater supply often causes damage elsewhere in a DC coupled amplifier as valves switch off and anode voltages rise to the full HT voltage.
- Grid bias supplies to output valves should **never** be fused because failure of this supply would immediately destroy the output valves.

For these reasons, valve amplifiers rarely have any fuses other than on the incoming mains.

Glass-bodied fuses should **never** be used to protect high voltage circuits such as AC mains. A short circuit causes the fuse to rupture instantly and vaporise, thus depositing a conductive metal film on the inside face of the glass, which continues to pass a small current, but heats the glass. If glass is heated sufficiently, it becomes a conductor in its own right, so the fuse has failed to protect the circuit. Fuses suitable for high voltage use have ceramic bodies filled with sand to prevent the creation of a continuous conductive film.

To prevent tampering, mains and high voltage fuseholders accessible from the outside of the chassis must require a tool to release the fuse. Although conceivably accessible by a strong fingernail, fused IEC inlets sidestep the problem by forcing removal of the incoming mains lead before access can be gained.

Class I and Class II equipment

Class II appliances have all hazardous voltages (> 50 V) **double insulated** from contact with the operator, and use a twin-core double-insulated mains lead. Double insulation requires two insulating barriers, one of which may be air, each independently capable of withstanding the shrouded voltages and protecting the user. It is possible to make a Class II appliance that has exposed metal, but rigorous testing is required to ensure that the appliance meets the full technical standard. A printed symbol consisting of two concentric squares signifies double insulation.

Class I appliances require only one layer of insulation from hazardous voltages, but this layer must be totally shrouded by a conductive layer **bonded** to mains earth via a low-resistance path (see later). It is far easier to make equipment that conforms to Class I than Class II, so amateur equipment should **always** be built to the Class I standard to ensure safety.

These classes of insulation do not merely apply to the appliance, but also to the mains cable from the wall socket. An earthed conductive sleeve such as coaxial braid develops voids when flexed and cannot be relied upon as a conductive barrier, so mains cables are invariably double insulated, leading to a thick external insulating sheath over their insulated conductors.

Earthing

Earthing is often the cause of hum problems, but once considered logically, there is no need for it to cause any hum problem whatsoever. Colloquially, the term "earthing" refers to the mains earth safety bond to the metal chassis, and also to the 0 V signal wiring, but the two are quite distinct.

Earth safety bonding

The three wires leaving a domestic outlet are line, neutral, and earth. Neutral and earth are connected together at the substation, or possibly at the electricity supply company's cable head within the house. This means that if line contacts earth, a fault current flows, determined by the **earth loop resistance**, which is the entire resistance around the loop, including the resistance of the line wires. See Figure 3.31.

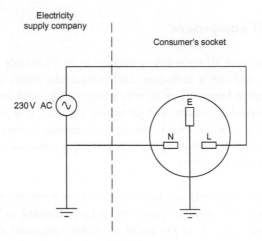

Figure 3.31
How an earthed chassis and a fuse protect against shock.

The purpose of the mains earth safety bond is to provide a sufficiently low-resistance path such that should the line conductor come into contact with the exposed metalwork (which would then become a shock hazard), the resulting line to earth fault current is sufficiently great to rupture the fuse **quickly**. Because the time taken for a fuse to rupture is proportional to the square of earth loop resistance (fuses have I^2t ratings), there is no such thing as an earth loop resistance that is too low.

Although exposed valves may superficially appear to conform to Class II, because the electrodes are insulated by a vacuum **and** the glass envelope, once the envelope is broken, the secondary layer of insulation also disappears. To ensure conformity, valves on the top of the chassis require enclosure by a perforated metal cover to meet Class I, or insulating barriers that meet Class II.

If we build an amplifier on a chassis with exposed metal, then the construction must conform to Class I, with all hazardous voltages insulated from and totally enclosed by the earth bonded metalwork. The ideal place for the earth bond is close to the entry of the power cable. The ideal earth bond would weld the incoming earth wire from the mains cable directly to the chassis, but this is rarely practical. A practical alternative takes the incoming earth directly to a mechanical soldered joint wire onto an M6 solder tag. The tag is then screwed to the chassis with a serrated washer either side of the chassis, and another serrated washer above the earth tag, followed by a flat washer (to prevent the tag rotating when the bolt is tightened), then secured with a locknut. The nut and screw should be firmly tightened with a large screwdriver and a spanner **after** soldering, otherwise the secure thermal bond to the chassis prevents the iron from heating the tag, resulting in a poor (high-resistance) joint. See Figure 3.32.

A thick over-rated mains cable minimises earth resistance to the bond point. Although it is permissible for 3 A rated equipment to have 0.5 Ω of resistance from the pin of the mains plug to the chassis (**not** measured directly at the bond point), reducing this resistance to <0.1 Ω, by using 2.5 mm² mains cable, reduces hum and maximises fuse current (minimising rupture time) in the event of a line to chassis fault, improving safety.

The preceding arguments apply to equipment that is directly powered from the mains, but valve pre-amplifiers often have remote power supplies. Nevertheless, the same arguments

Figure 3.32
A good earth bond maintains low contact resistance for the life of the equipment (this bond is 25 years old, yet contact resistance measures 2 mΩ).

can still be applied, so a substantial cable should extend mains earth to the pre-amplifier chassis, and the bonding technique should be the same. Likewise, mains motors on turntables should be earth bonded via their mains cable, and not via a flimsy pick-up arm lead loosely connected to a possibly indifferent mains earth.

Sometimes there will be conductors that could come into contact with high voltages but do not have a guaranteed electrical path to the mains earth bond, and examples are:

- The acoustically suspended motor on a turntable.
- A mains transformer or choke that is acoustically isolated because of its vibration.
- Any separate anodised aluminium panel supporting a mains connection, such as a front panel mains switch, or mains transformer on the baseplate.

Each of these must have a connection, such as a wire or a screw, via a serrated washer, electrically bonding them to the mains earth bond, but in the first two examples it is important that the wire not be so stiff so as to short-circuit the acoustical isolation, implying a loop or short helix of wire.

0 V system earthing and unbalanced signals

It is the 0 V signal earth connection to chassis that causes the hum due to loops between multiple earths, **not** the mains earth safety bond.

Hum loops are circuits within earth paths that can have hum currents induced into them by nearby mains transformers or by leakage currents caused by mains filters. The circuit inevitably contains resistance, so an unwanted voltage is developed, and if the circuit is common with the audio circuit, the unwanted voltage causes the audible hum. See Figure 3.33.

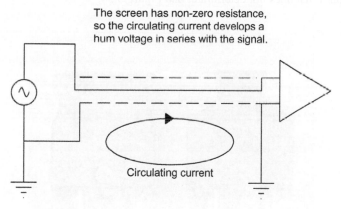

The screen has non-zero resistance, so the circulating current develops a hum voltage in series with the signal.

Circulating current

Figure 3.33
How a hum loop causes audible hum.

To remove the hum, the loop must be broken, and this could be done by disconnecting the earth wire within the mains plug of one piece of the equipment, **but this is extremely dangerous**. The loop should be broken by removing the 0 V signal earth bond to chassis within one or more pieces of equipment.

Breaking the 0 V signal earth to chassis bond

Fortunately, commercial equipment intended for an international market tends to be double insulated, so hum loops do not often occur, but a useful improvement on Class I equipment is to provide a **ground lift** switch or pluggable link that can make or break the 0 V signal earth to chassis connection at will on each piece of equipment. This method allows the optimum 0 V system earthing arrangement to be determined quickly and safely.

Switches certainly start with low-resistance contacts, but resistance rises as the contacts tarnish, especially if they are not cleaned by periodic use, as would be the case for a 0 V signal earth bond switch, so a pluggable link is far better. The best pluggable link is the **U-link** that was used by telecommunications companies as the interface between their equipment and the customer's equipment in private telephone exchanges. The British telecommunications U-link had a pair of linked 4 mm plugs on ½″ centres and provides a low-resistance path that does not deteriorate. See Figure 3.34.

Figure 3.34
A U-link intended for telecommunications use make an ideal removable 0 V signal earth to chassis bond.

The U-link plugs into two 4 mm chassis sockets, one insulated (connected to the 0 V signal earth) and the other uninsulated and bonded firmly to the chassis via serrated washers.

Ideally positioning the 0 V signal earth to chassis bond

We have seen that to avoid earth loops, there must be only one 0 V signal earth bond to chassis, and hence mains earth. We should now consider the optimum position for this single bond.

An amplifier amplifies the **difference** between its two input terminals. We tend not to think of the 0 V signal earth terminal as an input, but it is. In order to screen an amplifier, we surround it with a conductive casing/chassis connected to the amplifier's 0 V signal earth. Inevitably, there is capacitance from the chassis to mains. Similarly, there is capacitance from 0 V signal earth wiring to mains. Both capacitances return currents to the point where they are connected together. The further we are from the bond, the larger the unwanted voltage drop developed across the intervening (non-zero) impedance.

A moving magnet RIAA stage having a sensitivity of 2 mV at 1 kHz has a sensitivity of 284 μV at 50 Hz, so it cannot tolerate any interference voltage due to unwanted voltage drop. Thus, the optimum place to bond the 0 V signal earth to chassis is at the input of the RIAA stage. The bond wire should be as short and thick as possible to minimise its impedance and ensure that it is a good bond even at radio frequencies.

Audio interconnections between equipment

It is convenient to use alternating current for power because transformers allow power to be delivered at any convenient voltage or current, but the disadvantage is that there is always an electrostatic and magnetic leakage field that can induce 50/60 Hz hum in signal cables. Reducing a signal cable's susceptibility due to electrostatic fields is easy — just add an earthed conductive screen. Reducing susceptibility to magnetic fields is harder, and determined by:

- Loop area: Signals are currents that must flow from their source, through the load, and return to the source, requiring a loop comprising flow and return conductors. Unless the area enclosed by that loop is zero, when immersed in a magnetic field one conductor encounters a different flux to the other and unequal currents are induced. Hum is proportional to the **difference** in currents.

- Series resistance: If the two conductors had zero series resistance, there would be zero voltage drop along their length due to the induced currents, and no hum. If the two conductors had non-zero but equal resistance, equal longitudinal voltage drops would be developed, but there would be no potential difference between the conductors at the far end, and no hum. If the two conductors had non-zero and unequal resistances, the longitudinal voltage drops would be unequal and there would be a potential difference between the conductors at the far end, causing hum.

If the two signal conductors could be made coincident, loop area would be zero and they would encounter identical magnetic fields, inducing equal currents, and no difference current. Coincidence can be approximated by a tightly twisted parallel pair or coaxial construction (the screen's average position is coincident with the core).

Unfortunately, coaxial cables invariably have unequal core and screen resistances. As an example, BBC PSF1/2 video cable has a double-braided coaxial screen having a measured resistance of 5.5 mΩ per metre, but the core resistance measured 33.7 mΩ per metre. The potential difference between the two conductors due to the induced current "i" is:

$$v = iR_{core} - iR_{screen}$$
$$= i(R_{core} - R_{screen})$$

Thus, it is the **difference** between the resistances that is important, $R_{imbalance}$.

The induced current "i" must circulate the entire loop, whose resistance is dominated by the destination's load resistance. Alternating magnetic strengths are commonly specified in terms of volts per metre (especially at radio frequencies), so the potential difference between the two conductors due to imbalance becomes:

$$v = \frac{El}{R_{load}} \cdot R_{imbalance}$$

$$= El \cdot \left(\frac{R_{imbalance}}{R_{load}} \right)$$

where:

E = magnetic field strength (V/m)
l = length of cable (m)
$R_{\text{imbalance}}$ = screen resistance (Ω)
R_{load} = input resistance of destination amplifier.

This equation tells us that longer cables pick up more magnetic hum (no surprise there), and that it is the ratio between the imbalance resistance and loading resistance that is important. Moving coil cartridges are invariably loaded by quite a low resistance (typically $100\ \Omega$–$1\ \text{k}\Omega$), so to render magnetically induced cable hum insignificant, the imbalance resistance needs to be as close to zero as possible, and a screened twisted pair should connect from the pick-up arm to RIAA stage, not coaxial cable.

If a screened twisted pair is used to connect unbalanced signals (one leg connected to earth), it is important that the screen is connected to 0 V signal earth at the source end. The screen should be earthed at the source end because the source has low output impedance that can firmly define a voltage difference between the two output wires. One of these wires is connected to the screen of the output cable. The screen picks up radio frequency interference, which it superimposes onto the commoned signal wire. The radio frequency interference is also superimposed onto the other signal wire via the output impedance of the source, and if the source has a truly zero output impedance at radio frequencies, both output wires now have the full interference superimposed on them. This might seem undesirable, but at the amplifier end the input stage responds to the **difference** in signal between the two input wires, and therefore rejects the interference that is identical on both input wires.

If instead we connect the cable's screen to earth at the destination end, the induced interference picked up by the screen now has to travel down the entire length of the cable to the source before it can be coupled via the output impedance of the source to an inner wire. Once coupled, the interference then has to travel the entire length of the cable before it can arrive at the amplifier input. The interference signal on one wire has now had to travel twice the length of the cable, but has suffered the effects of series inductance and shunt capacitance, which form a low-pass filter. This means that one of the wires at the input to the amplifier has the full interference signal and the other has a filtered signal, resulting in a difference signal to which the amplifier is sensitive.

If a transformer is used on the interconnection between two pieces of equipment, its ideal position is at the receiving equipment's input, not the source's output. This is because a zero-impedance source forces induced EMFs in the cable to be the same even if one leg is grounded, but adding an output transformer raises source impedance. Provided the earth is switched as well as the signal, a single input transformer can break earth loops from a number of domestic sources, whereas the thicker wire needed to minimise an output transformer's output resistance requires a larger core to accommodate the same number of turns, making it more expensive, yet it breaks only one earth loop.

Internal earth wiring of amplifiers

Once within an amplifier, the 0 V signal earth path can either travel in the same way as it did between equipment, in which case, it is known as **earth follows signal** [1] or it can be **star earthed**.

Earth follows signal is the traditional method of wiring valve amplifiers, and is the easiest to do properly. The traditional method uses an **earth bus-bar**, which is a thick (typically 1.6 mm diameter, or 16swg) tinned copper wire 0 V signal earth connected directly from input sockets to the 0 V signal earth, or source, of the power supply (commonly the reservoir capacitor or, if fitted, the 0 V output of the regulator).

The repetition of "0 V signal earth" may seem pedantic, but it reminds us that the signal earth also carries power supply currents. This last factor is extremely important, since the power currents are often many times larger than their associated signal currents. Because individual stages amplify voltage differences referenced to their local earth, the earth bus-bar needs a low impedance to minimise the spurious voltages developed by power supply or signal currents passing through it. This last point is often referred to in EMC texts as **common impedance coupling** because it is the presence of a non-zero impedance common to two circuits that causes coupling between them. Thus, we should either minimise the impedance (bus-bar) or minimise the common circuit (star earth).

The brute force precaution of having a low-impedance earth bus-bar is not sufficient, and connections must be made to it in the correct order so as to avoid developing voltages in sensitive input circuitry (minimising the common circuit). The author remembers an RIAA stage that had been constructed to the Mullard two-valve design, but had considerable hum. The hum was cured by moving **one** wire 150 mm along the (1.6 mm) earth bus-bar.

The correct order from the input socket is: input circuitry (such as shunt capacitors, etc.), grid leak resistor, cathode bypass capacitor (if fitted), cathode resistor, any anode signal circuitry (such as equalisation), next valve's grid leak resistor, etc. If in doubt, think **current** rather than voltage. Decide whether you would be happy for a particular current to develop a voltage drop at a particular point along the bus-bar, and whether that voltage drop would then be amplified.

Depending on earthing arrangements, the earth bus-bar may need to be firmly bonded to the chassis via a solder tag. Traditionally, one of the screws retaining the input valve socket was recommended for the earth bond, but all of the rules regarding mains safety earths still apply, and a few milliohms of contact resistance is sufficient to cause an irritating hum, so an M6 or ¼″ solder tag that can be tightened down firmly onto the chassis via an intervening serrated washer is much better than the typical fastener that can pass through a B9A socket. If a valve socket has a central spigot it should be bonded to chassis as this reduces capacitance between valve pins.

To make a neat earth bus-bar, the thick wire needs to be straightened, and this is not a trivial task. The traditional way to achieve this is as follows:

Grip one end of the wire in a vice, and then grip the other end in a substantial pair of pliers, such as would be used for working on a car. The wire is than wrapped one turn round the jaws of the pliers, and the pliers are firmly gripped with both hands whilst one foot is braced against the vice. The wire is then firmly pulled until it can be felt to stretch, and **without moving the position of the pliers** is cut at the vice end. A kink-free length of wire results, and this can now be cut away from the pliers.

It should be realised that considerable force is required to achieve this result, and this can apply dangerous forces to your back if you do not position yourself correctly. If you are in any doubt as to how to position yourself, or have back problems, **do not attempt to use this method**.

Unfortunately, tinned copper wire oxidises over time, and becomes difficult or impossible to solder, but a beautifully shiny and easily soldered surface can be restored using metal polish intended for brass. A wipe with isopropyl alcohol removes the last traces of polish and enables perfect soldered joints. The polish trick would probably also work with tinned PCBs that have lain around for years before being populated but might remove silk screen lettering.

Rather than using a bus-bar of circular cross-section, flattening it leaves resistance unchanged but reduces inductance, yet we still need thickness to maintain rigidity, so a 1 mm × 10 mm copper or silver strip becomes a possibility. The very large cross-sectional area ensures low resistance, whilst the width lowers the inductance and makes soldering fractionally easier. Holes can be drilled into the bus-bar to match the component wire's diameter, or V notches filed at the edge to locate the component whilst soldering. The wetted iron should first be pressed firmly to the strip to pre-heat it, then scraped to touch the component lead, and solder applied to form the joint. A 200 W iron is needed, as a smaller iron takes so long to raise the joint to soldering temperature that heat has time to flow to distant components, damaging them. The strip bus-bar is an expensive and not terribly effective solution, but can be useful for power amplifiers where components tend to be quite large.

The ultimate expression of the earth bus-bar is the radio frequency earth plane, which is a two-dimensional conducting earthed surface to which earth connections are made (a wire is considered one-dimensional). This construction is now common on audio printed circuit boards as designers have realised how important radio frequency immunity is to audio circuitry. On a multilayer PCB, an entire copper layer can be used as an earth plane, which has low inductance because it is so wide, and therefore guarantees a good radio frequency earth to every point that contact is made. It is very hard to argue a good case for **not** using a PCB with earth plane for an RIAA stage.

Some classic valve amplifiers (such as the Rogers Cadet III) approximated to an earth plane by soldering to conductive posts pressed directly into the (steel) chassis, and using this as the 0 V signal earth. Neither this method nor a steel chassis is recommended for permanent construction, but tin plate biscuit or tobacco tins are ideal enclosures for proof-of-concept prototypes or test jigs because earth connections can be soldered to the "chassis" and sensitive sections can be quickly partitioned from one another if necessary with more tin plate. Beware that such thin metal snatches easily when drilled and leaves large sharp burrs that need cautious removal.

Star earthing is achieved by having a **single** earth point, usually bonded directly to chassis, to which all 0 V connections are brought individually, thus minimising their common impedance. Ideally, all the connections to this point are made with short leads to minimise inductance and interference pick-up (they're all aerials), but building an entire amplifier using this method can become messy. See Figure 3.35.

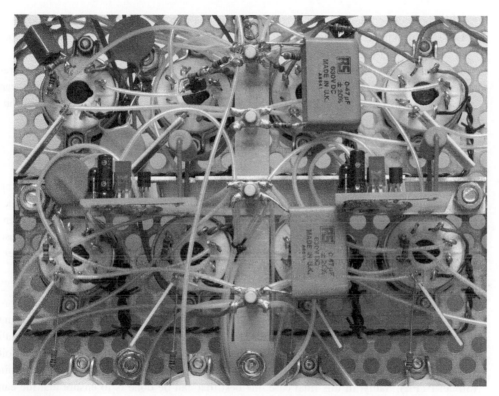

Figure 3.35
True star wiring uses many wires, but is manageable with care. Stars from top to bottom: 0 V signal earth, +270 V, +160 V, −335 V.

Because of the difficulty of making so many 0 V connections to a single tag, traditional constructions often used a combination of star and bus-bar earthing, with the input stage star earthed, and following stages earthed to the bus-bar.

Earthing capacitor cans

Traditional paper-in-oil and Soviet PTFE capacitors are hermetically sealed in steel cans, and there are two reasons why a deliberate connection might need to be made between such a can and chassis:

- If earthed, the capacitor's conductive can provides one of the safety layers required of Class I equipment. This requirement tends to apply to the larger painted steel can paper in oil capacitors used for power supply smoothing.
- The capacitor is being used for coupling the audio signal and must be screened to prevent hum. This is quite a rare requirement as it is more usual to leave the can unscreened in order to minimise stray capacitance to earth, but a physically large coupling capacitor might pick up sufficient interference as to make earthing essential.

The problem lies in making a long-term, low-resistance connection to the can.

The first type of capacitor tends to be quite large and usually requires a separate clamp that cannot be relied upon to cut through the paint and make a long-term gas-tight connection, particularly as the chassis flexes over the years and clamps loosen. The only guaranteed solution is to scrape a little paint off the rolled seam at the bottom of the can and use a soldering iron having a large tip to solder a wire quickly to the (usually quite thin) steel can, then bond this wire to the chassis in the usual way.

The second type of capacitor might be of a similar construction to the first (and therefore require the same solution), or it might have integral flanges welded to the can, allowing serrated washers to make a gas-tight joint. However, it is more likely that the can is small and soldering is impractical either because it is aluminium (most old small paper-in-oil capacitors) or the can is so thick that the required heat would damage the capacitor within (Soviet PTFE). Sadly, in this final case, the best that can be done is to ensure that the can and securing clamp are clean and bright before fixing.

Rectifiers and high-current circuitry

Although bridge rectifiers are commonly available, they invariably use epitaxial diodes and we might want to make our own bridge rectifier using better diodes. If we choose diodes that are in an insulated package (such as STTH512F, rather than STTH512D), then fitting them to an aluminium angle bracket (to allow heatsinking) is easy. See Figure 3.36.

Figure 3.36
Insulated diodes make bridge rectifier construction easier.

Some parts of an amplifier unavoidably carry high currents. In a capacitor input power supply, the loop from the mains transformer via the rectifier to the reservoir capacitor carries the capacitor ripple current, so it is essential that no connections are made to the 0 V signal earth **within** this loop because the large current would develop an interference voltage across the unavoidable wiring resistance. The two circuits are split at the reservoir capacitor. Assuming that the reservoir capacitor has screw terminals, it is best to connect the solder tag directly contacting the capacitor to go to the rectifier/transformer and the one further from it to go to the load. (This prevents the high currents in the capacitor/rectifier/transformer loop from developing a voltage in the external load circuitry.) See Figure 3.37.

Figure 3.37
Star connections to a reservoir capacitor reduce interaction between incoming and outgoing currents.

In order to minimise the electromagnetic field caused by the passage of transformer to reservoir capacitor currents, the capacitor/rectifier/transformer loop should ideally be a twisted pair to minimise its area, and be as short and thick as possible to reduce its resistance and consequent voltage drop.

A power amplifier's output stage is effectively supplied from a single capacitor, so output valve cathodes should be brought individually to the 0 V terminal to form a star. Similarly, individual wires from output transformers should be brought individually to the HT terminal to form a star. See Figure 3.38.

Figure 3.38
Star connections to this capacitor feeding the output stage minimise interactions between valves.

The output to the loudspeaker from a power amplifier is also a high-current loop, and additional connections to sense this voltage, such as global negative feedback, should be made very carefully. The ideal method is to connect a screened twisted pair to the output terminals, with the screen connected to chassis at one end only (to prevent loop currents flowing in the 0 V signal earth). This cable is then routed to the input stage, where one side is connected to the lower end of the cathode resistor, and the other is connected via a series resistor to the cathode (assuming cathode feedback). This method ensures that the feedback voltage is derived from the correct point and presented to the correct point. See Figure 3.39.

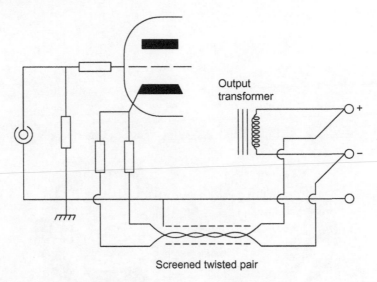

Screened twisted pair

Figure 3.39
Screened twisted pair derives the feedback from the correct point (amplifier output terminals) and prevents induction from transformers entering the feedback loop, then applies the feedback at the correct point.

Balanced interfaces

Almost all the previous wiring considerations were required because the amplifier amplified the voltage difference between two points, one of which was the 0 V power supply rail (which is invariably also connected to chassis and mains safety earth). If the connection to the 0 V supply rail could be broken or made irrelevant, interfering voltage drops due to hum loops or power supply ripple would also become irrelevant. One way to do this is to use a balanced interface between pieces of equipment, carrying the signal over a screened twisted pair, terminating in a connector having three pins, one for screen, two for the signal.

The traditional connector is the XLR (pin $1 = \mathbf{X} = $ chassis, pin $2 = $ signal **Line** $+$, pin $3 = $ signal **Return** $-$).

Balanced interfaces become more important as audio signals become smaller, cable lengths increase, or more signals are involved. Thus, a mixing desk in a control room receiving many microphone signals from an acoustically isolated remote studio invariably has balanced inputs. The one domestic signal that is a clear-cut contender for a balanced interface is the vinyl cartridge. Balanced interfaces always drive input circuitry designed to reject common-mode interference, whether it is a transformer or some form of electronic balancing, and because transformers are expensive, much ingenuity has been shown in developing electronic circuits that reject common-mode interference.

Although circuit design is outside the scope of this book, construction practicalities are very much within, so we must consider the XLR connector and its cable. The purpose of the cable screen is not simply to cover the twisted pair, but to extend an unbroken screen from one chassis to another. It is expected that interference currents will flow in the screen, so it must make a low-impedance connection to the chassis at each cable end, and that means minimising pigtails. XLR chassis connectors have a chassis tag near pin 1 and the two should be connected. If the amplifier needs a connection between chassis and 0 V power supply rail, then that is a separate connection.

Layout of small components within a chassis

There are various approaches for positioning the smaller components such as resistors and capacitors, ranging from one extreme to the other:

1. Position the components in a regimented manner in neat lines, and use neatly harnessed wiring to make the complex interconnections. The parallel wires in the harness inevitably increase stray capacitances, but this method was used in the Quad II power amplifier. See Figure 3.40.
2. Position the components in a regimented manner in neat lines, and either use exposed neat wiring (Mullard recommendations) to make the interconnections between components, or use less tidy but hidden wiring arranged for lower capacitance (Leak and Tektronix). These techniques reduce capacitances compared to method 1, but can take up significant space. See Figures 3.41 and 3.42.

Figure 3.40
The Quad II harnessed its wiring into a loom, greatly increasing stray capacitance.

Figure 3.41
This fine example of a Mullard 5-20 made by John Lavender was spotted at a radio fair.

3. Methods 1 and 2 had regimented rows of components enforced upon them by the layout of the tags on the boards. If that restriction is waived, then a small resistor no longer requires the same space as a large coupling capacitor, and space is saved. The hidden wiring becomes copper tracks and the board becomes a printed circuit board (PCB).

Figure 3.42
This 10 W amplifier is one of a pair used in an "old technology" TV monitor project and receives power from a central power unit rather than having its own supply.
Courtesy of Brian Terrell.

4. The preceding methods are two-dimensional. Interconnecting wire length can be further reduced if the assembly becomes three-dimensional. Components are now soldered directly to valve sockets and, if necessary, to nearby stand-offs, or perhaps a tag strip. Done well, this construction method offers low stray capacitance and leakage, but it needs considerable thought beforehand to keep the construction tidy and accessible for testing/maintenance. Tag boards, tag strips, and stand-offs are all useful ways of supporting components and joints insulated from chassis. See Figure 3.43.

5. At valve impedances, practical inductances are utterly insignificant except in the 0 V bus-bar and power supply decoupling capacitors, so good layout in a valve amplifier minimises stray capacitances, and the best way of reducing capacitance between wires or components is to cross them at right angles. Capacitance to earth is minimised by short leads. One approach that is not immediately obvious is that sleeved wires have higher stray capacitance than self-supporting bare wires because $\varepsilon_r \geq 1$. It is the combination of these requirements that forces component layouts that appear to have been thrown together by an avant-garde sculptor. We thus arrive at the surprising conclusion that a good component layout probably looks untidy, but the converse is unlikely to be true. See Figure 3.44.

Figure 3.43
A selection of tagstrips, tagboards, and stand-offs.

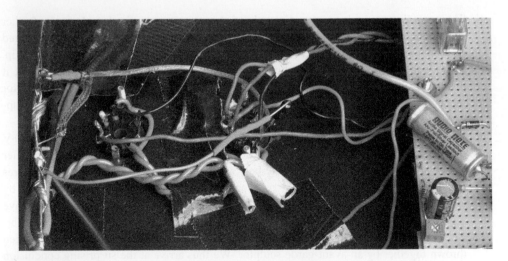

Figure 3.44
Bad practice: How many howlers can you spot?

This appalling example features a variety of gaffes:

- Gaffer tape: Gaffer tape is for bodging on stages and studio floors, not for putting inside electronics.
- Terrible soldering: As a single example, the thickest wire on the earth bus-bar was clearly soldered by an iron having an insufficiently large tip, resulting in poor solder flow and damaged insulation.
- Earth bus-bar: The earth bus-bar should be associated with the valves, not the input socket.
- If the input valve had been close to the input socket, screened lead would not have been needed to transfer the signal.
- Screws: The screws securing the valve sockets are too long. It's very difficult to install good heater wiring when your fingers keep hitting obstacles.
- Heater wiring: The wire is too thick, and stranded, and because the wire wasn't pre-tensioned before twisting, these factors resulted in a loose twist just where a tight twist was most important (near the small-signal valves).
- Diagonal components on matrix board: Diagonal components do not themselves cause problems, but they **do** indicate a poorly planned layout.

Whether the layout is on PCB or hard-wired, good layout requires considerable care, and thermal considerations must also be accommodated.

PCBs

PCBs have been mentioned several times, and the author uses them frequently, but they are not ideal for the novice because they are really a production method of construction, and it requires considerable confidence to design a theoretical circuit and commit it directly to a PCB. Whilst it cannot be argued that a PCB gives a thoroughly professional appearance to the finished project, the high resistance of 35 μm copper foil compared to a 1.6 mm-diameter bus-bar means that we cannot rely on brute force to minimise common impedance coupling, but must minimise common paths between circuits. Thus, although potentially the best from an EMC point of view, it is easy to make a PCB having poor layout, resulting in measurably inferior performance compared to hard-wiring. Whilst theoretical circuit design is important, it is only half the story, and as signal voltages and currents fall and operating frequency rises, layout becomes more important. Despite these caveats, making a single-sided PCB is not nearly as difficult as you might think, and it doesn't need specialist equipment.

Choosing PCB valve sockets

Obviously, it has to be the right kind — a nine-pin B9A is no use for a valve having an International Octal base, but there are other considerations:

- Ceramic bases are far less leaky than phenolic. This could be important in a low-noise, high-impedance application, but if leakage is crucial, use a chassis mounting PTFE socket and hard-wire — air is the best insulator. Incidentally, there is no point in buying an expensive low-leakage PTFE socket then degrading it by mounting it on a (comparatively) leaky FR4 PCB.
- A glazed ceramic socket can't absorb water, making it a better insulator. On the other hand, if it's a modern socket and glazed, it may have poor pin tension, ultimately leading to intermittent contact.
- Beware dirty unglazed ceramic. If NOS sockets have been stored loose, the tinning on the pins can rub off onto the (abrasive) ceramic forming a surface leakage path.
- Centre spigot: If you haven't had to route tracks across the middle of the socket, an earthed centre spigot reduces leakage currents and capacitance between pins. If you were forced to route tracks across the socket centre, choose a socket without a spigot, or carefully curl it into a smaller diameter with pliers to remove it neatly (quite easy to do).
- Long pins lift the valve out of the component forest, allowing better cooling. They also allow high g_m valves like E88CC/6922 etc. to oscillate more easily because they add inductance before the grid-stopper; you probably want short pins.
- Some sockets have 3 mm-wide pins (low inductance) necessitating an expensive narrow milled slot. The traditional solution simply drilled a 3 mm-diameter hole and accepted the resulting poor soldered joint. Try to use the type where the pins are cut down to ≈ 1.2 mm wide on the last 4 mm.

Many different types of PCB socket have been made, so it's worth searching for the most appropriate one for each application. See Figure 3.45.

PCB layout

The hardest part is working out the layout. Unfortunately, achieving a good layout is very much like learning to ride a bicycle — logical once you have experience, but tricky to describe to a novice because so many factors have to be taken into account simultaneously.

Figure 3.45
A selection of ceramic B9A PCB sockets. Note the dirty (and possibly leaky) ceramic on all but the second from right socket.

Points to remember are:

- If the circuit diagram was drawn well (connections in the right order), a good PCB will be laid out very much like the diagram.
- Most traditional leaded components (except valve sockets) fit a 0.1″ grid.
- You are working from underneath (the foil side), so if you use ICs, remember to work out their pin-out as viewed from underneath (IC manufacturers' diagrams give the pin-out viewed from above, whereas valve manufacturers give the view from underneath).
- A simple audio PCB should not require any links. If you need links, you probably haven't tried hard enough (or you used the autorouter in a PCB design package).
- PCB foil is thin (typically 35 μm), so it is vital to think about currents and ensure that you don't pass heavy currents through sensitive areas. Remember that you can always widen the track, or star tracks, to a reference point to minimise interaction between currents. See Figure 3.46.
- PCBs inhibit cooling, so try to put hot resistors or transistors near the edge where they can be bolted to a heatsink or can cool naturally. If you can't do that, raise them well clear of the board and give them space from other components to allow cooling air to flow round them.
- Don't put power valves on a PCB — it's just asking for thermal problems.
- When you put valves on a PCB, don't even attempt to do heater wiring using PCB tracks, it makes layout of the audio almost impossible, and heater wiring is far better done afterwards with twisted pair spaced away from the board so that it rests snugly against the chassis. See Figure 3.47.

Figure 3.46
Star earth and power on a PCB.

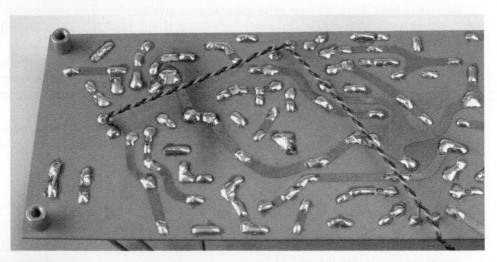

Figure 3.47
Heater wiring on a PCB is best achieved with a twisted pair.

- Surface mount resistors make excellent grid-stoppers, and really aren't that difficult to use, but in order that surface tension aids rather than hinders positioning, it's important that your PCB tracks are the same width as the component. You need the smallest tip on your iron and <0.5 mm silver-loaded solder. If not already tinned, tin the pads where the resistor will go using the absolute minimum of solder, position the resistor with tweezers, and with a very lightly wetted iron, reflow the solder at one end to form a badly soldered joint. Even using tweezers designed for surface mount components, it can be tricky to position the component flush to the board, so once reflowed, the author then uses the tips of the tweezers to press the component firmly to the board, and reflows again. Clean the iron tip, and solder the other end properly with a little fresh solder, then return to the first joint and solder it properly with a little fresh solder. It's actually more difficult to describe the process than to do it! If you accidentally apply too much solder (easily done), 2 mm-wide desolder wick easily removes the excess.
- Pre-amplifiers tend to need many earth connections, so a double-sided PCB with the component side used as an earth plane not only reduces common impedance coupling, but provides electrostatic screening. If you use a homemade double-sided board (without plated-through holes), radial leaded capacitors such as Wima FKP1 obscure their leads on the component side, so any earth connection requires its lead to be bent horizontally (without stressing the component) to solder directly to the earth plane. They don't, therefore, need a hole to be drilled through the board for this connection. Professionally made PCBs have plated-through holes, so the previous example can have holes at both ends.
- A power amplifier's driver stage usually has its valves showing on the top of the chassis, but PCB capacitors can sometimes be quite tall, forcing the PCB to be mounted so far below the chassis that a valve envelope barely peeps through the chassis, restricting cooling. In this instance, it may be better to mount the valve sockets on the foil side (beware that, from the point of view of the tracks, this reverses the pin order). Unfortunately, mounting the socket on the foil side is mechanically weak because removal of a valve immediately tries to tear tracks off the PCB. The solution is to choose a flanged PCB socket and use PCB mounting spacers of just the right length to press the flange firmly against the chassis plate. See Figure 3.48.
- If you are forced to have two nearby tracks that are hostile to each other (perhaps the input and output of an amplifier), you can always put an earthed track in between to guard them.

There are three ways of determining the layout:

Method 1: You obtain a pad of graph paper marked in 0.1″ squares. This is usually described as "10ths, ½ and 1 inch". You then sit down with a 2H pencil and a plastic

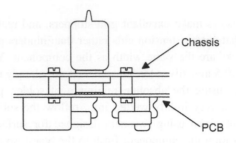

Figure 3.48
The chassis can help in supporting valve sockets soldered to the foil side of a PCB.

eraser and work out your layout. After much rubbing out and redrawing, you obtain an efficient layout, which you then copy onto a virgin sheet of graph paper.

Method 2: You use a familiar computer drawing program with "snap" set to 0.1″. After much shuffling and redrawing, you end up with an efficient layout and you struggle to persuade your printer to print the correct size. This process initially takes longer than Method 1, but allows for changes and copies later on.

Method 3: You hone your command of invective whilst learning to live with the nightmare idiosyncrasies and inconsistencies of a genuine PCB layout package and produce gerber files that you send to a PCB production house. In return for a fixed set-up cost and a variable unit cost, they make professional quality PCBs faithful to your files. Job done, but you'd better get every detail of those files right first time.

Making PCBs yourself

Whichever method you used, you now have a drawing the same size as your final PCB.

Some constructors, having produced their layout on a computer, print the layout onto a transparency. Ensure that the transparencies you use in a laser printer are suitable for photocopiers and/or laser printers, otherwise you will wreck your printer. Using PCB material freshly coated with UV-sensitive etch resist, you put the PCB and transparency into a UV light box and take a contact photograph of the layout which is later developed. Note that the drawing must be mirrored before printing so that the printed side is in contact with the board. If the printed side is not in contact with the board, the image on the board becomes defocused, leading to a poor PCB. Because this is a photographic process, exposure time is critical, as is the contrast ratio that your printer can produce on a transparency (usually very poor). Additionally, film emulsion deteriorates with time, which is why you needed fresh PCB material.

Alternatively, you could buy UV etch resist and spray the board yourself. If you buy etch resist, carefully check the "use by" date and reject it if there is less than a year remaining. Treat etch resist, developer and unused (coated) board like any other photographic film, and store them in a fridge to maximise their shelf life.

Correctly made photographically produced PCBs have a perfect finished appearance, but severely restrict your choice of PCB material unless you are prepared to coat boards yourself. Additionally, UV light boxes are not cheap, and practice is required to achieve correct exposure. Because of all these photographic problems, the author still uses an appallingly low technology method.

First cut the PCB material to size and clean the edges. Whether you use glass-reinforced plastic (GRP), commonly known as FR4 (Fire Resistant type 4), or synthetic resin bonded paper (SRBP), the best way of tidying the board edges is to rub them carefully against a fresh sheet of 160 grade silicon carbide paper held taut on a flat surface such as the lower table of a pillar drill. Once the edges are smooth, burrs can be removed by light dragging the entire edge at an angle against 320 grade silicon carbide paper. Smooth edges reduce water ingress that increases leakage.

Next, wash/scrub the foil using a scouring pad in hot water with plenty of detergent until the copper gleams. Stick the paper layout gently to the foil side. Use a scriber to mark all the component and fixing holes through the paper. See Figure 3.49.

Carefully remove the paper. Wash the board again under very hot water so that no trace of glue remains. Wash your hands thoroughly. Dry the board, and handle it by the edges only. You now have a board covered in dots. Sit down at a well-lit table, put some J.S. Bach on (greatly helps concentration), take a 00 paint brush, a pot of well-shaken enamel paint, and "join the dots". It is not necessary to be a painter of Michelangelo's calibre, just to have a steady enough hand to paint smooth curves that do not touch. Preparing the board for painting tenses the muscles, and you will find that you produce better work if the board is prepared the day before, so that you have a relaxed, steady hand. It is most

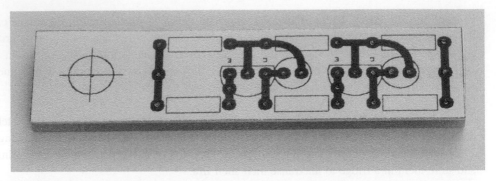

Figure 3.49
Marking "join the dots" holes through the paper template.

important not to touch the foil, as grease/sweat could impede the subsequent etching process, so an artist's mahl stick to support your wrist can be useful when painting large boards. See Figure 3.50.

Figure 3.50
This easily-made mahl stick prevents accidental contact with the PCB.

The author uses red paint (Humbrol 220) because it is easy to see errors against the copper backing. See Figure 3.51.

Figure 3.51
The PCB is painted and ready for etching.

If you are impatient, once painted, you will gently warm the board to dry the paint quickly. Although that worked well with proper oil-based paints, much of the water-based rubbish sold nowadays promptly spreads and joins all your tracks. Test your paint first to determine how it behaves before converting your electronic Sistine Chapel into an Impressionist disaster. If in the slightest doubt, leave it to dry naturally overnight.

Wash your brush carefully in white spirit followed by copious warm soapy water.

Etching

The following description is full of warnings/precautions, but provided that these are heeded, perfect etching will be achieved with an absolute minimum of excitement and domestic upset.

PCBs are etched using ferric chloride solution. Ferric chloride granules are available from all electronics factors – just add water. Bear in mind that it is intended for etching, so it is highly corrosive, and should be stored in a labelled strong PVC container that should be replaced every five years (an old container might burst and release a torrent of corrosive ferric chloride solution). Recycled fruit juice bottles are a possibility, but replace all original labels with new conspicuous ones and keep the container well from inquiring hands.

Assuming that you have a board covered with developed etch resist or dry paint, it is time for etching. Tear off a few squares of kitchen roll and keep them handy in case you accidentally spill a drop. Pour the etchant into the plastic etching tray to a depth of ¼″ (≈ 6 mm). If a spill occurs, immediately mop it up with the kitchen roll, then wash down with a wet cloth to dilute any remainder – speed is essential to prevent stains. Warm etchant works faster than cold (chemical reactions double in speed with each 10°C rise in temperature), so the author discreetly pops the filled etching tray in the kitchen microwave for a minute to noticeably warm it (perhaps 30–40°C) before etching.

Ease the board gently into the etching fluid foil side up, and rock the tray gently and continuously to keep the etchant on the move and clear etched debris from the surface of the board. After a few minutes you will see the board begin to etch around the edges. Continue agitating the fluid and watching. Depending on temperature, foil thickness and age of fluid, the process takes between 10 and 20 minutes. It is most important not to leave the board in the etchant for too long or it will undercut tracks beneath the resist.

Place your etching fluid container in the (unplugged) sink, turn the cold tap on, uncap the container, return the tray's fluid and replace the cap. It's at this point that you understand why some lunchboxes are better than others; the ideal box is thick (>2 mm) PVC with sharply radiused corners (<6 mm) and no external flanges. Not only does the right box allow you to return etchant to store without spilling a drop, but it inhibits wetting, so that virtually no etchant remains in the tray. If you should spill any etchant, the running water will immediately dilute it and the sink will not be stained. Put the etchant away, and run cold water gently (to avoid splashing) into the etching tray to dilute the remaining etchant and allow it to overflow until clean water results. You can now retrieve the board and closely inspect it for incomplete etching. See Figure 3.52.

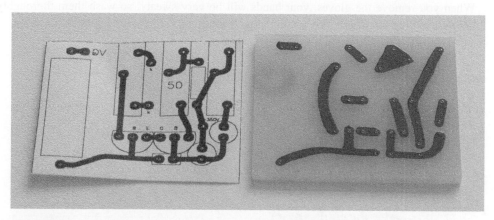

Figure 3.52
PCB etched.

Assuming etching is complete, the etch resist can be removed with chemical paint stripper, which is also nasty stuff:

- Dichloromethane (the active constituent of paint stripper) evaporates readily, so make sure the bottle top is fully tightened after use. Even better, store the sealed bottle within a sealable polythene bag to further reduce evaporation and maintain effectiveness.
- Paint stripper sold in tin cans eventually corrodes the can, starting at the crimped seam round the bottom, but as it oozes out it dries and seals the leak. You discover the problem when you wrench the can away and break the seal, releasing a tsunami of corrosive goo. The author stores chemicals in strong PVC boxes that act as a bund and confine any escape.
- If it's powerful enough to strip paint it's certainly not going to do your fingers any good, so wear rubber gloves. Washing-up gloves tend to be clumsy, but the disposable type used by medics are ideal and surprisingly easily available.

Pour the paint stripper generously all over the board, wait a few seconds until the paint crinkles nicely, then scrape it (and the paint) away using an old toothbrush under running cold water. Don't press hard with the brush or you will scrub paint into the board surface never to be removed. Don't allow paint to build up on the brush — periodically wipe it clean in the plug hole. Don't hesitate to apply fresh stripper rather than force to remove stubborn paint.

When you remove the gloves, your hands will be very sweaty, so wash them thoroughly before handling the board. The board is now ready to be drilled. See Figure 3.53.

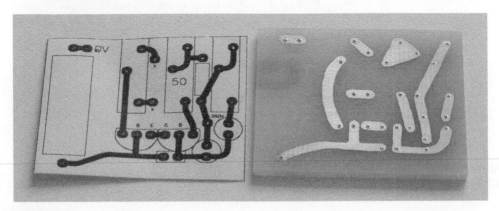

Figure 3.53
PCB stripped and ready for drilling.

Drilling

If you use FR4, you ideally need silicon carbide drills, but they are expensive and brittle. Ordinary high-speed steel (HSS) drills **do** work, but they blunt quite quickly. Typical hole size is 0.8 mm, but you may wish to measure components individually and drill holes sized for each component lead to ensure the very best possible solder joint. Small drills must be treated carefully and need very high drilling speeds (see Chapter 2).

Stuffing and soldering

Once drilled, the board can be stuffed. Component leads should be cropped short **before** soldering, otherwise the mechanical shock from cropping disturbs the soldered joint, leading to later failures. See Figure 3.54.

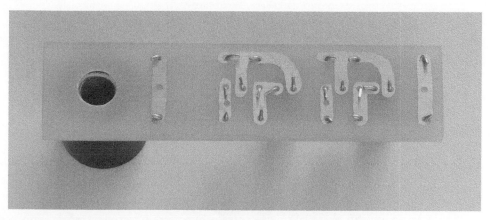

Figure 3.54
The board has been stuffed, leads cropped, and is ready for soldering.

Once soldered, the board should be defluxed (covered in detail earlier in this chapter) to minimise surface leakage currents. See Figure 3.55.

Although a board is a sub-assembly, there is no reason why it should not form part of a larger sub-assembly. In this example, each board is a constant current sink that ultimately controls the total anode current and balance of a pair of output valves in a push—pull stereo power amplifier. See Figure 3.56.

The aluminium bracket supports a pair of PCBs by their trimmer potentiometers, which have been positioned so that their shafts align with the perforated aluminium sheet to which this bias assembly will be fitted.

Modifying PCBs

Although it might seem that a circuit constructed on a PCB is set in stone, modification is perfectly possible — it just takes care.

Phenolic boards are far more fragile than glass-reinforced plastic boards (FR4 etc.), so decide carefully how you will remove a component before attacking the board.

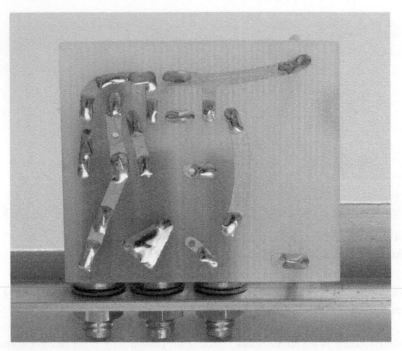

Figure 3.55
The joints have been soldered and defluxed, enabling checking of joint quality.

Figure 3.56
A pair of boards on a bracket form a larger bias sub-assembly.

Never attempt to salvage components. If you bent the wire over before soldering, the component will not come out simply by heating the solder and pulling from the component side. If you are lucky, desolder wick will remove all the solder, allowing flush cutters to snip away the bent wire, allowing the component to be pulled free. Alternatively, cut the component away from the component side, and use a sharp point such as a needle or a dedicated PCB rework tool to push each lead through whilst heating the solder on the track side. Or using the bend in bent nose pliers as a fulcrum allows controlled force to be applied to the component lead on the reverse side of the board, and the wire can be tugged through, although this tends to damage the side of the hole and pad where the wire was bent over.

If a wire fragment stands proud, it can easily be removed by a pair of fine tweezers. Never apply force to wire fragments — you could tear a track away from the board. Sometimes it helps to remove some of the solder from the track side before removing a wire fragment. Paradoxically, you may not want to remove all the solder because this makes it difficult to get the heat into the remaining solder that bonds the unwanted component to the track. Desolder wick is a safer option than a desolder gun — the recoil can kick tracks away.

The replacement component may not be the same length as the original. Don't attempt to impose the old hole spacing on the new component if they are different, use one old hole and drill a new hole to suit (but make sure that you don't accidentally drill into another track). Pass the wires through, and bend the wire from the blank hole so that it contacts the track you want, then solder it.

Sometimes you need to cut and remove a section of track. A sharp No. 10 scalpel blade easily cuts thin copper foil, so you don't need to murder the board. Make a cut either side of the section of track to be removed, tin the track, heat it with the iron, and being careful not to dig into the board, use the tip of the scalpel to ease the unwanted track away. If you are careful, it is possible to remove sections of track without anyone knowing you've been there.

Sometimes you need to add a track from one side of the board to the other. The standard industry technique is to use green (any colour would do, but green is used) wirewrap wire to make the link. Cut the wire to length and strip ½″ from each end. Tin each end to allow the stress of stripping to be relieved and shrink back the insulation, trim the exposed wire to 1−2 mm as appropriate for the joint, then solder. If the wire is long, a few small spots of cyanoacrylate glue can be used to secure it.

If you need to add circuitry and there's a blank area of board, it's surprising how neat a job can be achieved by drilling holes for component leads, and bending and soldering their wires together on the foil side in lieu of tracks.

Problems and solutions

The following section covers techniques that you may find useful but that didn't sit neatly in previous sections.

Identifying the outer foil of a capacitor

Electrostatic pick-up is proportional to the capacitance between the interference source and the affected part, so large objects like coupling capacitors are especially vulnerable. Because capacitors are invariably wound as a coil of foil, the outer foil screens the inner foil, and should ideally be connected to earth. Obviously, a coupling capacitor has neither end connected to earth, but an induced current develops a smaller voltage if the impedance to earth is minimised, so outer foils should always be connected to whichever end has the lowest impedance to earth, usually the output of the preceding stage rather than the following grid.

Unfortunately, the outer foil is not often marked, but it can usually be identified. See Figure 3.57.

The capacitor under test is connected as a reservoir capacitor to crudely rectified AC, and a strip of metal foil is wrapped around the capacitor. The metal foil is connected to the input of an oscilloscope or an amplifier. If the outer foil of the reservoir capacitor is connected to earth, very little interference will be picked up, but if it is connected to the unearthed output of the rectifier, the capacitance between it and the added foil will easily couple the higher harmonics of the rectified AC into the amplifier or oscilloscope. Thus, the outer foil can be identified by finding which connection couples the most interference.

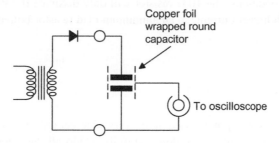

Figure 3.57
Identifying the outer foil of a capacitor.

Unfortunately, this method might not work for high voltage polypropylene capacitors because they are often made from two capacitors in series placed end to end, so they don't strictly have outer and inner foils.

High-current (>2 A) heater regulators that shut down at switch-on

Single-ended amplifiers cannot cancel heater-induced hum from directly heated valves in their output transformer, so one solution is to power the heater from regulated DC. The 5 A LM338 is the ideal three-terminal regulator for the job, but its comprehensive protection can sometimes be tripped by the (much lower) cold resistance of a valve heater. If this occurs, the solution is to add a resistor between input and output of the regulator to bleed current directly into the heater. At switch-on, the regulator shuts down, but the bleed resistor passes sufficient current to warm the heater, raising its resistance, eventually allowing the regulator to come out of shut-down.

The value of the bleed resistor is not critical, and setting it to pass 10% of the final current generally solves the problem. Thus, if we had a 10 V/3.25 A heater, and dropped 3 V across the LM338 under load:

$$R = \frac{V}{0.1 \times I} = \frac{3 \text{ V}}{0.1 \times 3.25 \text{ A}} = 9.23 \ \Omega \approx 10 \ \Omega$$

Under working conditions, the 10 Ω resistor will only dissipate 0.9 W, but at switch-on, it must pass a much higher current, so an aluminium-clad resistor bolted directly to the chassis is ideal.

Tektronix tag strips

Valve-era Tektronix equipment used some very nice ceramic tag strips that can be salvaged and reused, but it has to be done carefully. Undo all the screws holding the valve sockets in place so that the tag strips are not restrained by hidden wires to the sockets. The strips can be loosened from the reverse side of the chassis by firmly pressing the central white spigot so that it sits flush with its transparent surround (the author uses the square ends of his short car pliers), then use a pin punch or small screwdriver to gently push the white spigot all the way through. The strips plus wiring loom plus sockets should now pull away from the chassis, making it much easier to separate the tag strips. The clear spacer requires considerable force to remove from the chassis, but can be pushed through from the reverse side using the flat of a flat-bladed screwdriver — there will be a click as it releases. Silver-loaded solder must be used on these tag strips to prevent the plating peeling away from the ceramic, and it is better to cut components away than try to desolder them whole; once unsupported wires remain, they are easily desoldered.

Tarnished silver-plated stand-offs that won't solder

Silver tarnishes when exposed to air, so 30-year-old silver-plated stand-offs may not be solderable. However, dipping the tag into Goddard's Silver Dip (Tarn-X is the US equivalent) converts the tarnish back to silver, enabling soldering. As with all chemicals, Silver Dip should be handled with care, and only the silver-plated tag of the stand-off should be allowed to touch the liquid, but mounting them on a scrap of cardboard neatly solves this problem. See Figure 3.58.

Rinse the stand-offs in clean water, and dry them with forced hot air from a hairdryer. Although this technique is very effective, it is rather smelly, and the tags tend to tarnish quickly afterwards, so it is best to treat them just prior to soldering.

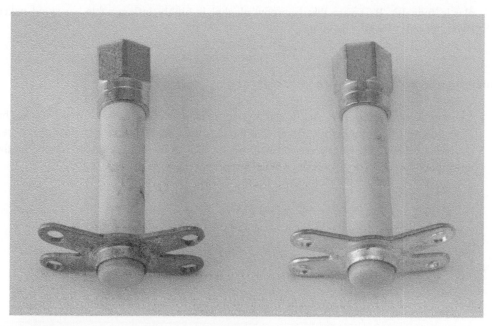

Figure 3.58
Silver-plated tags before and after dipping in "Silver Dip".

NOS valve sockets that won't solder

NOS valve sockets are likely to be at least 40 years old, and the (usually tinned) tags may be reluctant to solder – this seems to be a particular problem with Loctal and Septar PTFE sockets. Taking a wire brush to the pins is not a good idea because even a fine wire brush scratches the insulation supporting the tags and embeds those scratches with particles of (conductive) solder, making the base much leakier. One solution is to soak them in a jar of isopropyl alcohol for a day or two, periodically giving the jar a good shake. Really recalcitrant sockets can be periodically removed and brushed with a toothbrush until the tags become clean and shiny. Finish with a good rinse in clean alcohol, followed by clean water, and dry them with forced hot air from a hairdryer.

The alcohol quickly becomes dirty, but can be recovered by filtering it through a coffee filter paper supported in a stainless steel kitchen funnel (pure alcohol might damage some plastic funnels). Be aware that isopropyl alcohol is highly inflammable, so avoid naked flames or sparks whilst handling.

Enlarging the hole for the wire in a solder tag

You might think this is easy, but solder tags are made of brass, and brass has a nasty habit of snatching when you drill it. Given that the tags are small, you've only just discovered the problem, and you want to get on with wiring, this is the perfect recipe for slicing the tip of your fingers. The solution is a small reamer. See Figure 3.59.

Figure 3.59
Tapered reamer for enlarging small holes.

This is a very handy tool to keep with your wiring tools. Because it cuts a tapered hole, it needs to be used from both sides of the hole. It throws up a burr, but this isn't a problem, because when you fold your wires around the joint to form a mechanical joint before soldering, the high pressure at the burr makes an excellent contact even before you solder.

Securing wiring in place

Long runs of wiring often need to be pressed securely down to the chassis, or into the corner, but its flexibility (even if solid core twisted pair) causes it to lift. The following techniques can be useful:

- Bend a U-shape onto the end of a solder tag and use it to clamp the wiring from a convenient screw. Only really useful for a twisted pair.

- Hot melt glue. Beware that glue strings easily make more mess than the glue solves.
- If a number of wires can be laced together, the resulting rigid harness tends to stay wherever it is pushed.

Valves with multiple grid pins such as the 6C45, PC900

Multiple grid pins allow a UHF tuner head to be made with short wires no matter what the layout. This was important because at UHF the lead inductance of 6 mm of wire is significant. However, a useful audio trick is to fit a grid-stopper on each grid pin then connect them together; parallelling the grid-stoppers in this way reduces series resistance (and therefore noise) yet still damps the resonance.

Replacing broken valve pins

A valve socket's pins are inserted from the valve side, then retained by a twist or bend in the solder tag. McMurdo B9A and B7G pins are interchangeable and removable, so a B9A socket that's riveted in place can have pins replaced with pins taken from a B7G socket, or vice versa. This means that if a single pin is broken, it doesn't mean that the entire socket has to be replaced. Many circuits don't use all the valve pins, so if a wired pin breaks, replace it with an unwired pin from the same base.

Reference

1. S. Dove, Designing a professional mixing console: Part Six — When is a ground not a ground? *Studio Sound*, March 1981, pp. 56—60.

Recommended further reading

H.W. Ott, *Noise Reduction Techniques in Electronic Systems*, 2nd edn. Wiley (1988), ISBN 0-471-85068-3.
Slightly misnamed (its main focus is interference reduction), this was one of the first books to formally cover EMC and is still one of the best.

J. Goedbloed, *Electromagnetic Compatibility* (Phillips Research Laboratories). Prentice-Hall (1990), ISBN 0-13-249293-8.

Given that Phillips was a component manufacturer at the time, this excellent book contains plenty of theoretical considerations for component design and usage backed up by measurements.

W.C. Elmore and M. Sands, *Electronics*. McGraw-Hill (1949).

This book was born from the Manhattan Project, so it's no surprise to find that the electronics is concerned with counting nuclear pulses. Despite the seeming dissonance with valve audio, many of the practical problems are the same, particularly when trying to minimise noise in high-impedance circuits.

R.S. Villanucci, A.W. Avtgis, and W.F. Megow, *Electronic Techniques: Shop Practices and Construction*, 2nd edn. Prentice-Hall (1981), ISBN 0-13-252486-4.

A useful book from the early days of transistor audio containing useful tables of (US) wire gauges, etc.

SECTION 2

TESTING

CHAPTER 4

TEST EQUIPMENT PRINCIPLES

A fully equipped electronics laboratory has the entire gamut of test gear from spectrum analysers and oscilloscopes to DVMs, insulation testers, variable power supplies, signal generators, and component analysers. Sooner or later, you will have to bite the bullet and buy some test equipment. Your purchase might range from a tatty used moving coil meter found at an amateur radio fair to a shiny new oscilloscope. Either way, your money is a finite resource and you will want maximum bang per buck.

In this chapter, we will investigate how instruments work, enabling us to understand their inevitable limitations and putting us in a position to judge which features are worth paying for, which can be neglected, and how to get the best out of each instrument. We will also glance at uncertainties because they are key to deciding whether or not a measured result is significant.

The moving coil meter and DC measurements

All test equipment needs an indicator to display the results of its measurement. In the valve era, the choice was between a moving coil meter or cathode ray tube (CRT), and because CRTs and their necessary support circuitry were always expensive, the vast majority of test equipment used a moving coil meter. This has two significant consequences. Firstly, it is fashionable for modern studio and consumer audio equipment to be styled to look "retro" using hexagonal black bakelite knobs and round black meters having gothic pointers and cream scales. Secondly, there are plenty of second-hand moving coil meters available for peanuts.

Building Valve Amplifiers. DOI: http://dx.doi.org/10.1016/B978-0-08-096638-0.00004-7

How a moving coil meter works

If we place one magnet near another, they will leap together so that their unlike **poles** touch (North to South). Conversely, if we try to force like poles together (North to North, or South to South), they repel. When we pass a current through a wire, it generates a weak magnetic field, and if we wind the wire into a tight coil, this concentrates the magnetic field. The coil is then known as an **electromagnet** because electricity is used to produce the magnetic field. Unsurprisingly, the strength of the electromagnet is proportional to the product of the current and the number of turns (formalised as **magnetising force**, and expressed in ampere-turns).

If we were to place the electromagnet near a **permanent** magnet, it would be attracted or repelled with a force proportional to the current and, if we could measure this force, we would have indirectly measured the current. Moving coil meters use the force to extend a spring, and because the extension of a spring is proportional to the force (Hooke's law), a scale that actually measures extension can be **calibrated** in units of current.

It is quite difficult to make a low-friction **linear** bearing (one that moves in a straight line), but a low-friction rotating bearing is easily made. Unfortunately, although we did not mention it earlier, the attraction and repulsion between the electromagnet and permanent magnet are inversely proportional to the **square** of the distance between them, resulting in a scale cramped in some sectors but unnecessarily expanded in others. Ideally, we would like a scale with equal-sized graduations throughout its range. Fortunately, shaping the permanent magnet so that it applies a constant radial magnetic field at all positions of the rotating electromagnet produces a linear scale. See Figure 4.1.

We now have an instrument whose pointer deflection is directly proportional to the current flowing through the coil.

$$\theta = \frac{BANI}{k}$$

where:

θ = angular deflection
B = flux density in the gap (T)
A = area of coil (m^2)
N = number of turns in coil
I = current in coil (A)
k = spring constant.

Figure 4.1

Internal view of a moving coil meter (note the shape of the pole pieces around the coil to produce a radial magnetic field).

Thus, we can make the meter more sensitive by winding more turns (N) of finer wire on the electromagnet, using a more powerful permanent magnet (B), or weakening the spring (k). Increasing the coil's area (A) would not only require a proportionately larger permanent magnet to maintain flux density, but also increase the moving mass to be supported by the bearings, probably increasing their friction. In practice, it is difficult to make a robust meter that requires less than 50 μA for full-scale deflection (**FSD**), and 1 mA FSD meters are particularly common.

Measuring larger currents

A manufacturer could wind different meter coils for different currents, but it is cheaper to make a standard meter movement and adapt it for larger currents. Suppose that we need a 100 mA meter, perhaps for monitoring current in an output valve. Our meter movement still only needs its design current, perhaps 1 mA, to drive the pointer to full scale, so we must bypass or **shunt** 99 mA away from the delicate meter movement. See Figure 4.2.

Our problem is to determine the correct value of shunt resistance that will allow 100 mA of total current to be shared correctly between the meter (1 mA) and the shunt resistor (99 mA). Fortunately, now that we have stated the problem, the solution is quite easy.

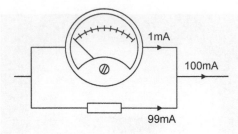

Figure 4.2
Converting a 0–1 mA meter to read 100 mA.

The meter's internal resistance and electrical sensitivity are often stated at the bottom of the scale plate, even if the scale itself is calibrated in foot/Lamberts per fortnight.

If we know the current passing through a known resistance, we can use Ohm's law to calculate the voltage developed across that resistance. As an example, suppose that our 1 mA movement has an internal (coil) resistance of 65 Ω:

$$V = IR = 0.001 \text{ A} \times 65 \ \Omega = 65 \text{ mV}$$

The shunt resistor is in parallel, so it sees exactly the same voltage as the meter, and if we want to shunt 99 mA, we simply use Ohm's law to find the required resistance:

$$R = \frac{V}{I} = \frac{0.065}{0.099} = 0.657 \ \Omega$$

Sadly, this calculation demonstrates two important points. Firstly, there is no hope of the required resistor being a standard value and, secondly, the resistor is such a low value that the resistance of its wires becomes critical. Fortunately, this is a sufficiently common application that dedicated meter shunts may be available, or a reasonably close value can itself be shunted by another resistor until the correct value is obtained. One solution would be to use five 3.3 Ω resistors in parallel (0.66 Ω), and trim this combination with a 500 Ω variable resistor in parallel.

Measuring small resistances ($<10\ \Omega$) accurately is difficult, so it is better to set up a test circuit to pass the required 100 mA (perhaps measured by a known accurate DVM), and finely adjust the shunt resistance until the meter reads full scale. Nevertheless, adapting a meter to measure different currents is always awkward.

What to do if your meter is unspecified

If you are unlucky, your chosen meter might **not** specify its sensitivity and internal resistance. Sensitivity can be found by placing it in series with a variable resistor in series with a calibrated DVM set to read current, all connected across a battery. Gently reduce the value of the variable resistor from 100 kΩ until the meter reads precisely full scale, then read off the current measured by the calibrated DVM. See Figure 4.3.

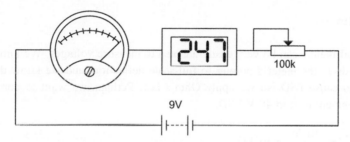

Figure 4.3
Determining the sensitivity of an unknown meter.

The internal resistance could be found using the resistance range of the DVM, but they sometimes sink rather a lot of current, and that might damage the meter under test. A safer method is to leave the variable resistor set to cause full-scale deflection, but move the DVM to measure the voltage across the meter under test, and use Ohm's law to calculate its internal resistance. See Figure 4.4.

As an example, perhaps the first test required 247 μA to attain FSD and when the DVM was used to measure meter voltage, 72.3 mV was seen across the meter at FSD:

$$R_{\text{internal}} = \frac{V}{I} = \frac{72.3 \times 10^{-3}}{247 \times 10^{-6}} = 293\ \Omega$$

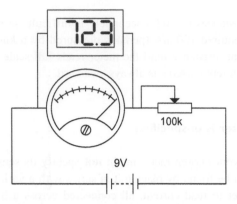

Figure 4.4
Determining the internal resistance of an unknown meter.

Measuring voltages

Moving coil meters can be very easily adapted to measure voltages. We know the required current to drive the meter's pointer to full-scale deflection, and we know the voltage that we want to cause FSD, so we apply Ohm's law. Perhaps we want to convert our 1 mA meter movement to read 40 V FSD:

$$R = \frac{V}{I} = \frac{40 \text{ V}}{1 \text{ mA}} = 40 \text{ k}\Omega$$

This calculation effectively states that a 40 kΩ resistor would pass 1 mA if 40 V were applied across its terminals. If we placed a perfect 1 mA meter in series with the 40 kΩ resistor, it would achieve FSD when 40 V was applied. You will notice that the final voltmeter has an FSD of 40 V and a resistance of 40 kΩ, or 1 kΩ per volt. It was common for moving coil multimeters to specify their loading resistance in this manner because it allowed the user to easily assess the loading the meter imposed on any given voltage range. See Figure 4.5.

But all meters have internal resistance, and a typical 1 mA meter has an internal resistance of 65 Ω, so we ought to subtract this from the series resistor to make the total resistance 40 kΩ:

$$R_{\text{series}} = 40 \text{ k}\Omega - 65 \text{ }\Omega = 39.935 \text{ k}\Omega$$

Figure 4.5
Multimeters often specify their sensitivity in Ω/V on their scale plate.

A freshly calibrated laboratory standard moving coil meter having a 5″ mirror scale could attain accuracy at FSD of ±0.5% or, to put it another way, there is little point in worrying about precisely trimming the series resistance by 0.16% when the meter itself contributes more than three times the error (we will see later that this factor of 3 is very significant). The author simply picks suitable theoretical resistor values from his stock of ±1% tolerance preferred values and doesn't even bother measuring them.

If the previous attitude offends your sensibilities, consider why we are fitting the meter in the first place. We very rarely measure to confirm a theoretical value – we generally make a measurement to display **aberrations** from the required value. In short, we want to know if the voltage/current is **not** what it should be. Taking dodgy action films as an example, meter scales in villain's lairs rarely show precisely calibrated scales; one section is generally marked green and the other red (perhaps subscripted "Danger" or "Evacuate" for the hard-of-thinking). Once the required value is observably wrong, the precision of the measurement becomes irrelevant. Distinguishing between the need to identify a fault condition rather than making a precision laboratory measurement is significant because it makes identifying fault conditions much cheaper.

Over-current protection for moving coil meters

It's a good idea to provide overload protection consisting of a pair of back-to-back parallel diodes across moving coil meters. A typical meter might drop 90 mV at FSD, so for best protection we need a diode that begins to conduct once that voltage is exceeded. Germanium diodes are habitually quoted to have a forward drop of 200−300 mV, so they would seem ideal, but they actually begin conducting far earlier than that, and an OA91 germanium diode passed 24.4 μA at 90 mV on test. See Figure 4.6.

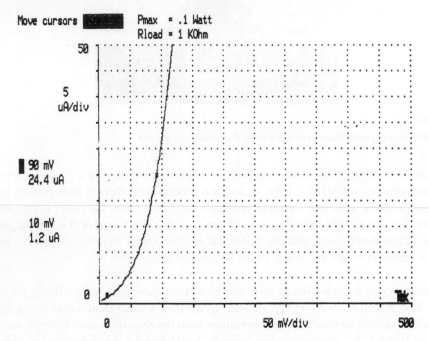

Figure 4.6
Note that this OA91 germanium diode begins conducting way before 200 mV.

24.4 μA might not seem a lot, but it would be a huge error if shunting a 30 μA movement. Marconi legitimately used germanium diodes across the meter in their TF2700 component bridge because it only had to indicate a null, so a non-linear diode current (that fell quickly to zero as the null approached) became an advantage.

The Shockley equation relates current to applied forward bias for a semiconductor diode:

$$I = I_{\text{s}} \cdot \left(e^{\frac{q_e V}{AkT}} - 1 \right)$$

where:

I = predicted diode current (A)
I_s = reverse bias saturation current (A)
A = empirical constant needed for practical diodes ≈ 2
q_e = electron charge $\approx 1.602 \times 10^{-19}$ C
V = applied forward bias (V)
k = Boltzmann's constant $\approx 1.381 \times 10^{-23}$ J/K
T = absolute temperature = $^\circ$C + 273.16.

The author measured a 1N4148 small-signal silicon diode's forward drop against applied current, then modelled it, adjusting the values of A (1.88) and I_s (2.7 nA). A tilt in forward drop at the highest currents was observed due to the additional voltage drop across the resistance of the bonding wires and bulk silicon, so 1.3 Ω of series resistance was added to the model, enabling a fit having <1% deviation from 5 μA to 25 mA. See Figure 4.7.

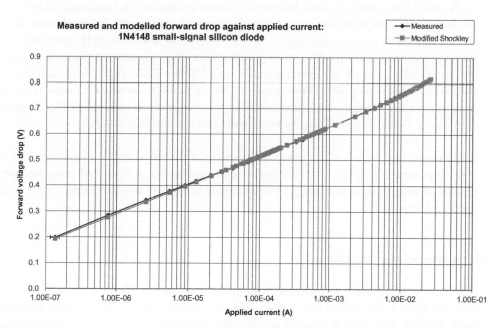

Figure 4.7
Semiconductor junctions conform closely to a logarithmic law (often over as many as seven decades).

Using the experimentally derived constants, a forward drop of 100 mV could be predicted to cause a diode current of 18 nA, which would equate to an entirely negligible measurement error of 0.07% even when shunting a 25 μA movement, demonstrating that a silicon diode is ideal for protecting moving coil meters.

Measuring output stage cathode current

It is often necessary to measure an output valve's cathode current. This can be done by inserting a 1 Ω resistor in the cathode circuit and measuring its voltage drop. Panel-mounting DVMs the size of a postage stamp tend to be 200 mV FSD, and 1 mA through a 1 Ω resistor produces 1 mV, so these meters can be connected across the 1 Ω resistor to give a reading scaled directly in mA.

We don't need to use a high-power 1 Ω resistor for current sensing, and it's actually a disadvantage to do so. The current sense resistor needs precision and fragility, so a 1 Ω ± 1% 0.6 W metal film resistor is ideal because it allows a reasonably accurate measurement, is low inductance, and might serve as a fuse in the event of catastrophic failure (fusible resistors are available, but typical tolerance is ±5%). Because DVMs include amplifiers and are so much more sensitive than moving coil meters, they are more difficult to protect. DVMs typically have 10 MΩ input resistance, so a 200 mV FSD DVM would draw 20 nA at FSD. A forward drop of 200 mV across a 1N4148 could be expected to cause a diode current of 160 nA, causing an intolerable error to a potential divider, but negligible error to a current divider provided that the meter was configured to read >2 mA.

Panel-mounting DVMs are not cheap, so it seems a shame to throw away their typical ±0.1% accuracy with a 1 Ω ± 1% current sense resistor. In a push–pull amplifier, the balance of output currents is much more important than their absolute value, so we need matched 1 Ω resistors.

Kelvin or four-wire connection

One-ohm resistors can't be measured reliably by most meters because the resistance of their test leads is typically 0.2 Ω, and variable contact resistance makes accurate measurement difficult. The solution when an accurate low-resistance measurement is required is to use a Kelvin connection. The principle is that a pair of leads known as "force" and "sense" connect to each end of the **device under test** (DUT), making a total of four wires from the measuring instrument. The significance of the four wires is that they eliminate lead resistance as a source of error because the sense voltage used for the meter's $V/I = R$ calculation is that across the DUT, not across DUT plus test leads.

The force wire connects to one side of the crocodile clip and the sense to the other, only becoming connected when the clip is applied across a conductor. When a current is forced down one of the force wires, its only return path is through the other force wire. Meanwhile, the sense wires feed a high-impedance voltmeter to measure the voltage dropped across the DUT. Because the meter is high impedance, contact resistance to the sense wire does not cause a voltage drop and the voltage measured genuinely is the voltage across the DUT. The forcing current is known, so the meter can accurately calculate the resistance of the DUT.

Kelvin connection is generally only provided on bench DVMs and the special leads and clips are an optional accessory, further increasing cost. Beware that if you interlock a pair of Kelvin crocodile clips having insulated outers, you will correctly see infinite resistance — test them on a thick conductor to verify correct operation of the meter and lead combination.

A cheaper Kelvin method of matching low-value resistors

Matched low-value resistors can be found by soldering a chain of them in series and applying a constant voltage from a regulated power supply across the far ends. We use a DVM to measure and record the voltage drop across each resistor then pick resistors having the same measured voltage drop. We are making a four-wire measurement because our power supply and its leads force the current, whereas our DVM and its leads sense voltage. Accuracy can be improved by:

- Rather than applying a constant voltage across the ends of the chain, we could drive a constant current using the DC constant current regulator described later in this chapter (see Figure 4.78), and power this combination from a regulated voltage, which assists in maintaining a constant current.
- Before making the first measurement, we adjust the forcing voltage or current to give a DVM reading that is just under that required to make it change range. For a 3½ digit DVM, 180 mV is a good voltage, because it would read 180.0 mV, whereas 210 mV would be a poor choice because it would read 210 mV, which is less precise (fewer significant figures).

Matching higher value resistors

High-value resistors ($>500\,\Omega$) having a stated $\pm1\%$ tolerance can be matched fairly well simply by measuring their resistance using a DVM. Matching $\pm0.1\%$ tolerance resistors is trickier. One way would simply be to buy a better DVM. But DVMs having $\pm0.05\%$ accuracy, or better, are expensive. A much cheaper way is to make a bridge. See Figure 4.8.

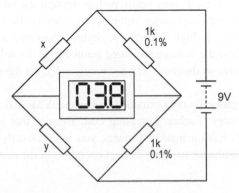

Figure 4.8
This simple bridge circuit enables precision comparison of resistors.

Although the bridge was actually invented by Samuel Hunter Christie, the circuit became known as a Wheatstone bridge because Sir Charles Wheatstone found so many uses for it. However, Wheatstone **did** invent the concertina, or accordion, which is perhaps less forgivable.

We have a pair of equal-value resistors in the right leg, and an upper resistor in the left leg for which we want to find a matching lower resistor. If the ratio of upper to lower resistor in each leg is identical, the meter will measure 0 V. If not, the imbalance will cause a voltage, which can be recorded. Resistors having equal imbalance voltage are matched to one another, but not necessarily to the upper resistor. This is a very sensitive test, and allows a cheap 3½ digit DVM on its 200 mV range to match resistors to $\pm0.01\%$.

If we want to match to a specific resistance, we need the upper and lower right leg resistors in the bridge to be perfectly matched. Using the previous method, we can find a pair of perfectly matched resistors, and substitute them into the bridge's right leg. We now insert our reference resistance in the bridge's upper left leg and test unknown resistors against it by placing them in the lower left leg — the smaller the reading on the DVM, the better the match.

Unless kept at a constant temperature, our matched resistors will drift in value, and not necessarily together. Thus, for long-term stability, the reference resistors would ideally be manufactured for low temperature coefficient, so they might be ±0.1% metal film, or even ±0.01% metal foil. Because the meter measures ratios, rather than an absolute voltage, power supply voltage is not critical, and the bridge can even be powered by a 9 V battery.

Comparative versus absolute measurements

The previous techniques for matching resistors were independent of the accuracy of the measuring instrument and simply compared two values. Although our comparison could be very accurate, we never knew absolute values.

We can only determine an absolute value by repeating the absolute physical measurement that defined the fundamental unit as an international standard or by comparison to a value referenced and traceable to that standard. Thus, any instrument calibrated in terms of absolute values **must** contain an internal standard to which it compares the external measurement. DVMs compare external voltages to an internal reference voltage that they treat as their standard (typically 200 mV for a 3½ digit DVM, 4.096 V for a 4¾ digit DVM, or 10 V for a 6½ digit DVM). Analogue component bridges often compare reactances to a 100 nF capacitor that they treat as their standard.

Obviously, measuring a higher voltage such as 150 V requires a scaling (attenuation) factor before comparison can be made to a 4.096 V reference, whereas measurement of a smaller voltage such as 42 mV requires a different scaling (amplification) factor. Similarly, a component bridge's range switches the scale of the external component's reactance to enable comparison with the internal reference. Scaling always adds further errors, so you will find that instrument specifications quote **basic accuracy** on the instrument's unscaled range.

Precision and accuracy

Because any instrument's measurements rely on the long-term stability of its internal standard plus any scaling factors, quantifying their deficiencies enables us to understand and communicate the quality of our measurements. We first need two very distinct definitions:

- **Precision:** This is the quality with which a number is described. As an example of two expressions of the same frequency measurement, 4.433 MHz has poor precision, but 4,433,618.75 Hz has excellent precision. Note that it is the number of significant figures that is important, **not** the number of digits after the decimal point. Having said that, once a measurement is expressed logarithmically (such as dB), it **is** the number of digits after the decimal point that is important, and not the number of significant figures.
- **Accuracy:** This is the quality of the actual measurement and is often described as the deviation of the measured value from the accepted value. As an example, many experimenters have measured the charge of the electron, so an accepted value is $1.6021765 \times 10^{-19}$ coulomb, whereas the number that appears on the display of our $10 + 2$ digit scientific calculator as a derived result from our measurements might be $1.601376521 \times 10^{-19}$, which although more precisely described is not especially accurate (-0.05% error).

The example illustrates a common paradox — it is very easy to produce a result that simultaneously has good precision but poor accuracy. We can therefore make the following important statement:

> *The precision with which a number is expressed should reflect the accuracy with which its measurement was made.*

But how do we know the accuracy of our measurement? It was easy when measuring the charge of the electron because this is a measurement that has been made many times and we could compare our result against an accepted value. But we often make measurements that have no accepted value, yet need to judge their accuracy.

Errors and uncertainties

Errors and uncertainties are quite distinct. An error is a single human mistake (possibly later correctable), whereas uncertainty is variation of a real-world measurement, often due to more than one cause, and those individual causes may not necessarily be identified or quantified.

Examples of errors are:

- **Misreading or transferring data incorrectly:** A decade resistance box was set to 4.578 kΩ, but the value entered into our calculations was 4.587 kΩ. If we are lucky, we later find that we recorded the value correctly in our laboratory notebook but calculated from the incorrect value, making this a perfectly correctable error.
- **Making the measurement at the wrong point (DC):** We intended to measure all voltages with respect to a valve's cathode, but actually measured with respect to 0 V. Although this is a correctable error, we lose accuracy compared to measurement from the correct point because our corrected values become the result of two measurements and two uncertainties (see later).
- **Making the measurement at the wrong point (AC):** We inadvertently connected to a noisy 0 V point when making a noise measurement, adding hum and interference to the measured signal. It is very difficult to recover reliable data from this scenario and the only real solution is to repeat the measurement correctly.

Examples of uncertainties are:

- **Analogue meter:** We watch the meter needle closely and observe that between one measurement and the next it has moved a fraction of the distance between two scale divisions. The distinction between the two measurements might be that we have deliberately changed the frequency of the signal whose amplitude is being measured, or we might have made no deliberate changes but observe that the needle moves with time. Either way, although the change is observable, it might not be quantifiable, leading to an uncertainty as to the measurement's true value.
- **Digital meter (DC):** We measure a DC voltage that is specified to be 2.000 V yet the last digit on our DVM flickers between 6 and 8, with 7 being the most common value.
- **Digital meter (AC):** We play the 5 cm/s 300 Hz sine wave on a vinyl stereo test record and measure signal voltages leaving the RIAA stage in order to determine left to right channel balance with a view to making a gain adjustment that matches them. We have deliberately selected a DVM capable of measuring the amplitude of a 300 Hz sine wave accurately, yet our expected value of 500 mV is not quite stable on the second digit, let alone the third.

Each of the previous uncertainty examples has an observable uncertainty associated with its reading, and if it is possible to quantify that uncertainty, then we should include it as part of the recorded measurement. Thus, we would note that the DVM's measurement of

2.000 V was 2.007 ± 0.001 V. Even better, we might comment on that uncertainty, perhaps noting that it appeared to be a cyclic variation (suggesting induced mains hum) rather than random variation (noise).

Often, we make a measurement that seems stable but we have no immediate indication as to its accuracy. Suppose that we made a series of measurements of the same parameter using the same instrument, perhaps at one-minute intervals, and logged them for a week. We plot the results against time and see a drift in the results, but was this due to variation in the parameter or drift in the measuring instrument?

Manufacturers of good quality instruments include information in their accompanying manuals that allow uncertainty estimates to be made. The manual for the Agilent 34410A DVM states that its alternating voltage measurements of sine waves 10 Hz−20 kHz have associated uncertainties of 0.02% of reading plus 0.02% of range plus 30 μV for frequencies below 1 kHz, and further states that these uncertainties should be summed algebraically. Thus, a reading of 0.51634 V of a 300 Hz sine wave made on this instrument's 1 V range would have a total uncertainty of $(0.51634 \times 0.0002) + (1 \times 0.0002) + 0.00003 = 330\,\mu$V, so we could state that the measured voltage was 516.34 ± 0.33 mV and this expression of uncertainty could be plotted directly as error bars on a graph. Alternatively, we might prefer to express the uncertainty as a proportion expressed as a percentage, in which case:

$$\text{Percentage proportionate uncertainty} = \frac{\text{Uncertainty}}{\text{Measurement}} \times 100\%$$

$$= \frac{330\ \mu V}{516.34\ mV} \times 100\% = 0.064\%$$

So we would specify 516.34 mV $\pm 0.064\%$.

In audio we generally prefer to express results in dB, and the uncertainty is habitually expressed slightly differently, requiring a modified calculation:

$$\text{Uncertainty (dB)} = 20\log\left(\frac{\text{Uncertainty} + \text{Measurement}}{\text{Measurement}}\right)$$

$$\text{Uncertainty (dB)} = 20\log\left(\frac{516.34 + 0.33}{516.34}\right) = 0.0055\ \text{dB}$$

It would be most unusual to specify a voltage as 516.34 mV $\pm$ 0.006 dB, but had the measurement been one in a series of relative measurements, perhaps $+0.23$ dB referenced to the 1 kHz measurement, then it would be quite usual to specify the relative measurement and its uncertainty as being $+0.23$ dB ± 0.006 dB.

Note that there is no point in specifying an uncertainty to a precision of better than one-tenth of the instrument's accuracy, and whilst many instruments can measure to an accuracy of 0.1 dB, very few can measure to 0.01 dB, so specifying any audio measurement to better than 0.01 dB is nonsense, hence the truncation of the calculated uncertainty from 0.0055 to 0.006 dB.

Combining uncertainties

A result will often be derived from a number of measurements, each having an associated uncertainty, and fully calculating that derived result's uncertainty generally involves more work than deriving the result. We rarely need to undertake rigorous uncertainty calculations, but a very brief introduction to such calculations is useful because it reveals some useful design tips.

It will not surprise you to learn that uncertainties always sum. Individual uncertainties might fundamentally derive from the same root cause, in which case they are said to be **correlated** and their uncertainties sum algebraically (implying that the uncertainties in the Agilent 34410A we saw earlier came from a single root cause). More commonly, uncertainties are **uncorrelated** (entirely unrelated) and their uncertainties add vectorially (power summation).

When we add or subtract two values, each having an associated uncertainty, we must sum the absolute values of the uncertainties. Thus, for correlated uncertainties:

$$(5.50 \pm 0.11 \ \text{V}) + (4.20 \pm 0.07 \ \text{V}) \quad = 9.70 \pm 0.18 \ \text{V}$$
$$(5.50 \pm 0.11 \ \text{V}) - (4.20 \pm 0.07 \ \text{V}) \quad = 1.30 \pm 0.18 \ \text{V}$$

But if the uncertainties were uncorrelated (as is far more usual):

$$(5.50 \pm 0.11 \ \text{V}) + (4.20 \pm 0.07 \ \text{V}) = 9.70 \pm 0.13 \ \text{V}$$
$$(5.50 \pm 0.11 \ \text{V}) - (4.20 \pm 0.07 \ \text{V}) = 1.30 \pm 0.13 \ \text{V}$$
$$\left(0.13 = \sqrt{0.11^2 + 0.07^2}\right)$$

Note that subtracting two similar results from another leads to a result having a larger proportionate uncertainty — and this is why the statement was made earlier that deriving a result from two measurements degrades accuracy compared to a single measurement at the correct point.

It is often convenient to express an individual uncertainty as a proportion such as a percentage. But if so, it must be converted back into an absolute value before it can be summed when two results are combined by addition or subtraction. Thus:

$$= (5.50 \ \text{V} \pm 2.0\%) + (4.20 \ \text{V} \pm 1.7\%)$$
$$= (5.50 \pm 0.11 \ \text{V}) + (4.20 \pm 0.07 \ \text{V})$$
$$= 9.70 \pm 0.13 \ \text{V}$$
$$= 9.70 \ \text{V} \pm 1.3\%$$

This particular example has a final proportionate uncertainty that is smaller than the individual uncertainties, and this property is exploited in the following example of ten 10 MΩ 1% tolerance resistors connected in series:

$$R_{\text{total}} = R + R + R + \ldots = 10 \times R = 10 \times 10 \ \text{M}\Omega = 100 \ \text{M}\Omega$$

But the total uncertainty is:

$$\text{Uncertainty} = \sqrt{100,000^2 + 100,000^2 + \ldots} = \sqrt{10 \times 100,000^2} = 316,228 \ \Omega$$

So the final resistance is 100 MΩ ± 0.32% — better accuracy than each original resistor, and this is because some individual uncertainties would have been positive and others negative, tending to cancel. Admittedly, those ten resistors occupy more space than a single resistor, but if we were already forced to connect resistors in series to obtain a higher voltage rating (perhaps in a high voltage probe), then the improvement in accuracy would be most welcome. Further, the shunt capacitance of the composite resistor is one-tenth that of each individual resistor — sometimes it really is worth connecting multiple resistors in series.

When two values are combined by division or multiplication, the resulting uncertainty is found by expressing individual uncertainties as proportions and summing those proportions. Thus, for the uncorrelated uncertainties, we sum the proportionate uncertainties vectorially:

$$\frac{23.1 \ \text{V} \pm 1.7\%}{5.50 \ \Omega \pm 2\%} = 4.20 \ \text{mA} \pm 2.6\%$$

$$(5.50 \ \Omega \pm 2\%) \times (4.20 \ \text{mA} \pm 1.7\%) = 23.1 \ \text{V} \pm 2.6\%$$

$$\left(\sqrt{0.02^2 + 0.017^2} = 0.026 = 2.6\% \right)$$

If we wish, we can work directly in percentages:

$$\sqrt{2\% + 1.7\%} = 2.6\%$$

There are two rules of thumb when combining two uncorrelated uncertainties:

- If one uncertainty is three times the other, it dominates.
- If one uncertainty is less than one-tenth of the other, it is negligible.

At first sight, these rules might not seem all that useful, but variations on them run through audio. The three times rule tells us to concentrate our efforts on reducing the larger uncertainty, because a small reduction of the smaller uncertainty will be unnoticeable. The one-tenth rule is applied subliminally when considering random noise. The white noise produced by one valve is certainly uncorrelated with the noise produced by another, so if a cascade of identical amplifiers is made, provided that the signal entering the second valve has more than ten times the amplitude entering the first, the proportionate noise is less than one-tenth, and the second stage's noise is negligible (0.04 dB change in signal to noise), so we concentrate our efforts on quietening the first stage.

Why worry about uncertainties?

The most likely reason for needing to determine uncertainties is that the results of the experiment are due for scientific publication and we must provide evidence to the world that a measurement's value is sufficiently accurate to be believable. Thus, if the claim is made that a neutrino has travelled faster than light (overturning contemporary physics), the arguments centre on whether the uncertainties are really as small as claimed, or whether they should be larger, allowing the superluminal speed to become a measurement error still within expected uncertainties. While the theoretical physicists debate uncertainties in the coffee bar, the experimental research students crawl in the dust beneath the rig checking plugging.

In the early stages of writing the fourth edition of *Valve Amplifiers*, the author used a pair of DVMs to perform an Ohm's law experiment to test the proposition that the resistance of some resistor constructions was not constant with applied voltage. Thus, while one meter measured the voltage drop across the resistance, the other measured the current through it. Both meters reported directly to a computer, account was taken of the loading caused by the voltmeter's finite input resistance, and connections were carefully made. When plotted against voltage, rather than having a constant value of resistance, the (carbon) resistor showed a clear voltage dependence. See Figure 4.9.

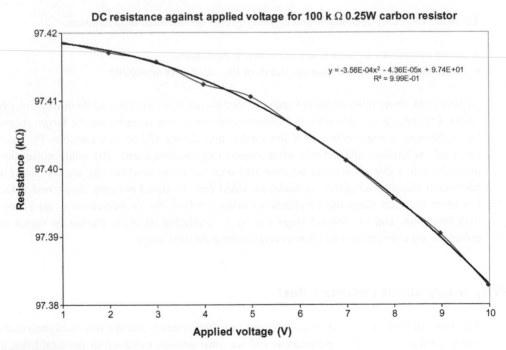

DC resistance against applied voltage for 100 k Ω 0.25W carbon resistor

$y = -3.56\text{E-}04x^2 - 4.36\text{E-}05x + 9.74\text{E+}01$
$R^2 = 9.99\text{E-}01$

Figure 4.9
This series of resistance measurements suggests voltage non-linearity.

Excited, the author measured some more resistors. But then the author rooted through the less interesting sections of the Agilent manual to find the uncertainty specifications, added those to his calculations, and produced a new graph adding those uncertainties as error bars. See Figure 4.10.

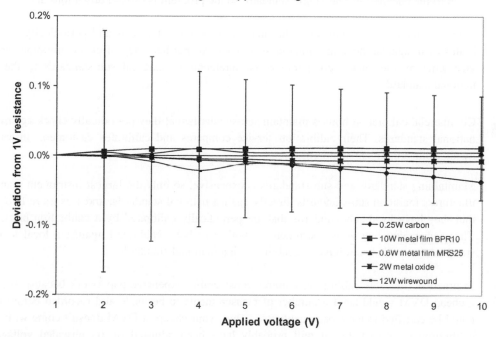

Figure 4.10
Calculating and displaying uncertainties shows that the voltage non-linearity suggested in Figure 4.9 is well within uncertainties.

As can now be seen, the measured resistances **might** have voltage dependence, and the carbon resistor **might** be the worst, but all the measured deviations were well within the expected uncertainties, so nothing firm could be said. (A later measurement of harmonic distortion proved conclusively that the carbon resistor's resistance was not constant with applied voltage.)

Calibration

If I measure a voltage with my meter and deem it to be 1.000 V $\pm$ 1 mV, then you bring your meter halfway round the world and measure that same voltage, you expect your meter to register ≈ 1.000 V, not 1.200 V. The only way that such standardisation can be achieved is if all measuring instruments have calibration traceable to an agreed single international standard. Obviously, it is impractical to compare each and every DVM right down to the cheapest to that single standard, so the problem is broken down into tiers.

Each country has an organisation that maintains national standards and periodically calibrates them against the single international standard. Further, they offer a calibration service whereby they make comparative measurements of external sub-standards to their national standard.

Commercial calibration houses maintain sub-standards that they periodically check against national standards. Their calibration service compares and calibrates customers' instruments against their sub-standards.

Maintaining standards and sub-standards is expensive, so only the largest instrument manufacturers maintain sub-standards directly against national standards, and it is more likely that they have local sub-standards that are periodically calibrated by a calibration house. Thus, when you buy your instrument, it will have been calibrated against a local sub-standard that will/should have traceability to international standards.

In practice, traceable calibration requires considerable expensive paperwork because even a cheap DVM would need a number of measurements to be made and recorded before it could be certified to have been calibrated. Thus, your cheapest DVM doesn't come with a calibration certificate, but it will probably have been adjusted on its unscaled voltage range to agree with a traceably calibrated DVM, ensuring that if it measures my 1.000 V, it is likely to agree to within a tolerable error.

Calibration is rather like a motor vehicle's mandatory safety test — the test only warrants performance at that precise instant. We make assumptions about how we expect performance to deteriorate with time and usage that set the calibration interval — often annual. We also make the important caveat that calibration becomes void the instant the instrument is physically abused or its covers removed (hence those tamper-evident calibration stickers).

Temperature changes cause materials to expand and contract, possibly resulting in relative movement that leads to calibration drift, so an instrument left permanently switched on in a stable temperature is more likely to maintain its calibration than a portable instrument taken out into the field (perhaps literally) and switched on and off for each measurement. If we needed to maintain calibration of the field instrument to the same performance as the laboratory instrument, we would probably require shorter calibration intervals.

Calibration could simply mean comparing and recording readings. An example might be a decade resistance box that would return from calibration accompanied by a certificate detailing the measured resistances of principal settings, but with no internal changes having been made.

Incidentally, **never** buy/use an unknown decade resistance box without first checking it. Fortunately, it is not necessary to check every possible resistance. Plug your DVM into the box using short thick leads (low resistance) having 4 mm plugs at each end, set the DVM to measure resistance, set all the box controls to zero and check that the meter reads $\approx 0\,\Omega$ (checks lead and switch continuity). Now rotate through the lowest resistance range and check that all values are within the tolerance stated on the box. Reset the range to zero and check the next highest range. Repeat until you run out of ranges. If your resistance box has come from teaching (and most have) a few resistors will have been damaged by overheating caused by students treating them as dummy loads. It's not expensive or hard to replace those (usually wirewound) resistors with modern equivalents, but once you know that they need replacement you can haggle over price. It's harder to damage decade capacitance boxes, but students are ingenious, so check each of their capacitances (and especially associated D) too. Good quality decade capacitance boxes use polystyrene capacitors to minimise the fall in capacitance with rising frequency, but inferior boxes using polyester capacitors have recently been made and should be avoided.

More sophisticated instruments have internal controls enabling matching to an external standard. Broadcasters often lock significant frequencies to atomic standards, and the UK's Droitwich transmitter that transmits BBC Radio 4 on long wave has a frequency of 198 kHz and the (now obsolete) PAL colour sub-carrier had a specified frequency of 4.43361875 MHz $\pm$ 1 Hz, but the BBC locked both frequencies to rubidium oscillators that were typically accurate to <0.01 Hz. Analogue colour televisions necessarily phase-locked an oscillator to the PAL colour sub-carrier, so this provided a domestic source of known frequency referenced to atomic accuracy. Thus, the author would occasionally tune an analogue receiver to the BBC and measure its PAL colour sub-carrier oscillator using his eight-digit frequency counter, then adjust the counter's CAL control until it read correctly, thereby ensuring that subsequent measurements were far more accurate than he could ever need.

The significance of adjusting controls at calibration to match an external standard is that we assume that the control does not subsequently move, causing calibration drift. If the instrument lives permanently on a shelf, vibration is minimal and controls are unlikely to move, but an instrument thrown in the back of a Land Rover and driven at speed down a dirt track suffers far more vibration. Unfortunately, frequent calibration and attendant adjustment also loosens controls, making the instrument more likely to lose calibration. Rather than periodically wading in and adjusting all internal controls in a bid for constant accurate calibration, bench instruments in a stable environment are more likely to maintain their accuracy if they are left alone until they are provably in error either from long-term drift or a fault. The problem is how to decide when an instrument's error is no longer tolerable.

Once you have more than one instrument capable of measuring a given parameter it is likely that one instrument will be better than the others, so this can be treated as your standard and all other instruments checked against it. Thus, 3½ digit DVMs could be checked against a 6½ digit DVM in the certain knowledge that any drift of the 6½ digit DVM would be far too small for the inferior meters to measure, so it could be deemed to be a local standard. Of course, this presumes that you have not abused the 6½ digit DVM, so it's useful to be able to check the standard. School physics laboratories traditionally had a Weston standard cell and these often appear in electronics junk shops for peanuts. If you see one, snap it up and use it as a voltage standard, but beware that:

- Weston cells have an output resistance of $\approx 1 \text{ k}\Omega$, so loading this with the standard $10 \text{ M}\Omega$ DVM input resistance causes an observable voltage drop on a 6½ digit DVM (so select its high-impedance mode or compensate).
- The EMF of a Weston cell has predictable temperature dependence, so for most accurate results you need to know its temperature and compensate.
- The cell may have been made prior to the definition of the volt used by the DVM manufacturer, so don't necessarily believe the cell's labelled voltage.

The easiest way to deal with all the previous points is to measure and record the Weston cell's voltage at a recorded temperature, then periodically check that it is substantially unchanged at other times when the temperature is roughly the same.

Similarly, a few 0.01% resistors are useful for checking calibration of resistance measurements. Standard capacitors tend to be rather large and rare; standard inductors are even larger and rarer.

Remember that there is no point in fretting about a 3½ digit DVM displaying a 0.2% error if the sum of its uncertainties on that range is ±0.5%. We define an erroneous measurement as being one that exceeds the calculated uncertainties when compared against a known good measurement.

Calibration becomes crucial the moment you need to question quality. It would not be a good idea to threaten withholding payment over a quality issue then have to recant upon discovering that your test equipment was out of calibration or that your test methodology was flawed.

Meters and AC measurements

A moving coil meter can only respond to DC, so to measure AC, a rectifier must be added to convert the AC to DC. But AC waveforms can be any shape, and we are not always able to view them on an oscilloscope, so we need a means of describing important parameters. Once we know the different ways of describing AC, we can choose the rectifier that delivers the measurement we need.

Peak voltages

One convenient way to specify complex waveforms is by peak voltages. See Figure 4.11.

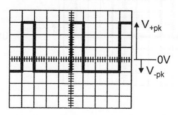

Figure 4.11
Measuring peak voltages relative to 0 V.

Note that because the waveform is asymmetric about 0 V, it is necessary to specify V_{+pk} and V_{-pk} separately. In this example, the vertical scale is 1 V/division, so $V_{+pk} = 3$ V and $V_{-pk} = -1$ V.

If required, V_{pk-pk} could be specified. This might be measured directly as 4 V_{pk-pk}, or it could be derived from the previous measurements:

$$V_{pk-pk} = V_{+pk} - V_{-pk} = +3 \text{ V} - (-1 \text{ V}) = 4 \text{ V}$$

V_{pk} and V_{pk-pk} are particularly useful measures when testing a stage for overload because the numbers directly correlate with the numbers predicted by loadlines on a valve's anode characteristics, making it easy to check theory against practice.

Mean level

It is often useful to know the average or mean level (either voltage or current) of an AC waveform. Since the waveform is assumed to be repetitive, the mean level is determined by finding the mean of one cycle of the waveform.

Conventionally, to find a mean, we sum individual data and divide by the number of items of data. Because an oscilloscope display is a graph of voltage against time, we sum voltages multiplied by their duration (producing area) and divide by the total duration of the waveform's cycle. See Figure 4.12.

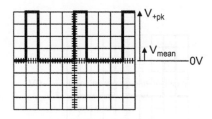

Figure 4.12
Mean voltage can be intuitively seen because it is proportional to the **area** under the waveform. Equal areas above and below 0 V would force $V_{mean} = 0$ V.

Note that although this waveform has exactly the same **shape** as the previous one, $V_{+\mathrm{pk}} = 4$ V and $V_{-\mathrm{pk}} = 0$ V. Over one cycle:

First unit of time: $V = +4$ V
Second unit: $V = 0$ V
Third unit: $V = 0$ V
Fourth unit: $V = 0$ V

We want to find the mean of four data items, so:

$$V_{\mathrm{mean}} = \frac{4 \text{ V} \times 1 + 0 \text{ V} \times 3}{4} = 1 \text{ V}$$

We have calculated the mean level over one cycle of a waveform, and assigned it a fixed value, which implies that it is unchanging. If the waveform is repetitive, each cycle is the same as the last, so the mean level must also be constant from one cycle to the next, and it must therefore be a DC voltage (or current).

Any repetitive waveform may be considered to have an **AC component** riding on a constant level of DC known as the **DC component** (V_{mean}).

When a single triode stage distorts, it produces predominantly second harmonic distortion, and a side-effect of this is that it changes the mean level of the signal. A cathode biased stage typically uses a resistor bypassed by a capacitor, and the capacitor integrates the audio signal to produce a voltage that is partly determined by the mean level of the amplified audio. The significance of this is that if we had an accurate DVM, but didn't have a means of directly measuring distortion, we could estimate or compare distortion by measuring the change in the DC voltage across the capacitor with and without the signal applied.

Later, we will find that mean level crops up in all sorts of odd places.

Power and RMS

It is often necessary to be able to determine the heating power that a waveform would dissipate in a resistor using $P = V^2/R$ or $P = I^2R$. For DC, this is simply a matter of measuring the appropriate quantities and performing the calculation.

For AC, we need a means of specifying amplitude such that the previous power equations give the correct power. This new value is known as "root of the mean of the squares" due to the calculation method; it is invariably abbreviated to RMS or V_{RMS}.

For this example, we have reverted to our original waveform, which we could now describe as having a 4 $V_{pk\text{-}pk}$ AC component, but zero DC component. Because the DC component is zero, the waveform has dropped down the oscilloscope screen so that part of it swings negatively. See Figure 4.13.

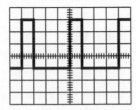

Figure 4.13
RMS voltage: No intuitive understanding is possible simply by viewing the waveform.

We calculate V_{RMS} in a very similar way to V_{mean}, but we first square each voltage, find the mean (of the squares), and finally take the square root of this mean:

$$V_{RMS} = \sqrt{\frac{3^2 \times 1 + (-1)^2 \times 3}{4}} = \sqrt{3} \approx 1.732 \text{ V}$$

Fortunately, we rarely have to calculate V_{mean} or V_{RMS} manually because a digital oscilloscope is really an application-specific computer in disguise, so it is ideally suited to performing hideous number crunching in the twinkling of an eye. See Figure 4.14.

The slight discrepancy in V_{mean} and V_{RMS} measured by the oscilloscope compared to the calculated values simply reflects the author's inability to set the controls precisely on his rather cheap (but perfectly adequate) function generator.

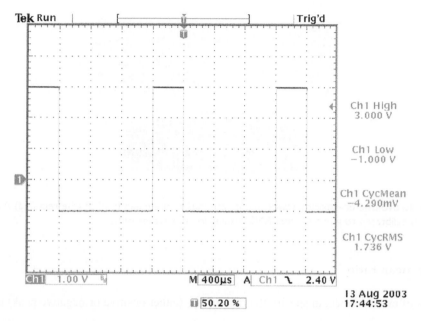

Figure 4.14
Oscilloscope automated measurements. Wonderful!

Some waveforms such as the sine wave are used so often that conversion factors have been calculated for determining V_{RMS} when V_{pk} is known. **For the sine wave only**:

$$V_{RMS} = \frac{V_{pk}}{\sqrt{2}}$$

The most obvious use of V_{RMS} is when determining the output power of an amplifier from a measurement of voltage across a known load resistance, but since we are only concerned with undistorted output power, a mean reading meter calibrated RMS of sine wave is perfectly acceptable. A moving coil meter responds to the mean value, so the waveform is rectified and the original waveform is **assumed to be a sine wave**, before a correction factor is applied. See Figure 4.15.

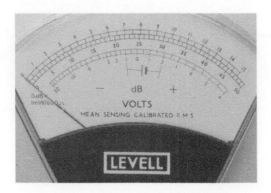

Figure 4.15
This AC microvoltmeter explicitly states that it responds to the mean level of (the rectified) AC waveform and is calibrated to give a correct RMS reading for sine waves only.

Peak to mean ratio

It is often convenient to specify the ratio of V_{pk} (either positive or negative peak) to V_{mean}:

$$\text{Peak mean ratio} = \frac{V_{pk}}{V_{mean}}$$

Human speech is often stated to have a high peak to mean ratio, although strictly the mean amplitude of any audio waveform is zero (otherwise it would contain DC), and the less familiar term **crest factor** is more appropriate.

Crest factor

A waveform composed mainly of high-amplitude narrow spikes is difficult to measure, so this property is defined:

$$\text{Crest factor} = \frac{V_{pk}}{V_{RMS}}$$

The full significance of crest factor will become apparent when we investigate dedicated audio test sets and the measurement of noise, but it is also useful when specifying power ratings of tweeters.

Unlike a cheap DVM, a **true RMS** DVM does **not** assume a sine wave and apply a conversion factor; instead, it actually calculates V_{RMS} for the waveform from first principles. Although you and I can calculate V_{RMS} for **any** waveform, true RMS DVMs are somewhat more limited, and they specify that limitation by stating the maximum crest factor that can be tolerated whilst accurately determining V_{RMS}. As an example, the author's Fluke 89 IV DVM can compute RMS accurately provided that the crest factor is $\leq 3 \times$ FSD whereas his rather better Agilent 34410A can cope with a crest factor of 10 provided that the signal waveform (including its harmonics) has a bandwidth of <300 kHz.

Although DVM manufacturers make a big fuss about true RMS, there is only one time when a low bandwidth (<1 kHz) DVM needs the facility, and that is when measuring alternating voltages containing noticeable distortion and we are truly concerned about the precise heating effect. Sadly, the preponderance of loads incorporating rectification and smoothing mean that mains voltage invariably has peaks tending towards flat tops. See Figure 4.16.

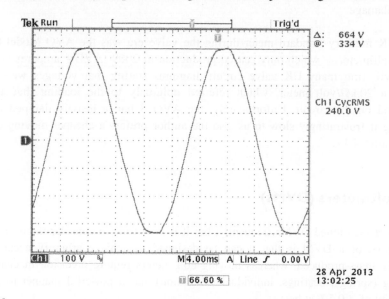

Figure 4.16
This waveform is typical of the author's mains voltage and contains 1.8% THD, but 5% is not uncommon.

Note that the oscilloscope's true RMS measurement (CycRMS) states 240 V, and for an undistorted sine wave, we would expect to see 679 V_{pk-pk}, yet the cursors show 664 V_{pk-pk}, which is 2.2% low, yet this mains waveform contained a comparatively low 1.8% THD (the author regularly sees 5% THD on mains). The significance is that mains voltage having flat tops causes DVMs that measure peak voltage and apply a sine wave conversion factor to read low, so we should ideally use a true RMS meter when measuring valve heaters powered from AC.

If we decide that we want a particularly accurate DVM, or one that can report to a computer, we find that true RMS is included whether we want it or not. However, if the DVM has wide bandwidth (>100 kHz), then we **definitely** want true RMS because it enables noise measurements (see later).

Speed of measurement

Despite the prevalence of DVMs, moving coil meters remain useful. A moving coil meter flicks quickly to its reading, and this might warn you to switch off sufficiently quickly to avoid damage.

The UK industry standard meter during the valve era was the AVO Model 8 moving-coil multimeter − which only ceased production in 2009 (AVO = amperes volts ohms). You will find many UK valve circuit diagrams stating that voltages were measured using a 20 kΩ/volt meter, which referred obliquely to the loading that an AVO 8 imposed on the circuit. Unfortunately, the AVO 8 had a heavily damped movement, making it frustratingly slow to use, so the author prefers a cheaper moving coil meter. See Figure 4.17.

Digital voltmeters (DVMs)

We have mentioned DVMs frequently, but it is now time to consider them in detail. The advantage of a DVM is that it can be designed to achieve arbitrary accuracy yet be cheaply mass-produced, whereas moving coil meters require precision mechanical engineering (springs, bearings, individual calibration) and a powerful magnet to achieve an accuracy of ±0.5% at best.

Figure 4.17
This cheap moving coil multimeter responds quickly, so it is still useful!

Despite the fact that we don't really know what time **is**, we can measure it stunningly accurately very cheaply. (You would be irritated if the digital watch given away with a can of engine oil had an error of one minute in a week, yet this corresponds to an error of <10 parts per million.)

Thus, whenever possible, digital measuring systems convert the measurement of the required parameter into a measurement of time. Unfortunately, if time is used to **make** the measurement, we must inevitably **take** time to make that measurement – usually by counting the pulses of a master clock. Thus, the 0–2 V range of a 3½ digit DVM might correspond to counting from 0 to 2000 pulses, and displaying that count scaled by an appropriately positioned decimal point. More significantly, if we wanted a 4½ digit DVM, it would need to count 20,000 pulses, which would either take ten times as long or require a master clock ten times faster.

We are now in a position to compare DVMs. The natural attitude is to say, "I want maximum accuracy with minimum cost, and I'm not really bothered about measurement time." Expensive experience forces the author to state that you **are** bothered about measurement time. It is not sufficient for the DVM to make the measurement once and stop, it must make the measurement repeatedly. As an example, we might want to set total output stage anode current to precisely 140 mA by adjusting a variable resistor. When we adjust the resistor, anode current changes instantaneously, but it takes time for the DVM to make the new measurement that reflects this change, by which time we may have overshot our adjustment. Remember that we usually measure to check that a parameter **isn't** wrong. If it **is** wrong, we must minimise the time that it is wrong, to minimise the smoke – so the addition of measurement time to human reaction time might risk damage to a set of output valves costing £200. You can buy a good DVM for £200, yet **one** measurement might recoup that cost. Are you still unconcerned about measurement time?

A DVM is fundamentally a precision analogue to digital converter (**ADC**) preceded by switchable attenuators and amplifiers. **All** ADCs have the following relationship:

$$\text{Cost} \; \propto \; \text{Sample rate} \times \text{Number of bits}$$

In terms of a DVM, the sample rate is the number of measurements per second, and the number of bits reflects the basic accuracy of the DVM – best on the unscaled direct voltage range.

Although manufacturers are usually open about uncertainties and price, they might not mention measurement speed. In an effort to beat the price/performance equation, handheld DVMs often add an "analogue" bar graph display to their digital display. The bar graph display is not accurate, but it **is** fast. Because any bar graph display must be composed of discrete segments, it is usually supported by auto-ranging electronics to allow it to display **small** changes. The upshot is that the display faithfully displays instantaneous **changes**, but the precise value is unknown.

In short, although bar graphs are very useful for revealing unstable parameters, nothing beats a fast accurate measurement, and this is one of the fundamental differences between cheap and expensive DVMs.

Successive approximation ADCs and comparator clatter from DVMs

Although early DVMs measured using a charge/discharge ramp timing technique, a successive approximation ADC is more likely in later DVMs. Successive approximation ADCs determine the digital word that corresponds to the analogue quantity by setting the word's most significant bit (MSB) and a comparator tests whether its input voltage is above or below the word. If it is above, the MSB is left at 1; if not, it is set to 0. Having done this, the process is repeated with successive bits until the least significant bit has been tested and set. The significance is that every time we use a DVM to make a measurement, we connect a comparator to our test point. Comparators invariably draw a current pulse from their input as they switch, and a 3½ digit DVM needs at least a 9 bit ADC, so the comparator is exercised nine times every time it samples. Thus, DVMs inject bursts of interference into the circuit they measure at their ADC sample rate.

The situation is not quite as bad as it first appears because the comparator is invariably buffered and all DVMs have input amplifiers and attenuators that tend to attenuate comparator clatter, but some still leaks through. Mostly, the problem does not make itself apparent, so it tends to be forgotten, but if we try to make a noise measurement whilst simultaneously monitoring a direct voltage or current, we may inadvertently find ourselves measuring DVM comparator clatter. Always sweep an oscilloscope through a wide range of time base settings to check that noise waveforms do not contain a periodic waveform.

Obviously, the solution to eliminating DVM comparator clatter from a noise measurement is to disconnect the DVM. Less obviously, we need to disconnect **both** leads – particularly when using mains-powered bench DVMs. The reason is that although disconnecting one lead prevents the DVM from making a measurement at the test point, its ADC continues sampling, and a mains-powered DVM inevitably has capacitance to earth that completes the circuit to our (almost invariably earthed) device that allows even a single test lead to inject comparator clatter.

Choosing a DVM

Some DVMs may not measure current, but this is not a great loss, since to measure current you must break a wire and subsequently reconnect it. Others may measure capacitance, but they may not be particularly accurate and tend not to measure small capacitances very well, so you may feel that it is better to put money aside towards a second-hand component bridge.

A few DVMs are the size of an (overgrown) pen, allowing you to read the display without looking away from the position of the probe tip — very useful when probing quick but fragile lash-ups.

One very useful feature on a DVM is a resistance range that is guaranteed not to switch on semiconductor junctions, because this allows you to measure resistors in circuit without semiconductors upsetting the measurement. Some early but good quality DVMs **didn't** have this facility because restricting the applied voltage to 200 mV would have made it difficult to achieve an accurate measurement. DVMs whose resistance test doesn't trip diodes generally have a separate diode test range that measures the forward drop (often up to 5 V, so useful for checking LEDs). Thus, if you see a cheap DVM at a fair, the presence or absence of the diode test range tells you whether it can measure resistance without switching on semiconductor junctions.

Summarising, there is no such thing as a single perfect meter, and each type has its advantages and disadvantages. If you are forced to have a single good DVM, you need a fast one that probably includes a bar graph to allow changing values to be clearly seen. Once you start testing, you will quickly discover that you can't have too many meters, so it's worth having different types. At the last count, the author had seven DVMs and six analogue meters.

Oscilloscope principles

Whether you use a 5 GHz digital confection complete with all possible bells and whistles, or a 5 MHz valve dinosaur found buried within the strata of rubbish in your uncle's shed, the fundamental principles are the same. An oscilloscope is simply an electronic graph drawing machine that plots graphs of voltage (vertical, or "Y" axis) against time (horizontal, or "X" axis).

There is a great deal of software available for converting a computer having a sound card into a clumsy oscilloscope. Forget it. A 192 kHz sound card has a maximum theoretical bandwidth of only 96 kHz, and you need far more than that.

Analogue oscilloscopes use a CRT (cathode ray tube) to produce the display, which requires perhaps 10 kV of EHT to accelerate electrons towards the phosphors. High-voltage circuits are always fragile, and oscilloscope EHT supplies are no exception. Although many parts are generic, the step-up transformer and voltage multiplier chain are invariably specific to the instrument, so future availability is always questionable, and you should judge the value of a second-hand oscilloscope by assuming that it will die irreparably within three years. Paradoxically, valve oscilloscopes are usually more serviceable than their silicon progeny and their display tubes may use a large-diameter electron gun that achieves sharper focus at reduced EHT voltage − improving reliability.

Not only is an oscilloscope likely to be the first major piece of test equipment that you buy, it will almost certainly be the most expensive − so you need to understand exactly what you're buying.

Analogue oscilloscopes

Although production of analogue oscilloscopes has ceased, their operating principles are an excellent introduction to the (far more complex) digital oscilloscope, and although analogue anomalies **can** be detected using a good digital oscilloscope, if available, an analogue oscilloscope may be quicker and easier.

An analogue oscilloscope uses a cathode ray tube (CRT) to draw the graph by repeatedly tracing a beam of high-velocity electrons onto the fluorescent phosphor screen to form a visible trace. The CRT is a large valve having a pair of focusing anodes (collectively known as an electron lens) that shape the electron stream from cathode to final anode into a beam that is brought into focus on the fluorescent phosphor coating behind the tube face to form a sharply defined spot. The phosphor coating is not conductive, but current flows because the electrons in the beam strike the phosphor with sufficient velocity that a low-velocity secondary electron is emitted in addition to the photon of light. The low-velocity electron is easily attracted to the concentric final anode, and thus a circuit is formed. The spot is scanned repeatedly across the tube face by applying a ramp waveform to the horizontal deflection plates, and the persistence of the phosphor coating coupled to the persistence of the eye produces a continuous display.

In order to produce a stationary trace, **all** oscilloscopes have three main blocks that have to be adjusted correctly. See Figure 4.18.

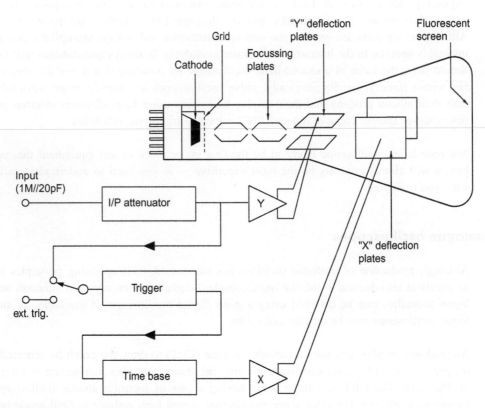

Figure 4.18
Analogue oscilloscope block diagram.

The "Y" amplifier

The **"Y" amplifier** is responsible for controlling vertical deflection and positioning of the beam. Adjustment of the relative DC potentials on the "Y" deflection plates **shifts** the beam. Adjustment of the input attenuator controls the sensitivity of the deflection − commonly labelled in **volts/div**. This control uses a 1, 2, 5 sequence because it gives equal spacing on a logarithmic graph and allows the oscilloscope to adapt smoothly to the natural world. (Currency uses this logarithmic sequence for the same reason: 1p, 2p, 5p, 10p, 20p, 50p, £1, £2, £5, £10, £20, £50.)

There is often a **fine** control (sometimes called **variable**, or **vernier**) that allows the vertical scaling to be finely adjusted to make a signal conveniently fit the screen. You probably can't now measure amplitude, but an analogue oscilloscope is primarily for displaying a waveform's shape rather than measurement.

All oscilloscopes also have a choice of input **coupling**, marked **AC**, **DC**, or **GND**. See Figure 4.19.

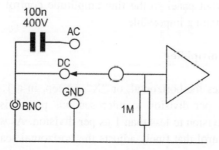

Figure 4.19
Input coupling to the "Y" amplifier.

- **DC:** All signals, including DC, are passed to the deflection plates. This is the preferred mode of operation as it does not attenuate low frequency signals, and therefore does not distort square waves, but it may not be practical when investigating small audio signals in a valve circuit.
- **AC:** DC is blocked by a capacitor (usually only rated at 400 V); this is useful for investigating the AC conditions of a circuit separately from the DC conditions − such as faultfinding.

- **GND:** This connects the input of the oscilloscope to ground/earth. If the trace is then moved to a convenient reference point, when returned to DC coupling, the movement of the trace from the reference allows the applied DC to be measured easily. (Using an oscilloscope to measure DC might seem like using a sledgehammer to crack a nut, but it responds faster than even an analogue meter, so it can be handy when faultfinding.)

Most oscilloscopes allow signal polarity to be **inverted**; this can be simply for user convenience, but it is more common to use it for matching signals between two channels. The oscilloscope is calibrated by connecting both channels to the same signal. One channel is then inverted, and the two channels are **added** or **summed**. Adding one signal to its inverse is the same as subtracting, and because the two signals are identical, the result should be zero. In practice, the oscilloscope always has small gain discrepancies, so the fine amplitude controls are adjusted for a null, indicated by a straight line. If the oscilloscope's inputs are now disconnected from one another and individually connected to test points, a straight line only results when the amplitudes of the two test points are equal. This is a very handy test for stereo channels; one signal is applied to both channels, then comparisons are made at all points along the circuit to ensure that channel balance is acceptable. Beware, however, that the final generation of analogue oscilloscopes used digital control from the front panel so the fine amplitude control was not infinitely variable, making exact gain matching impossible.

The time base and "X" amplifier

The **time base** provides the horizontal, or "X", sweep, in a 1, 2, 5 sequence with a basic range of milliseconds per division to microseconds per division — better oscilloscopes range from 10 s per division to less than 1 ns per division. As with the "Y" amplifier, there is often a variable control that finely adjusts the horizontal scaling to make the waveform conveniently fit on the screen, but the horizontal sweep is now uncalibrated, so absolute time measurements are no longer possible.

The time base control adjusts the frequency of a precision ramp generator. Because the oscilloscope display tube uses electrostatic deflection plates to deflect the beam by an amount proportional to applied voltage, the combination of the two produces a sweep having horizontal deflection proportional to time.

Depending on the tube type and accelerating voltage (a higher voltage produces a brighter and more sharply focused display, but the faster electrons are harder to deflect), the deflection plates require $\approx$200 $V_{pk\text{-}pk}$ to deflect the beam from one side of the tube face to the other. A dedicated high voltage "X" amplifier very similar to the "Y" amplifier is required to deliver the sweep voltages required by the tube.

If the sensitivity at the input of the "X" amplifier is made the same as that at the input of the "Y" amplifier, the "X" amplifier's input can be switched to accept either the ramp from the time base generator, or Ch2 (leaving Ch1 to go to the "Y" amplifier). This is known as **"XY" mode**, and because it is so cheap to provide, all oscilloscopes offer it. "XY" mode allows the display of **Lissajous** figures – which are virtually useless, but were loved by producers of early "James Bond" films. See Figure 4.20.

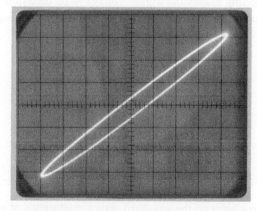

Figure 4.20
A Lissajous figure showing a perfectly straight line (rather than this slight ellipse) would indicate zero phase error.

The previous Lissajous figure compared the voltage waveform across a choke with the current waveform through it. Because the phase between current and voltage is zero at resonance, and changes very sharply around it, testing phase is a very sensitive way of determining resonant frequency. When the two waveforms are perfectly in phase, the ellipse changes to a straight line, and this powerful example is the sole reason for elevating the modern status of Lissajous figures to virtually useless.

Triggering

If we were to allow the time base to sweep randomly with respect to the input signal, we would see a mess of moving patterns on the screen. We need to sweep the electron beam in such a way that it draws each trace perfectly overlaid onto the previous trace, thus producing a single trace of the input waveform. If instead of letting the time base sweep continuously, we only allowed it to begin a sweep at the instant that a particular feature of the input waveform had been identified, this would **synchronise** the time base to the input signal and force the traces to overlay.

The **trigger** identifies the significant part of the input signal's waveform both by its absolute level and whether that voltage is rising or falling. The simplest form of **edge** triggering is **AC**, which triggers the time base each time the input waveform passes through zero. **DC** or **normal** triggering allows the trigger voltage to be determined by the user adjusting the **trigger level** control. Whether in AC or DC mode, trigger **slope** can be chosen to be "+" (rising edge) or "−" (falling edge). See Figure 4.21.

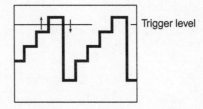
Trigger level

Figure 4.21
The trigger identifies a unique point on the waveform using a combination of voltage and slope.

In this example, the trigger level has been set so that the oscilloscope trigger fires on the top step of the staircase waveform. But the waveform goes **up** the step, then **down**, so if the trigger fired on both edges, we would see two traces, one shifted horizontally from the other. To avoid this problem, trigger slope enables us to select whether the trigger fires on the rising or falling edge.

Note that analogue oscilloscopes generally produce a sweep that is slightly wider than the display, so the start of the sweep at the trigger point and the end of the sweep are not seen unless deliberately shifted onto the screen.

The previous example had the time base set to display more than one cycle of the waveform, yet as described, the oscilloscope would only manage to sweep part-way across the screen before being triggered again, and starting a new sweep. To prevent the problem of truncated sweeps, once the trigger has fired, it is **guarded** until the time base has completed the sweep and is ready to begin the next.

If the time base is only allowed to sweep if the trigger has been fired, it is known as **normal** triggering. However, the disadvantage of normal triggering is that if you haven't been able to trigger the oscilloscope, you don't have a display, so you don't know why you failed to trigger. To circumvent this problem, oscilloscopes have **AC auto** triggering (occasionally known as **bright line**), which automatically triggers the time base after each sweep (**free run**), or is triggered when the trigger waveform passes through 0 V. Although this mode of triggering does not guarantee a locked trace, it makes it easier to determine which controls need to be adjusted to approach the required display. Once a reasonable trace has been achieved, it's usually worth switching to normal triggering for the most stable and controllable display.

The signal may contain interference that we want the trigger to ignore. If we were looking at a small signal in a pre-amplifier, it might well contain high frequency noise, and if we were to allow that noise to reach the trigger, it would make the display of our wanted signal unstable. (Remember that noise voltage is proportional to the square root of bandwidth, so a 20 MHz oscilloscope sees 30 dB more noise than we can hear in the 20 kHz audio bandwidth.) For this reason, manufacturers include **HF reject** (high frequency reject), which is typically a ≈ 30 kHz low-pass filter at the input to the trigger. Conversely, our small signal might be riding on some mains hum, and we might be able to avoid triggering on the hum using **LF reject** (low frequency reject, which is a high-pass filter ≈ 80 kHz), but we will see a much better way in a moment.

The trigger needs a signal from which to trigger. **Internal** triggering picks off a signal channel after the attenuators, and is typically marked Ch1 or Ch2. **External** triggering may come from another input connector (sometimes hidden round the back) or from **AC line**, which is the mains supply.

Suppose that you are investigating an amplifier, and have applied a 1 kHz sine wave from an oscillator to its input. By definition, signal amplitudes within the amplifier change as you move from one test point to another, probably requiring adjustment of trigger level if internal trigger is selected. However, if you take the oscillator's trigger output to the oscilloscope's external trigger input, and trigger from this signal, the display will always be correctly triggered, no matter where you probe, and no matter how much noise, equalisation, or distortion has been added to your signal. External triggering is almost always better than internal, so it is a good idea to get into the habit of using it whenever possible.

If you select AC line triggering when investigating ripple from linear power supples, the trace will always be locked. Incidentally, if you suspect that you are seeing mains hum, but are not sure, if the trace becomes stationary when you flick the trigger to AC line, mains hum is confirmed.

Bandwidth

The single most important specification for a motorcycle is its engine capacity, and the counterpart for an oscilloscope is bandwidth. And just like the motorcycle, an oscilloscope's bandwidth is invariably prominently displayed next to the manufacturer's name.

Strictly, bandwidth is defined as the difference in frequency between the high frequency and low frequency roll-offs where the response has fallen by 3 dB. Thus, a VHF tuner might define the bandwidth of its 10.7 MHz intermediate frequency amplifier as being 10.85 − 10.55 MHz = 300 kHz. Because all oscilloscopes respond to DC (which is 0 Hz), it is sufficient to specify an oscilloscope's bandwidth simply by quoting the high frequency roll-off.

Twenty megahertz is the slowest oscilloscope that you could buy today, and this is just fast enough for audio. But if the ear can only hear to 20 kHz, why do we need 1000 times as much bandwidth? The first clue lies in the definition of bandwidth. A 20 MHz oscilloscope displays a 20 MHz sine wave not at its correct amplitude, but attenuated by 3 dB. A well-focused analogue trace is just capable of displaying a 1% amplitude error, so we would ideally like the limited bandwidth of our oscilloscope to contribute a smaller error, perhaps 0.5%, at the highest frequency whose amplitude we want to measure amplitude correctly. If the oscilloscope's transient response has been optimised, its high frequency response corresponds to that of a single CR network, and we find that for 0.5% error we need $f_{-3\,dB}$ to be ten times higher. In other words, our 20 MHz oscilloscope introduces noticeable amplitude errors above 2 MHz. (It was precisely this 10:1 ratio that justified the otherwise strange existence of 60 MHz oscilloscopes − the European PAL analogue television signal had a video bandwidth of 5.5 MHz.)

Having established that our 20 MHz oscilloscope is only accurate to 2 MHz, this is still way beyond 20 kHz, so why do we need this bandwidth? Unfortunately, prototype audio power amplifiers do not always behave as they should, and it is not uncommon to find them oscillating between 1 and 2 MHz, so our oscilloscope needs to be able to display such oscillation perfectly in order that we can see it, and do something about it. There's an old saying that you can be neither too rich nor too slim. As far as oscilloscopes are concerned, you can't have too much bandwidth. (Actually, you can, and this is why >100 MHz oscilloscopes invariably include a switchable 20 MHz low-pass filter for removing noise from low frequency waveforms.)

The other parameter that goes hand in hand with bandwidth is the fastest time base speed. The ideal display contains one or two cycles of the waveform. Taking a 60 MHz oscilloscope as an example, the **period** (duration of one cycle) is:

$$T = \frac{1}{f} = \frac{1}{60 \times 10^6} = 16.7 \text{ ns}$$

Oscilloscope screens generally have ten horizontal divisions, so to display two cycles of 60 MHz, we need a time base that sweeps five divisions in 16.7 ns. In other words, it needs to be able to sweep at 3.3 ns/div. In practice, very few 60 MHz oscilloscopes were able to sweep as fast as this, and the fastest calibrated sweep of the author's cheap and cheerful (analogue) Tektronix 2213 was 50 ns/div, so it could only adequately display 4 MHz, despite being described as a 60 MHz oscilloscope.

For faultfinding audio circuitry that includes high g_m valves or power FETs (that can oscillate at tens of MHz), we would ideally like a time base that can sweep at 10 ns/div or better. The next step up from 60 MHz is a 100 MHz oscilloscope such as the 1970s vintage Tektronix 465 or 1980s vintage Tektronix 2445, which (if working) can sweep at 20 ns/div and has many extra features that we now need to investigate.

Be warned that the otherwise very desirable Tektronix 2465B oscilloscope (and possibly others from the 24xx series) is not repairable if a deflection amplifier's output transistor fails. The deflection amplifier is a hybrid, but the hybrid manufacturer collapsed, so there are no supplies of replacement amplifiers. The problem can be spotted by applying a signal (perhaps the probe calibration signal on the front panel) having a displayed amplitude of one large division and using the vertical position control to move it up and down the screen — if amplitude changes with position, the vertical hybrid has failed. Check the horizontal amplifier by displaying a signal such that ten cycles occupy the width and sweep horizontal position — if period changes with position, the horizontal hybrid has failed.

Bells and whistles

Bells and whistles allow detailed examination of more complex signals, but these refinements still fall into the basic three blocks of: "Y" amplifier, time base, and trigger.

The "Y" amplifier

Frequently, we need to investigate more than one signal at a time, and compare relative timings or voltages. To do this, we add extra input attenuators and amplifiers (known as channels), then switch sequentially between them at the input to the final "Y" deflection amplifier. See Figure 4.22.

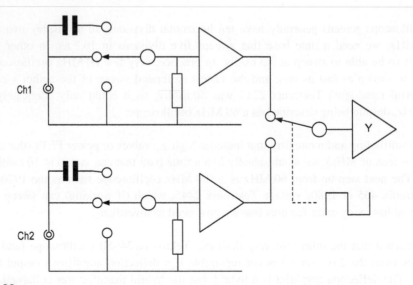

Figure 4.22
Adding channels to an oscilloscope simply requires additional attenuators, input amplifiers, and an electronic switch.

Early oscilloscopes required you to manually choose between **alternate** or **chop** modes to switch between the channels. Alternate mode alternates entire sweeps between Ch1 and Ch2, and is best for fast time base settings (<2 ms/division), but below that, chop mode is better because chopping between the channels during each sweep avoids irritating flicker at slow time bases. Later oscilloscopes automatically switch between alternate and chop at a time base setting matched to the persistence of the display tube's phosphor.

Really old oscilloscopes were dual beam, using two entire sets of vertical electronics and two pairs of "Y" deflection plates one above the other within the display tube. Unsurprisingly, this complication made them more expensive than the dual channel alternative, so they weren't a commercial success. Dual beam oscilloscopes can be identified by the fact that one channel can't sweep to the top of the screen, whereas the other can't sweep to the bottom — this is not a fault, but a natural consequence of the stacked "Y" deflection plates.

Although most analogue oscilloscopes have two channels, some can display four channels at once, but the screen becomes rather cluttered, and the display dims. (Brightness at a given point is proportional to the number of electrons striking that point, so if the electron beam has to be shared between four traces instead of one, each trace receives only a quarter of the electrons and therefore a quarter of the brightness.) Surprisingly, the more significant limitation is that four channel oscilloscopes might not have a separate triggering channel, so the fourth channel ends up being used as the external trigger, and your extra money actually bought a three-channel oscilloscope (possibly with limited "Y" attenuators on Ch3 and Ch4 only suitable for logic signals), so check this very carefully — and remember to check the back of the oscilloscope, because the external trigger may be hidden there.

Cursors added at the "Y" amplifier's input produce horizontal or vertical lines on the display that can be moved at will by the user, and are the modern development of crystal calibrator **pips** that were sometimes included in very early oscilloscopes and radar. Because cursors are digitally generated, the count that determines their position can be displayed as a number, and this number can be scaled by "Y" amplifier setting or time base setting to give a read-out directly in terms of time or voltage. Cursors have three valuable advantages:

- Without cursors, you measure amplitude (or time) by counting squares and multiplying by the amplitude or time per division — which is tedious, and prone to error.

- Because the phosphor trace is on the inside glass face of the CRT and the ruled **graticule** (scale) is on the outside, the absolute position of your eye changes their relative position. On **internal graticule** oscilloscopes this **parallax** error was eliminated by painting the graticule lines on the **inside** face of the CRT (before applying the phosphor coating).
- Measurement referred to a ruled scale assumes that CRT deflection is linear, that deflection amplifiers are linear, and that the time base ramp is linear. In practice, none are perfectly linear, but these errors cancel because cursor position is distorted by the same amount, although input amplifier and attenuator errors remain.

The addition of cursors converts an oscilloscope from an indicator of waveform shape into a measuring instrument. See Figure 4.23.

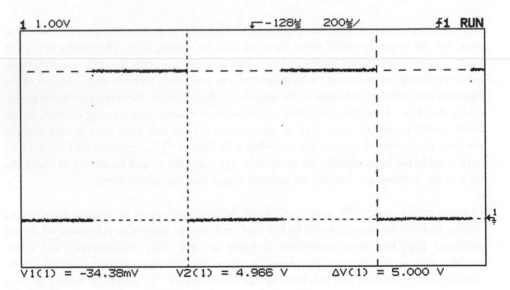

Figure 4.23
Cursors enable quick and accurate measurement of voltage or time.

The time base

Although we argued earlier that a time base that could go fast was best, a fast time base is rather like looking through a telescope — you can see in great detail, but you don't necessarily know where you are looking. High-magnification astronomical telescopes solved the problem by fitting a low-power sighting telescope to the barrel of the main instrument. Once the sighting telescope had located the region to be investigated, the observer switched to the main instrument. Similarly, fast oscilloscopes use their **main** or "**A**" time base to find the region on the waveform to be investigated, then the **delayed** or "**B**" time base is engaged to give the detailed view. Since position across a sweep is time, adjusting the **delay** before firing the "B" time base determines which part of the waveform is to be investigated in detail. See Figure 4.24.

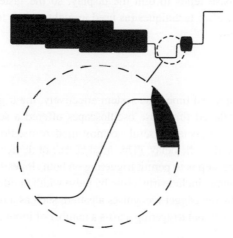

Figure 4.24
Delayed time base enables detailed examination of any part of this complex waveform.

To help the user, analogue oscilloscopes have two intermediate modes between the "A" and "B" time bases. **Intensified** mode adds a **bright-up** to the main display, which highlights where the "B" time base will sweep. Once the correct region has been chosen by adjusting delay, **mixed** mode can be engaged, which alternates between "A" time base and

"B" time base, allowing fine delay adjustment with confidence. Finally, the "A" display can be switched off by selecting "B" time base only, resulting in an uncluttered display of the precise region of interest.

Delayed time base is intended for investigating high frequency detail buried in a low-frequency waveform, so it is particularly suitable for investigating:

- Ringing at the leading edge of square waves caused by incorrectly terminated audio transformers.
- Ringing at the leading edge of square waves caused by poor compensation of a global feedback loop.
- Interference spikes caused by rectifier switching.

Using delayed time base tends to dim the display, so the fastest analogue oscilloscopes used CRT image intensifier techniques (as used in night sights) to make the display brighter, but at increased cost.

Triggering

In order to use the improved time base system effectively, the triggering must become more sophisticated. Early delayed time base oscilloscopes offered a separate trigger for the "B" time base, but this isn't especially useful, so most modern oscilloscopes simply run the "B" time base immediately after the delay. (This must be one of the few whistles to be discarded!) A seemingly retrograde step is to permit triggering on both slopes – very useful when investigating jitter. Other features include triggering by pulse width or edge slope, derivations of pattern triggering whereby the trigger recognises a pattern such as a particular sequence of serial logic pulses, or combinational triggering across a number of input channels.

Hold-off is a very useful facility that allows the trigger to be guarded by an adjustable time, so that only the intended transition triggers the oscilloscope. This facility is particularly useful for persuading a simple oscilloscope to synchronise to the data words in the AES/EBU serial digital audio data stream sent to an external DAC.

Digital oscilloscopes

Analogue oscilloscopes applied an amplified input waveform directly to the CRT, so it needed the same bandwidth as the input signal; 20 MHz CRTs were easily and cheaply made, and even 100 MHz wasn't too much of a problem, but >400 MHz was much more expensive, so the 1 GHz Tektronix 7104 was a very rare beast.

Digital oscilloscopes solve the CRT bandwidth problem by divorcing the wide bandwidth of the input waveform from the display. They do this by converting the analogue waveform into digits, storing them in memory, and reading them out at a (much slower) rate chosen to suit the display. Thus, when digital oscilloscope manufacturers refer to bandwidth, they mean the **analogue** bandwidth of the input attenuator and amplifier section before the analogue to digital converter (ADC). Not only does storing and displaying the waveform at a slower rate mean that an extremely cheap display is perfectly suitable for a 5 GHz oscilloscope, but it also means that a faster time base is easily achieved, allowing even the first generation 100 MHz HP54600B to sweep at the ideal 2 ns/div, enabling two cycles of 100 MHz to fill the screen.

To make an informed choice when considering the purchase of a digital oscilloscope, we must investigate some of the murkier areas of digital principles about which some oscilloscope manufacturers are distinctly coy.

The ADC (analogue to digital converter)

Only digital oscilloscopes have an ADC, commonly known as the **acquire** block, and its principles of operation need to be understood to obtain good results.

Sampling and quantising

An analogue signal can change continuously in both time and amplitude. Digital oscilloscopes take the analogue signal and break it up in time (**sampling**) and in amplitude (**quantising**). Having broken the signal, they convert it to a stream of binary digital numbers that describe that signal.

When an analogue signal is quantised, there are always errors because analogue amplitude cannot be perfectly described by a finite number of quantising levels. See Figure 4.25.

As an example, the 100 MHz Tektronix TDS3012 has 2^9 (512) quantising levels, which means that when amplitude falls between two levels, there is a maximum error of $\pm 0.1\%$

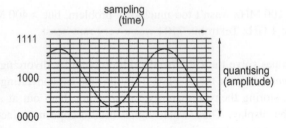

Figure 4.25
Quantising breaks amplitude into levels; sampling breaks time into intervals. The two operations are entirely separate and can be applied in either order.

(this argument assumes an ideal ADC). Given that ±0.1% quantisation error is trivial compared to the ±2% analogue uncertainty in the input attenuator and amplifier section, you might feel that using nine bits is extravagant in terms of ADC and memory cost. There are two justifications for this bit depth:

- 512 quantising levels map directly (with a small overlap) onto an LCD display having 480 vertical pixels. This was a worthwhile advantage at a time when LCD displays were firmly linked to the (525 line) NTSC television standard that had 475 active lines and a 4:3 aspect ratio, conveniently served by a 480×640 pixel display ($480 \times 4/3 = 640$).
- Because the quantising errors are so small, a trace can be expanded by a factor of 10 **after** capture, allowing investigation of details without quantising errors becoming too noticeable.

We have seen that quantisation (amplitude) errors are almost insignificant, but sampling (time) errors are far more of a problem.

Nyquist, and equivalent sample rate

The Nyquist criterion states that a repetitive waveform can be correctly reconstructed provided that the sampling frequency is greater than double the highest frequency to be sampled. Practical considerations usually increase this frequency slightly, so the digital audio on compact disc needed a 44.1 kHz sample rate even though its audio bandwidth was limited to 20 kHz. This simple theory implies that a 100 MHz oscilloscope requires an ADC that samples at 200 MS/s (megasamples per second), yet the first generation HP54600B could only sample at 20 MS/s, whereas its usurper the TDS3012 sampled at 1.25 GS/s. Why did these two 100 MHz oscilloscopes have such wildly different sample rates?

Equivalent sample rate

The HP used a technique known as **equivalent sample rate** (termed **repetitive sample rate** by LeCroy), and it works as follows. We assume that the input signal is repetitive, perhaps a 100 MHz sine wave. Sampling at 20 MS/s means that we sample at 50 ns intervals, but an entire cycle of 100 MHz only lasts for 10 ns, so we miss four entire cycles completely and sample a single point on the fifth cycle. This doesn't sound very useful, but if we were to delay sampling after the next trigger point by a time equivalent to one-hundredth of the screen width, our next sample would be at a slightly later point on the waveform, and would therefore plot a different voltage. If we keep on incrementing the sampling time delay after trigger by one-hundredth screen width intervals, we eventually build up an entire waveform that appears to have been sampled at a much higher rate. Thus, if we had set our time base to 2 ns/div (to display two cycles of 100 MHz), the sweep would display 20 ns, and if we had set our delay increments to give 100 points, we would appear to be sampling every 0.2 ns, giving an equivalent sample rate of 5 GS/s. Very clever, yes?

The problem with equivalent sample rate lies in the assumption that the input waveform is repetitive. A general rule of thumb for oscilloscopes is that you need at least ten points in a cycle to give a reasonable approximation to the shape of that waveform. But if your waveform is a one-off pulse, the ADC must take a genuine ten samples in one cycle, so a sample rate of 1 GS/s is needed to give a reasonable approximation of a 100 MHz sine wave in one-shot mode. Thus, the TDS3012 **did** have a one-shot bandwidth of 100 MHz, but the 20 MS/s sample rate of the earlier HP54600B meant that it had a one-shot bandwidth of only 2 MHz.

The significance of the previous comparison is that many digital oscilloscopes are not entirely transparent about their sample rate specifications.

287

Contravening Nyquist, and aliasing

When we apply a modulating frequency (f_m) that is slightly higher than half the sample rate (f_s) to an ADC, the frequency of the input signal is misinterpreted as a low frequency, and this phenomenon is known as **aliasing**. See Figure 4.26.

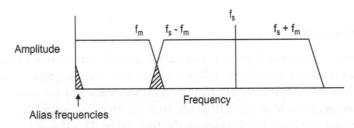

Figure 4.26
Once $f_m > f_s - f_m$, alias frequencies begin to crawl up from 0 Hz.

When audio or video leaves the analogue transducer and is converted to digits, the ADC is preceded by a low-pass **anti-aliasing filter** set to a maximum of slightly less than half the sampling frequency. In this way, aliasing is eliminated, but as we will see in a moment, digital oscilloscopes **cannot** incorporate an anti-aliasing filter, so they remain susceptible to this problem.

Sample rate, record length, and time base speed

Remember that a digital oscilloscope samples the incoming signal and writes the resulting data into memory. Later, the memory is read to the display at a convenient rate. Suppose that we want to display two cycles of 100 MHz, and we have chosen an ADC that can run at 1 GS/s. Each cycle generates ten points, and we have two of them, so we generate twenty points across the screen. Now suppose that we want to display two cycles of 50 Hz. Each cycle lasts 20 ms, so the total sweep lasts for 40 ms. In that time, sampling at 1 GS/s, we generate 40,000,000 points, requiring a lot of quite fast memory. Unlike computer memory, we don't need to be able to access data points in this memory in a totally random fashion, so there are various fudges, collectively known as **demultiplexing**, that permit the use of slower (much cheaper) memory. Nevertheless, oscilloscope manufacturers charge heavily for memory.

We should ideally have enough samples in memory to allow the trace to be expanded without generating visible gaps on the display. If the display was 640 pixels wide, 10,000 samples would allow us to expand by a factor of 10 without producing visible gaps. Thus, we need a total **record length** of 10,000 samples or points, usually abbreviated to 10 kpt. See Figure 4.27.

Waveform record

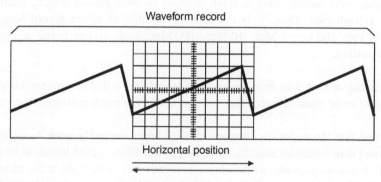

Horizontal position

Figure 4.27
The horizontal position control moves the display window back and forth across the waveform record.

The screen has ten divisions, so we can find the maximum permissible sample rate using:

$$\text{Maximum permissible sample rate} = \frac{\text{Record length}}{10 \times \text{time}}$$

Note that sample rate is now **totally dependent** on the amount of memory, so if we set the time base to 10 µs/div, our hypothetical oscilloscope having a 10 kpt record length would be forced to drop its sample rate from 1 GS/s to 100 MS/s. This isn't a problem, because even if we required ten points per cycle, this sample rate and time base setting could correctly capture 1000 cycles at 10 MHz, and we would never be able to distinguish this number of cycles across the screen, so the display would be **identical to an analogue oscilloscope**. It might seem odd to choose 10,000 points across the screen when we know that we will have to **decimate** (discard samples) to 640 pixels, but 10,000 produces nice round numbers for the sample rates generated by the 1, 2, 5 sequence of the time base.

Because sample rate varies with time base setting, we cannot precede the ADC with an anti-alias filter, and this means that aliasing is always a possibility. If we want to record a trace over 1 s, then spot a sharp glitch and examine it in detail, we must run the ADC at a much faster rate, requiring a great deal more memory, and this is why oscilloscope marketing puff brags about record length. Beware that, at some settings, the display may occupy only a small fraction of total record length, leading to slow display refresh rate. Thus, it is common to be able to select record length, perhaps 1 kpt, 10 kpt, 100 kpt, 1 Mpt, 10 Mpt, 100 Mpt, with 10 kpt being a good general-purpose setting.

Summarising, at slow time base speeds, it is record length that determines sample rate, and the need to avoid aliasing means that you can't have too much maximum record length.

Be warned that the major oscilloscope manufacturers currently work to a different business model than computer manufacturers and do not allow record length to be extended by a simple hardware upgrade, so the choice generally has to be right at the moment of purchase. However, some manufacturers offer the option of upgrading record length post-purchase via a software key. What they're doing is applying economies of scale by making all their oscilloscopes physically the same (containing the maximum amount of memory), and selling variants having restricted access at a reduced price. They're rather coy about this practice, perhaps fearing that a full explanation might lead to emotive arguments about whether oscilloscope prices genuinely reflect production and development costs or what the market will stand.

Features unique to digital oscilloscopes

We have seen that a high sample rate is crucial, but that it can only be maintained at slower sweep speeds by having sufficient record length, and that analogue bandwidth becomes a secondary consideration. Worse, because digital oscilloscopes are really computers with knobs on, they currently obey Moore's law (speed doubles every

18 months), and depreciate quickly, so any oscilloscope having a maximum real-time sample rate of less than 100 MS/s (appropriate for 20 MHz analogue bandwidth) is effectively junk. Despite these caveats, digital oscilloscopes have many features that simply **cannot** be provided in an analogue oscilloscope, so it is time to look at their advantages.

Negative time

Because digital oscilloscopes acquire, store, and display the input signal continuously, they can show events **before** the trigger point, allowing them to look backwards in time. By definition, this is impossible for an analogue oscilloscope.

Usable slow time base settings

At very slow sweep speeds, an analogue CRT no longer produces a graph, but a decaying swept spot, making the display very difficult to interpret. At these slow sweep speeds, digital oscilloscopes may refuse to trigger using auto-triggering and free-run (known as **roll mode**), which produces a non-decaying display akin to an analogue chart recorder. Alternatively, proper (normal) triggering restores a triggered display that updates periodically. Either way, sweep speeds of 10 s/division become possible, which is very useful for observing slow events like heater warm-up times.

As an aside, to the author's knowledge, the obsolete HP54600B family has an unusual quality shared by no other oscilloscope. If armed with its optional memory brick, it can store 100 traces, but more importantly it can recall them all simultaneously to the screen. If the traces were expected to be completely different, this wouldn't be useful, but if displaying heater warm-up times (expected to be similar) it is extremely useful — sufficiently useful to make this dinosaur worth rescuing from a skip.

Capturing and displaying infrequent events

Analogue oscilloscopes produced trace brightness proportional to the time spent by the beam sweeping overlaid traces per second, so if the oscilloscope wrote infrequently, the trace dimmed, perhaps to imperceptibility. By contrast, even a single event written into digital memory can be written to the display repetitively to give a perceptible trace.

Colour

The colour of the trace on an analogue oscilloscope was determined by the required persistence of the phosphor screen coating, so the trace produced by Ch1 was the same colour as that produced by Ch2. Digital oscilloscopes with monochrome displays were made, but proper digital oscilloscopes have a colour display, allowing each channel to have a distinct colour, which means that instead of using the top half of the screen for Ch1 and the lower for Ch2, both traces can occupy the full height of the screen because they are easily identified.

The previous point is more important than at first appears. The screen's "Y" axis corresponds directly to the codes leaving the ADC (ranging from 0 to 256 for an 8 bit ADC). If we only use half the screen for one channel (0−128), we have effectively thrown away one bit of ADC resolution, so we should always use the full height of the screen to maximise accuracy. Of course, the same argument could be applied to an analogue oscilloscope, but the crucial significance is that reduced trace height degrades the accuracy of digital automated measurements. Once we have two waveforms each occupying the full height of the screen, they can only be easily distinguished if they are different colours, so a colour display becomes a necessity rather than a luxury.

Storing and exporting traces

The digital advantages mentioned so far pale into insignificance compared with the ability to store traces for later comparison. When fettling an amplifier, and testing the effect of changing a component, it is extremely useful to be able to compare "before" and "after", and this is the reason for internal **reference** memories that allow a trace to be stored and later recalled to the screen.

Most oscilloscopes can export traces, but first generation digital oscilloscopes sometimes required an add-on module, proprietary software on the PC, and perhaps an uncommon port (such as GPIB) at the computer. Each of these extras costs money, and data transfer required the oscilloscope to be directly connected to a specific computer, which may be inconvenient. Such requirements greatly reduce the value of early digital oscilloscopes, so if you are considering buying one, check very carefully how it exports traces and make sure that the file format does not require dedicated software.

Far more powerful and more recent digital oscilloscopes encountered a somewhat different problem in that their record length was sufficiently deep that trace files were simply too large to fit on any portable media available at the time of their manufacture, so these oscilloscopes transfer their data via a network port. This ties the oscilloscope to a bench, but at least not to a particular computer. However, be aware that it costs more to interface to a GPIB port than an Ethernet port.

The ideal oscilloscope has a USB socket for a memory stick, so you simply store a trace and take the stick to any computer at your leisure, but even here there can be a problem. Oscilloscopes are occasionally limited to the amount of memory they can address, so it would be unfortunate to find that although the oscilloscope was working perfectly, you couldn't buy a memory stick small enough for it to talk to.

The most common way to export an oscilloscope trace is as a graphic image that includes the graticule and any on-screen information such as cursors or automated measurements. However, it is also possible to export an oscilloscope's trace as raw sampled data (usually in comma-separated-variable ".csv" format), allowing post-analysis on a computer. Beware that even short record lengths (10 kpt) generate large files and that spreadsheets written for 16 bit operating systems are limited to 65,536 data entries, making it essential for the oscilloscope to allow explicit control of record length.

Note that few oscilloscopes can return an exported trace to their screen. Thus, if you think you might need to later compare a trace, save it to one of the internal reference memories as well as exporting it.

Data decimation: sample, peak detect, high resolution

Rather than continually adjusting the ADC's sample rate, it is simpler to leave it running at maximum rate and discard or select data as necessary — this process is known as decimation. We saw earlier that at slow time base speeds, finite record length forces the oscilloscope to lower the rate at which it writes to memory, so decimation is frequently necessary, but there are various ways in which it can be achieved.

If data rate needs to be reduced by a factor of 8, every eighth sample could be selected and sent to memory, discarding the remaining seven samples, and this is known as **sample** mode.

As more and more data samples are discarded from sample mode, it becomes progressively more likely that an important event (perhaps a glitch) that occurs between samples will be missed. **Peak detect** mode captures transients by looking over the time frame needed to select a pair of samples and retaining the two ADC samples having maximum and minimum amplitude. In this way, both positive or negative peaks are detected and their results drop into a nearby memory location. See Figure 4.28.

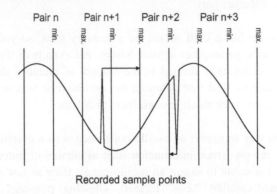

Figure 4.28
When sample rate falls, a fast event could fall **between** recorded samples, so peak detect captures maxima and minima and records them as a data pair over two samples.

The fact that peak detect causes events to be recorded with a slight time error and gross distortion is not nearly as serious as missing the event. Peak detect alerts the user that a more detailed investigation is necessary, and is invaluable for finding the cause of otherwise insoluble faults, such as those caused by random spikes on the output of a faulty switched mode power supply. Peak detect always makes signals look noisier than sample mode.

Sample mode selected individual samples at regular intervals without regard to intervening amplitudes, whereas peak detect selected a data pair having maximum and minimum amplitudes, but both techniques simply selected samples whose amplitude was unmodified. The very first Phillips CD players used 14 bit digital to analogue converters but over-sampled them by a factor of 4 to achieve 16 bit resolution, and digital oscilloscopes can use the same technique. Rather than selecting a single sample from "*n*" data points, we could calculate an average value across those data points, thereby oversampling and achieving resolution below the ADC's least significant bit. Because oversampling averages across a number of samples, it attenuates uncorrelated signals such as noise, but this might be precisely the detail you need to see, so oversampling is always offered as a selectable option known as **high resolution**.

Averaging and noise

Because a digital oscilloscope effectively stores the data resulting from each sweep as a data base, it can explicitly perform mathematical calculations such as **averaging**. The oscilloscope displays a repetitive waveform, so each sweep across the screen draws a trace very similar to the one before it. Provided that the oscilloscope is triggered from the repetitive signal, averaging across traces (usually termed **waveforms**) retains the repetitive element but random noise averages towards zero. Averaging is typically adjusted in binary logarithmic steps (2, 4, 8, etc.), perhaps from two to 512 traces. The reduction in noise is proportional to $\sqrt{n}$ so averaging over 512 traces reduces noise by 27 dB, but at the expense of increasing the time required for the display to stabilise.

Averaging is a very powerful method of extracting signals seemingly buried in noise, but when the repetitive element is obscured by noise, external triggering is required. As an example, mains hum buried in noise can easily be detected and measured by triggering the oscilloscope from AC line, then invoking averaging.

A crosstalk test might inject a square wave into one audio channel from a signal generator, then look for crosstalk on the power supply or another audio channel, and trigger the oscilloscope from the generator's trigger output.

The value of averaging and external triggering really becomes apparent when we precede the oscilloscope with an external pre-amplifier in an effort to measure signals at the microvolt level. The author has an ancient Levell TM3A 3 MHz microvoltmeter whose most sensitive range is 15 μV FSD, but its self-noise is 5 μV. However, the meter has an output before its rectifier that can be taken to an oscilloscope, effectively making the meter a variable gain pre-amplifier. Thus, when the author needed to measure the amplitude against frequency response of an LC filter having an anticipated 120 dB voltage range from 10 Hz to 1 MHz, the following set-up was ideal. See Figure 4.29.

The meter's range switch had 10 dB steps, but was switched in 20 dB ($\times$10) steps because this allowed the probe calibration menu on the oscilloscope to be adjusted by a matching factor of 10. In this way, when the meter added 40 dB of gain, the oscilloscope's voltage probe menu was set to 1/100, allowing the oscilloscope's automated measurement system to reflect the signal at the input of the meter. The significance is that when the meter's amplifier gain was set to 1000 and the oscilloscope was registering 1 μV/division, averaging could be invoked to significantly attenuate the inevitable noise generated by the

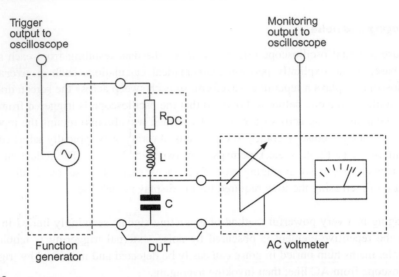

Figure 4.29
This test set-up exploits the digital oscilloscope's averaging function to measure synchronous signals below the AC voltmeter's self-noise.

meter's high-gain amplifier, and automated oscilloscope measurements below 1 μV became possible. Note that this noise reduction technique only worked because the signal to be measured was synchronous with the external trigger.

Beware that not all oscilloscopes allow their probe configuration to be set to divide (the Rigol DS1052E doesn't), so check first if you think you may need this facility.

Although averaging operates across samples widely separated in time rather than across adjacent samples, its result is a single sample, so it is invariably grouped in the user menu with the decimation techniques of sample, peak detect, and high resolution.

Display modes: envelope, persistence, and vectors

Envelope mode never deletes a sweep, so successive sweeps are displayed over one another in the manner of a multiple exposure photograph. Although the display becomes completely distorted, making it impossible to discern a waveform, it enables maximum voltage excursions to be measured easily, and this is particularly useful for determining how much overload capability is required in a given stage. We simply apply music for as long as we like, and measure the maximum vertical excursion. Conversely, data communications engineers use envelope for measuring time jitter (horizontal excursion) on their data waveforms.

Displaying a sweep indefinitely is not always ideal, so envelope mode is often adjustable in 2, 4, 8 steps to limit the number of sweeps for which the results of a given sweep are stored and displayed, and this enables the appearance or disappearance of infrequent pulses to be monitored.

All displays (and the human eye) have a property known as **persistence**, whereby a gradually decaying visible image remains even though its cause has gone. This might seem to be a defect, but it can be quite useful because it assists in seeing momentary events. Persistence in an analogue oscilloscope was pre-determined by the choice of display phosphors, and was therefore unchangeable.

Digital oscilloscopes can mimic persistence electronically and controllably. The way that this is done is to treat the array of display pixels as a three-dimensional database where the third dimension at each pixel is brightness. Inferior digital oscilloscopes simply place a "1" or a "0" in each pixel on each trace and this means that infrequent events are displayed with exactly the same brightness as frequent events, leading to a confusingly cluttered display. However, if brightness is stored as a 4 bit number at each display pixel, it can be incremented by one each time the waveform hits that pixel (perhaps from 0100 to 0101). If a calculation is applied simultaneously that gently decrements that number over time, pixel brightness can be made proportional to that number, and we have recovered the analogue advantage of persistence. Tektronix call this technique **digital phosphors** and it allows a digital oscilloscope having sufficient sample rate at all time bases to produce a display almost indistinguishable from a top-quality analogue oscilloscope, but with all the digital advantages. This truly is the best of both worlds, and was the deciding factor in the author's purchase of a TDS3032.

The key difference between persistence and envelope is that envelope mode switches pixels off abruptly whereas persistence causes the brightness of older pixels to gently decay. See Figure 4.30.

Modulation of pixel brightness using a 4 bit code allows brightness to be changed over a 16:1 range, and more bits would allow a larger range. The limits are determined not only

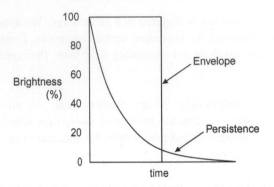

Figure 4.30
Oscilloscope brightness against time: persistence (exponential decay) vs. envelope (abrupt switch-off).

by maximum and minimum pixel brightness, but also by incident light upon the display, so it becomes difficult to better 256:1 range (8 bits). An alternative approach uses false colour to denote how often a pixel has been overwritten, and thereby how frequently an analogue event has occurred, potentially accommodating a much larger range, although display interpretation has to be consciously learned.

Because an oscilloscope samples and quantises, it plots points on a screen of virtual graph paper. It is conventional to join the points on graph paper, and oscilloscopes can also perform this interpolation function. Most oscilloscopes allow the user to choose whether the journey from one point to the next is made by linear interpolation or by sin $(x)/x$ curve, so this feature is usually known as **vectors**. However, there is an implicit assumption in joining the sampling points that the curve really did move smoothly from one point to the next. Finite sample rate means that this assumption may not be true, particularly at slow time bases, so most oscilloscopes can be switched to points (often described as **dots**) only.

Automated measurements

A typical display on an oscilloscope is likely to contain one or two cycles of a repetitive waveform. Each cycle is thus a histogram composed of many vertical bars, and measurements or calculations that involve maxima, minima, or areas under one cycle of the graph can easily be made. Provided that a digital oscilloscope can trigger and capture the waveform without overload, it should be able to apply the V_{RMS} or V_{mean} calculation principles explained earlier in this chapter to **any** waveform with accuracy limited solely by sample rate and the number of quantising bits.

We saw earlier how to manually calculate V_{RMS} over exactly one cycle, but its full significance will now become apparent. Oscilloscopes typically offer V_{CycRMS} and V_{RMS}, and the difference is that the first measurement calculates over the first cycle after trigger, whereas the second calculates over the entire trace, which is almost certainly not an integer number of cycles. Thus, V_{CycRMS} is the calculation we must use when we display one or two repetitive cycles (perhaps of a sine wave), and we should only resort to V_{RMS} if displaying a waveform that never repeats itself within the displayed trace. You can prove the preceding statement to yourself by displaying a sine wave, requesting both measurements, adjusting time base, and noting that the discrepancy between the two measurements reduces as more cycles of the sine wave are displayed because the non-integer remainder of displayed cycles becomes a progressively smaller fraction of the total.

The most severe test of the V_{RMS} and V_{mean} measurement statement is to measure V_{RMS} of bandwidth-limited white noise, so the author compared his Tektronix TDS3032 and Rigol DS1052E oscilloscopes (V3.41 and V00.02.06 firmware respectively). The oscilloscopes were first tested using a 1 V_{RMS}/1 kHz sine wave, which they both measured correctly ($<$1% error). The Agilent 33220A generator was then switched to produce noise having 1 V_{RMS} amplitude and 10 MHz specified bandwidth. Both oscilloscopes had their time bases adjusted to allow a 50 MS/s sample rate (sufficient not to suffer aliasing on a 10 MHz signal). Within the limits of observation (by definition, any measurement of random noise must be unstable), the TDS3032 gave readings centred around 1.06 V_{RMS} (2% high compared with the 1.039 V_{RMS} $\pm$ 2 mV stated by an R&S RTO1012), whereas the DS1052E gave readings centred on 1.72 V_{RMS} (4.7 dB high). The DS1052E has switchable record length between 2.5 kpt and 1 Mpt, so the author tried the longer record length and slowed the time base to maintain the 50 MS/s sample rate, resulting in a more stable noise reading centred on 3.34 V_{RMS} (10.5 dB high).

Further investigation of the DS1052E revealed that its V_{RMS} calculation was incorrect (although it gave the right values for sine and square waves). A 5 V_{pk+} rectangular pulse 50 μs wide was applied at a frequency of 1 kHz. Using cursors, V_{pk+} seen by the DS1052E was 5.04 V, and V_{pk-} was -40 mV. Calculating from first principles using these numbers gives $V_{RMS} = 1.13$ V, but the oscilloscope stated 1.33 V (18% high). Looping the same waveform through the TDS3032 gave $V_{pk-pk} = 4.96$ V and

$V_{pk-} = -20$ mV using its cursors, and it stated that $V_{CycRMS} = 1.103$ V (1.105 V expected, so 0.2% low). No doubt the Rigol problem can be fixed with a firmware update.

The huge practical advantage of automated measurements is that they are live, and track waveform changes, whereas cursors have to be manually matched to each measurement.

DC nulling or self-calibration, and warm-up time

Traditional analogue oscilloscopes were designed and constructed for minimum DC drift so that their 0 V baseline was stable even in the face of changing temperature and range switching, but this was expensive. Digital oscilloscopes approach the problem in a different way and store individual offsets for each range so that (with no input) as the voltage/division control is switched, the trace does not wander from 0 V, and absolute DC offset performance can be traded for increased bandwidth. However, when temperature changes, the stored offsets become inappropriate and the 0 V baseline drifts. More significantly, any automated amplitude measurement (such as V_{pk} or V_{RMS}) referenced to that drifted 0 V baseline becomes incorrect, and this is why some oscilloscopes usefully offer an explicit standard deviation measurement (σ, sometimes termed RMS_{AC}) that effectively removes the baseline and any associated error. The warning sign of a drifted 0 V baseline is that with no input connected, the trace is not aligned with its vertical marker. See Figure 4.31.

The oscilloscope must have achieved a stable internal temperature (powered for at least 20 minutes) and input leads removed before its automated DC nulling or self-calibration routine is invoked. In general, the better any instrument's accuracy, the longer its warm-up time, so the Agilent 34410A 6½ digit DVM requires two hours warm-up time to meet its specifications.

Fast fourier transform (FFT)

The Fast Fourier Transform is a mathematical tool that allows data captured in the time domain to be displayed in the frequency domain. Put simply, although the vertical axis is still amplitude, it is now plotted against frequency, rather than time, and the oscilloscope has been converted into a spectrum analyser. This is extremely useful for investigating distortion harmonics. See Figure 4.32.

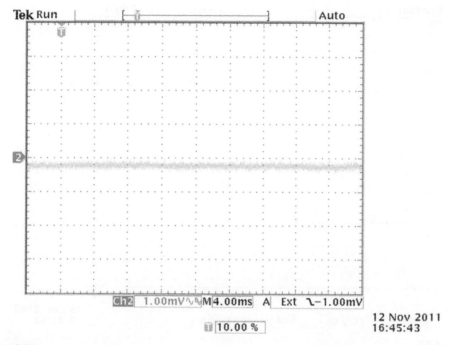

Figure 4.31
If the channel's 0 V marker isn't aligned with its trace (no input), automated measurements will be incorrect and oscilloscope self-calibration is needed.

The FFT is an immensely powerful tool, but it has limitations. In converting from time to the frequency domain, the mathematics of the FFT makes the assumption that the waveform to be analysed repeats itself periodically. This assumption may seem trivial, but it has **major** repercussions.

If we captured a **single** cycle of the waveform, and drew it around a circular drum (like a mechanical seismograph) so that the end of the cycle just met the beginning, then by rotating the drum we could replay the waveform ad infinitum and reproduce our original signal. Unfortunately, unless we have captured exactly one cycle, no more, no less, there will always be an amplitude step when we attempt to loop the recorded cycle back to itself on replay. However, if we capture more cycles on our drum, the step occurs proportionately less frequently per cycle, so capturing 1000 cycles reduces the error by a factor of 1000, at the cost

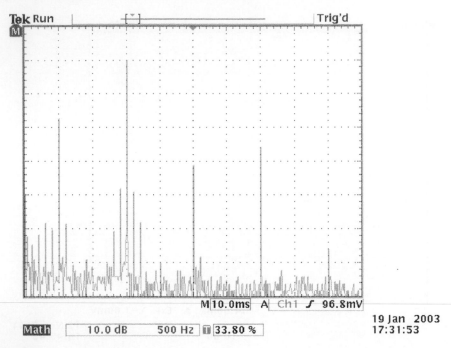

Figure 4.32
This FFT shows the amplitudes of the second to sixth harmonics of a push–pull output stage. As expected, the odd harmonics are higher amplitude than even, but more interestingly there are 50 Hz spikes around the second and third harmonics caused by intermodulation with a poor power supply.

of needing 1000 times the record length, making an oscilloscope's record length even more important if FFT is to be contemplated. Of course, if we have the freedom to adjust the frequency of the waveform to be analysed, we can enforce the capture of an integer number of cycles, avoiding the step, and this **synchronous FFT** has minimal skirts around peaks.

Another way of reducing the step is to force periodicity by applying a **window** to the waveform record. In this context, a window is a variable weighting factor that multiplies the values of the samples at the ends of the waveform record by zero, but applies more weighting (≤ 1) to samples towards the middle. Since any number multiplied by zero is zero, this forces the end samples to zero, and allows the waveform record to be repeated without steps. See Figure 4.33.

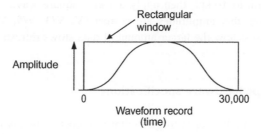

Figure 4.33
Shaping the window reduces the effects of periodicity violation compared to a rectangular window.

Because windowing distorts the waveform record, it must distort the results of the FFT calculated from that record. Windowing either spills energy from high-amplitude bins into adjacent bins, producing skirts around frequencies having high amplitude, or it changes bin amplitudes. (Because the process of sampling broke time into discrete slices, the results of an FFT must produce frequencies in discrete slices, and these are known as **bins**.) All windows are therefore a compromise between frequency and amplitude resolution.

A window that does not modify sample values is known as a rectangular window (because a plot of its shaping over the entire waveform record is rectangular). Because the rectangular window does not modify sample values, it cannot cause spreading between bins, and it offers the best frequency resolution, but peak amplitudes are likely to be in error because of periodicity violation. Conversely, although the **Blackman–Harris** shaped window has poor frequency resolution, it minimises peak amplitude errors, making it useful when measuring individual distortion harmonics.

In the same way that averaging reduced noise on an amplitude against time display, averaging can significantly improve the signal-to-noise ratio of an FFT display, but with the same cost of slowing measurement speed.

Although all digital oscilloscopes offer FFT, very few do it well. We have seen that reducing the effects of periodicity violation mandated a slow time base to capture many cycles of the fundamental, but avoiding aliasing requires a high sample rate, so the two combine to require a large waveform record containing a great deal of data to be processed. Many oscilloscopes simply don't have adequate processor power for that data. Test the FFT by setting record length to 10 kpt, then apply a 1 kHz square wave, set the display to span 0 Hz−20 kHz, check that amplitudes are correct (V, V/3, V/5, V/7, etc.) then increase record length to assess how the larger amount of data slows refresh rate.

Key digital oscilloscope hardware specifications

The main drive for oscilloscopes to go digital was cost − removing that horribly expensive CRT and its attendant expensive circuitry (EHT generator, high voltage "X" and "Y" deflection amplifiers). The fact that the first digital oscilloscopes were pretty poor (short record length, slow refresh rate, appalling ergonomics) didn't matter at the time because engineers had plentiful access to excellent analogue oscilloscopes, but that is no longer the case, and a digital oscilloscope is now likely to be our only oscilloscope.

The previous section investigated (mostly software) features common to all digital oscilloscopes, but we should now consider whether the hardware is good enough to support those features and acceptable at the price.

Screen size and ADC resolution

Analogue oscilloscopes typically had a usable screen size of 8 cm by 10 cm because electrostatic deflection CRTs had fixed proportions, and the typical depth was already 40 cm because of the length of the CRT, so a larger screen simply wasn't feasible. Liquid crystal displays (LCDs) impose no such depth penalty, so modern oscilloscopes have proportions more akin to a handbag. The seminal Tektronix TDS3000 series (introduced in 1998, now on its "C" variant, and much imitated) uses an LCD screen having a 6.5″ diagonal and 640 by 480 pixels, with a 10 cm by 8 cm area available for waveform display from its 9 bit ADCs that was clearly intended to be directly comparable to an analogue oscilloscope.

Larger oscilloscope screens are now available, potentially allowing us to see more detail in the waveform provided pixel size remains small. However, most cheaper oscilloscopes do **not** maintain pixel size, so ignore the hype about screen diagonal and look instead for the fine detail regarding display pixels, ADC bits, and whether oversampling can be invoked (high-resolution mode). 240 vertical pixels driven by an 8 bit ADC is half the vertical resolution of 480 pixels driven by a 9 bit ADC, and making the screen diagonal 8.5″ rather than 6.5″ is not going to magically improve it.

Interestingly, 12 bit oscilloscopes are beginning to appear (LeCroy), and such an ADC would ideally drive a vertical screen resolution of 4096 pixels ($2^{12} = 4096$).

Maximum record length

Depending on manufacturer, record length might be specified not in terms of the memory always available to each individual channel, but by the total amount of memory shared between all channels. Thus, 2 Mpt of record length might be stated in bold type at the front of the marketing puff, but the fine print could reveal that this is total memory, and that when you have all four channels running, each has only 500 kpt.

A mixed signal oscilloscope is a combination of a logic analyser and an oscilloscope, so it might have 16 digital channels and two analogue. A record length specification might again quote 2 Mpt total, but in the fine print note that this is when running a single digital channel (1 bit only), so running a single analogue channel (8 bit) would have access to a far less impressive 250 kpt, two analogue channels 125 kpt, and if both analogue channels and all 16 digital channels were running, only 62.5 kpt would be possible.

When comparing oscilloscopes, read record length specifications very carefully to make sure you understand what you are being offered.

Maximum sample rate

One popular way of increasing an oscilloscope's maximum sample rate is to **interleave** identical ADCs. Thus, to double speed, we use two ADCs each run at their maximum speed, one delayed by the required sample period. The principle is extensible, so 128 ADCs would increase speed by a factor of 128, but there is a downside. If we use 128 ADCs to achieve a 5 GS/s oscilloscope sample rate (perhaps matched to 500 MHz attenuator bandwidth), each ADC's sample rate falls to 39 MS/s or, to put it another way, we now have a 39 MHz clock whose fundamental and harmonics are well within the oscilloscope's analogue bandwidth, risking crosstalk into the signal to be measured. Conversely, if we had a true 5 GS/s single ADC, its clock would lie outside the 500 MHz analogue bandwidth. Beware also that interleaving forces successive samples to be measured by different ADCs, potentially increasing differential non-linearity due to mismatched DC offsets between ADCs. Provided that the oscilloscope is used purely in the time domain, the previous two issues should be inconsequential, but when the FFT is invoked to produce a graph of amplitude against frequency, internal clock spuriae and ADC linearity become more significant.

Just like record length, the ADC can be shared between channels. More usefully, we could consider the acquire block to be a single ADC and memory that is connected sequentially to each analogue input channel. Thus, when connected to one channel, all the memory and the maximum ADC sample rate are available to that channel, but if all four analogue channels are being used, each channel is only allocated a quarter of the memory and a quarter of the ADC's sample rate. Although it maximises use of the available resources to allocate the ADC in this way, very few engineers like the idea of sample rate changing just because another channel has been switched on or off.

Although the Tektronix 3000 series had a generous ratio of 10 between analogue bandwidth and maximum sample rate (100 MHz: 1 GS/s; 300 MHz: 2.5 GS/s; 500 MHz: 5 GS/s), a factor of 5 (100 MHz: 500 MS/s) is more usual on later designs, although it may be disguised. Thus, an oscilloscope might proudly have 1 GHz, 10 GS/s emblazoned on its front panel, but careful reading of the specification reveal that this is when only one channel out of a pair is used, and that when both are used, 5 GS/s is the maximum. Again, read the specifications very carefully.

Refresh rate

The screen's refresh rate has to be fast enough (>80 Hz) for flicker not to be visible, but if the oscilloscope's memory was only refreshed at this rate we might miss momentary glitches. Ideally, we would like every waveform capture to be followed by another at the first occurrence of the user's trigger conditions being satisfied, and for all of this data to be written to display memory. In practice, some triggers and subsequent acquisitions may be missed because the oscilloscope's memory system is not ready to accept data. Although refresh rate can be measured easily enough in terms of waveforms per second, it is hard to specify on a data sheet because it depends on time base setting, record length, triggering, and waveform shape, so vague descriptions like "up to" and "typical" are used. It is rare for an oscilloscope to reveal its refresh rate in use, which is a shame because indication allows the user to iteratively adjust settings to maximise it if necessary. Be assured that a cheaper oscilloscope is likely to save cost by having slower memory and busses that degrade its refresh rate.

Although difficult to specify, a good subjective assessment of a basic oscilloscope's refresh rate may be quickly made using orchestral music recorded in stereo without data compression. Using analogue audio, connect L to Ch1 and R to Ch2 (or vice versa), select

XY mode, select AC auto-triggering, and observe the resulting Lissajous figure. A perfect display would be identical to that produced by an analogue oscilloscope, resembling a continuous ball of wool smoothly pulsing in size and intensity without any jumps, glitches, or discontinuities. Unlike an analogue oscilloscope, a digital oscilloscope is forced to trigger and capture input waveforms before displaying them sequentially as discrete Lissajous figures, and because music is essentially random to an oscilloscope, really basic oscilloscopes fare very badly on this test due to their poor refresh rate.

Processor power

Oscilloscope manufacturers compete to offer the most record length because it's an obvious specification, but 1 Mpt is a lot of data, and any calculation using all of that data (such as FFT, bus decoding, or event searching) either requires a powerful processor or hardware acceleration of specific tasks. Thus, Agilent have hardware acceleration of bus decoding and Rohde & Schwarz of FFT. The reason that the Tektronix 3000 series has survived so long is that it is a very well balanced design having superb ergonomics and a short (10 kpt) record length that is easily handled by its processor.

Ergonomics

A dozen key specifications fully captured an analogue oscilloscope's performance, but although a digital oscilloscope's hardware and software features may be specified, it is essential to see how the combination is presented to the user – you need to try an oscilloscope before you buy it. Poor ergonomics are not confined to cheap oscilloscopes – as more and more capabilities are added, it becomes increasingly difficult to present them in a coherent manner.

As analogue oscilloscopes became more complex, they necessarily acquired an increasingly bewildering array of hard-wired knobs and switches. Digital oscilloscopes reduce front panel complexity by assigning groups of controls to multi-function knobs or resorting to drop-down menus. Thus, an oscilloscope might have a pair of vertical sensitivity and position controls that have to be assigned to control a particular channel. Some users hate this space-saving solution because it's easy to inadvertently adjust the wrong channel, so the Tektronix 4000 series upwards has dedicated vertical controls for each channel, yet cursors still need to be assigned to a channel – a slight inconsistency. Alternatively, the Rohde & Schwarz RTO series has multicolour LEDs under its shared controls that warn the user which channel the control will act upon. Vertical controls are used a lot, so make sure you're happy with the solution offered.

We have seen that record length is a key parameter, especially when using FFT, yet the user might not have explicit control. For any displayed waveform, there is always an optimum record length. A record length that is too long may slow screen update rate, but one that is too short may cause fine detail to be missed on an amplitude against time display or aliasing on an amplitude against frequency display. Further, as record length increases, the apparent effect is to capture more random noise, perhaps obscuring detail in the (repetitive) waveform. Thus, some oscilloscopes adjust record length with every time base setting, simply allowing the user to bias the algorithm between optimising resolution or sample rate. Whilst this strategy is fine for amplitude against **time** displays, using the FFT requires explicit control of record length and sample rate, and if a waveform is to be exported as a csv file, even 10 kpt is a lot of spreadsheet points. Check that you can enforce a known and constant record length when necessary.

Applying signals to the oscilloscope

Having acquired our oscilloscope, we need a means of applying signals from our circuit.

Input capacitance and voltage dividing probes

The most sensitive setting on a typical oscilloscope is 1 mV/division, and the standard oscilloscope input resistance is 1 MΩ. This is almost sensitive enough for a microphone, so we must use a screened lead to avoid picking up hum from all the (nearby) mains wiring. Typical coaxial screened lead has a capacitance of ≈100 pF/m, so we might use a 1 m lead having 100 pF of capacitance to connect the anode of an ECC83 or 6SL7 high-μ triode common cathode gain stage to our 20 MHz oscilloscope. Under typical operating conditions, this stage is likely to have an output resistance of 60 kΩ. Unfortunately, the capacitance of the lead and output resistance of the amplifier form a low-pass filter having a $-3\,\mathrm{dB}$ frequency of:

$$f = \frac{1}{2\pi CR} = \frac{1}{2 \times 3.14100 \times 10^{-12} \times 60 \times 10^3} \approx 26.5 \ \mathrm{kHz}$$

We cannot make useful audio measurements with this filter in the way. We cannot change the output resistance of the amplifier, so we must reduce the loading capacitance.

One way of reducing the capacitance might be to cut the lead shorter, perhaps to 10 cm, which would reduce the capacitance by a factor of 10 to 10 pF, and would move the filter frequency to 265 kHz, which is safely out of the audio range; 10 cm is somewhat less than the width of one line of text in this book, so it would be quite difficult to work with a lead this short.

Further, the oscilloscope itself has input capacitance, typically 10–30 pF, and the precise value is usually stated on the front of the oscilloscope next to the input connector. See Figure 4.34.

Figure 4.34
Most oscilloscopes state input loading next to their input socket.

The solution to the capacitance problem is to make a parallel connected pair of potential dividers, one resistive, one capacitive. See Figure 4.35.

The 1 MΩ resistor is the standard oscilloscope input resistance, and this is in parallel with the oscilloscope's input capacitance and the lead capacitance. Together with the 1 MΩ resistor, the additional 9 MΩ series resistor forms a 10:1 potential divider. The cunning part is the addition of the capacitor across the 9 MΩ resistor. There are two ways of looking at this capacitor's function:

- Potential dividers: The 9 MΩ/1 MΩ resistive potential divider has a loss of 10:1. We can also make potential dividers from capacitors, but because a capacitor's reactance is inversely proportional to its capacitance, a 10:1 capacitive divider requires the upper capacitor to be one-ninth the value of the lower capacitance.
- Time constants: We can view the upper resistor/capacitor combination as a high-pass time constant, and the lower resistor/capacitance combination as a low-pass time

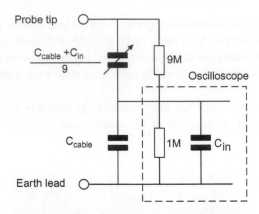

Figure 4.35

An oscilloscope probe is essentially two potential dividers in parallel. The complete circuit is contributed partly by the oscilloscope, partly by the probe.

constant. If the time constants are equal, their filtering effects cancel out, and the network has a flat frequency response.

Looking into the input of the circuit, at DC the capacitors are an open circuit, so we see the two resistors in series, giving an input resistance of 10 MΩ. At high frequencies, the capacitors are short circuits, so we can neglect the effect of the parallel resistors, and we see a pair of capacitors in series:

$$C_{\text{series}} = \frac{C_1 C_2}{C_1 + C_2} = \frac{13.33 \text{ pF} \times 120 \text{ pF}}{13.33 \text{ pF} + 120 \text{ pF}} = 12 \text{ pF}$$

At the expense of oscilloscope sensitivity, and by adding a resistor and capacitor at the far end of our lead, we have reduced input capacitance by a factor of 10 from 120 to 12 pF. **Oscilloscope probes** always state their input capacitance, either in the data sheet that came with the probe or on their connector. See Figure 4.36.

Because different oscilloscopes have different input capacitances, we need to be able to adjust the upper capacitor to match the oscilloscope. This adjustment is often a small screw in the side of the probe. See Figure 4.37.

Figure 4.36
Some probes state their essential data on the connector housing.

Figure 4.37
Probes are often equalised by a small screw in the body.

The adjustment is set by connecting the probe to the 1 kHz square wave provided by all oscilloscopes on their front panel, and often called **Cal**. The oscilloscope is set to display one cycle of the square wave, and the capacitor adjusted to give the flattest, squarest, leading edge. See Figure 4.38.

Generic ×10 probes are cheaply available for oscilloscopes having <250 MHz bandwidth. Faster oscilloscopes require specifically matched probes to minimise in-band response ripples, and such probes are typically ten times the price of a universal probe. Note that connecting a 500 MHz probe to a 500 MHz oscilloscope combines the two responses, so response at the probe tip is −6 dB at 500 MHz.

Although not a problem at audio frequencies, charging an oscilloscope probe's input capacitance through its ground lead inductance (roughly 0.75 nH per mm) causes significant ringing >10 MHz. Measurements within switched-mode supplies are particularly prone to error from this problem, requiring minimised ground lead lengths, so faster

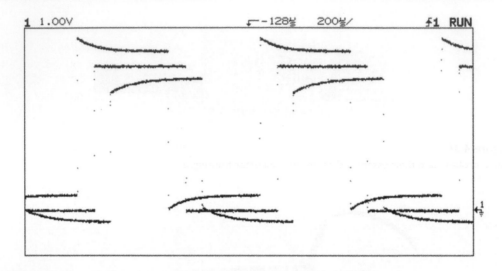

Figure 4.38
Correct probe equalisation produces a square leading edge (middle trace), rather than overshoot or rounding.

probes allow their ≈ 150 mm flexible ground lead plus crocodile clip to be replaced by a stiff ≈ 50 mm probe.

Active probes

Although our example **passive** voltage dividing probe reduced input capacitance from 120 to 12 pF, and 4 pF has been seen, it is possible to do better. **Active** voltage probes incorporate an amplifier close to the probe tip, reducing input capacitance, and because they incorporate an amplifier, sensitivity need not be lost. Active voltage probes are often **differential**, responding to the difference between their two floating input terminals rather than between one terminal and ground. There are two distinct classes of differential probes: high voltage and low voltage.

High voltage (usually >1 kV) differential probes tend to be quite low bandwidth (≤ 100 MHz), have long unscreened input leads terminated by large crocodile clips, and a large box between their input leads and the oscilloscope. Beware that just because the probe is differential, it is not **isolated** unless explicitly stated by the manufacturer. For example, the Pico TA044 7 kV probe is not isolated, but its 10 MΩ//10 pF input imped-ance to BNC earth from each leg means that its earth current is negligible at mains fre-quency and the distinction is moot. However, the probe might quite reasonably be used for investigating the final stage of a 10 MHz transmitter, in which case the BNC earth current via the 10 pF capacitance would certainly not be negligible, which is why the probe manu-facturer insists that the BNC be earthed.

It is perfectly possible to make a true isolated probe. The input signal can be converted to digits and transferred via an opto-coupler (which provides the high voltage isolation) before being converted back to analogue and sent to the oscilloscope. The floating ADC may either be battery powered or powered from the oscilloscope via a switched-mode supply incorporat-ing a transformer having a suitably high isolation voltage. All this complication adds cost.

Low voltage differential probes are specifically designed to maximise bandwidth at the probe tips and >1 GHz probes are common, so they minimise input capacitance (typically <1 pF) and their very short input leads (typically <50 mm) minimise lead inductance. Fast transistors tend to have low voltage ratings, and over-voltage protection adds crucify-ing capacitance, so the maximum voltage that can be tolerated before destruction occurs might be only ±8 V. Despite their frightening fragility and cost, a low voltage differential probe is the most accurate way of assessing in-circuit behaviour >1 MHz, especially within switched-mode power supplies.

Current probes determine the current in a wire by measuring the magnetic field it pro-duces, allowing current to be measured without breaking the circuit. They achieve this by directing the magnetic field through a coil of wire forming the secondary of a transformer, so unlike all other probes, passive current probes have a low frequency bandwidth limit, and it's surprisingly high − typically 15 kHz − and even worse, their cores are easily sat-urated by low frequencies or DC, so they're almost useless for audio (they're intended for investigating currents in switched-mode power supplies).

Wide-bandwidth **Hall effect** current probes still need a transformer to capture high-frequency information, but use their Hall effect device to generate a low frequency nulling waveform that they apply to the transformer (preventing core saturation at DC and low frequencies), then combine that nulling waveform with the transformer's waveform to deliver DC to 50 MHz bandwidth. Although useful, a new Hall effect current probe is compact but expensive, whereas previous generation probes required a bulky external amplifier, usefully reducing their price on the second-hand market.

Interestingly, the Tektronix P6302 Hall effect current probe plus its AM503 electronics has a sensitivity of 1 mA per division when the associated oscilloscope is set to the recommended sensitivity of 10 mV per division, making it ten times as sensitive as its later TCP202 counterpart. If the probe feeds a digital oscilloscope, the increase in noise caused by increasing oscilloscope sensitivity to 1 mV per division can be offset by invoking averaging, increasing sensitivity by a factor of 10 for repetitive signals. And if you don't terminate the cable between amplifier and oscilloscope, you can gain a further factor of 2 in sensitivity. (Not terminating the cable would probably be legitimate because the noise gain is now so great that you would already be limited to low frequency signals.)

All probes are expensive, but active ones especially so, and oscilloscope manufacturers typically charge £1000 per GHz of bandwidth per voltage input (so differential probes cost double). Moral: Look after your probes — the author returns his probes to their box immediately after use.

Probe interfaces

Early oscilloscopes required the user to remember to multiply the volts/division appropriately when a probe was used. The first step towards automation replaced the fixed marker outside the oscilloscope's volts/division control with a pair of LEDs, one being positioned appropriately for $\times 1$ probes and the other for $\times 10$ probes. A $\times 10$ probe had an additional pin outside the BNC's outer ring, and when it contacted a corresponding ring at the oscilloscope, the 11 kΩ resistor between them tripped a comparator, toggling the $\times 1$ LED off and the $\times 10$ LED on.

Oscilloscopes having automated measurement systems need to know when the signal has been applied via a probe, so the resistor/pin scheme originated in the 1970s still survives. But a current probe should advise an oscilloscope appropriately, and active probes require extra pins for power, leading to the birth of the oscilloscope probe interface. Early interfaces added pins as necessary and were purely analogue, but later interfaces exchange digital metadata such as probe internal parameters, serial number, and software version.

Although it is undoubtedly convenient for a probe to automatically configure the oscilloscope, oscilloscope manufacturers rarely support more than their proprietary interface. Because active probes are so expensive, they engender interface loyalty, making laboratories reluctant to change oscilloscope brand (presumably, exactly as the oscilloscope manufacturer intended). However, third-party active probes **are** available. See Table 4.1.

Table 4.1

Advantages and disadvantages of third-party oscilloscope probes

Advantage	*Disadvantage*
Typically half the price	Slower to use and perhaps prone to error
Output is on a standard BNC, so it can be connected to any instrument (not necessarily an oscilloscope), via an adapter if necessary	because scaling, coupling, and termination all have to be configured manually
	Requires external power supply
Sensitivity can be increased by adding an amplifier between probe and oscilloscope (and consequent noise reduced by invoking averaging)	

Specialised probes tend to be used far less frequently than standard $\times 10$ probes, making the slight inconvenience of a third-party probe tolerable for a halving in price, and its greatly increased flexibility is a bonus.

Transmission lines and terminations

Another way of connecting to an oscilloscope is via a 50 Ω **transmission line**, terminated by 50 Ω at either end, and this is why faster oscilloscopes have a 50 Ω switch on their input coupling. If you accidentally operate this switch with a $\times 10$ probe connected, your signal will disappear. The other hazard with the 50 Ω **termination** is that the resistor has a very low power rating, so it can easily be burnt out if a large signal is inadvertently applied directly to it.

Fortunately, transmission line effects do not become apparent until the cable length becomes a significant proportion of the length of one wavelength of the frequency travelling down the cable. Since the velocity of propagation down most cables is two-thirds the speed of light ($c \approx 3 \times 10^8$ m/s), this means that one wavelength at 20 kHz occupies 10 km of cable. We can completely ignore transmission-line effects in analogue audio – although until the early 1980s the BBC used an unamplified music line from Bush House (London) to their Daventry transmitter 110 km away, so transmission-line effects could presumably have been observed at audio frequencies on this cable.

Search coils

All transformers and inductors leak magnetic flux into their surroundings and it is often useful to assess this leakage. The simplest method uses an air-cored search coil connected via a length of coaxial cable to the oscilloscope and set to terminate the coil and cable with 50 Ω.

The search coil has inductance and forms a low-pass filter in conjunction with cable capacitance and oscilloscope input capacitance:

$$f = \frac{1}{2\pi\sqrt{LC}}$$

Search coil sensitivity is proportional to the number of turns, but if we want to investigate high frequency flux, the equation shows us that we must minimise coil inductance, which is proportional to the number of turns squared, so search coils that are useful at high frequencies are insensitive. Thus, we need a variety of quickly interchangeable coils, and should be prepared to make another for a specific purpose if existing coils seem unsuitable. Fortunately, the coils are so easily made that this isn't a problem. Coils can be wound on standard coil formers, but any plastic tubing from about 6 to 12 mm diameter will do, as 80 turns seems to give ≈ 38 μH whether on a 6 or 12 mm former. The essential is to be able to fit it securely to a BNC connector, either by soldering to the pins of a dedicated coil former or by glueing to tubing. See Figure 4.39.

Figure 4.39
A family of five search coils (L to R): 10, 38 (two constructions), 150, 650 μH.

Be warned that the neat coil with BNC glued into it was much harder to make than the others, yet technically no better. The (phenolic) tubing was cut to length and drilled 0.3 mm (easier than you'd think) to allow the wire to pass inside, coil wound (much harder than you'd think), coil sealed with clear varnish, leads soldered to the BNC, and the BNC glued into the former using epoxy. Although omitted for clarity, this coil could be made much more robust by an outer addition of heatshrink sleeving.

The author used 0.16 and 0.2 mm enamelled copper wire, but wire gauge isn't critical. The key parameter is the number of turns. See Table 4.2.

Table 4.2
Comparison of search coils

Turns	Likely inductance (μH)	Typical use
325	650	Mains transformers, chokes
160	150	Output transformers
80	38	
40	10	Switching supplies

When you wind a coil, measure its inductance, label it, and if the inductance seems wildly inappropriate, wind another with a different number of turns. Provided that you have chosen a coil of appropriate inductance, used a short connecting cable, and terminated it in 50 Ω, the shape of the waveform seen on the oscilloscope will be representative of the signal within the circuit. It is usually best to start with the lowest inductance coil first (to ensure observation of high frequency signals) then work up in inductance (and increasing sensitivity) if high frequencies need not be observed. The oscilloscope almost always needs to be set to its highest sensitivity, so an external pre-amplifier can be useful.

Unsurprisingly, these search coils do not allow absolute measurements, but they do allow useful comparisons to be made:

- Encapsulated mains transformers and chokes can have their coil orientation found by noting that maximum coupling will be obtained when the search coil is coaxial with the axis of the test device's coil. Drive the winding having the most turns from a signal generator and search. The signal leaving the search coil is always small and mains hum could be confused with the leakage field of interest, so it is useful to off-set the signal generator's frequency a little from mains frequency (perhaps 70 Hz) and externally trigger the oscilloscope from the generator — allowing any signal rolling through the stationary display to be discounted as mains hum. Once orientation is known, a mains transformer or choke can be aligned on the chassis for minimum interference.

- Very occasionally, the leakage inductance of a mains transformer can resonate with rectifier capacitance, producing bursts of perhaps 200 kHz — a search coil can identify this leakage flux and compare the before and after effect of adding snubbers.
- Switching supplies frequently radiate high frequency interference, and a search coil can identify where it's worst and which way it's pointing. Further, it can be used to test the efficacy of interposed magnetic shielding.

The intention is that the coil should respond primarily to the magnetic field rather than electric field, and this is largely ensured by the 50 Ω termination, which forms a high-pass filter in conjunction with any inadvertent coupling capacitance, and such capacitance is minimised if single-layer coils connect the near end of their coil to the BNC centre pin and their far end (which will be closest to the test device) to the BNC's earth pin.

Oscillators and dedicated audio test sets

If you have an oscilloscope, you need an oscillator, but an oscillator is also useful for supplying an external source of AC to an analogue component bridge so that components can be tested at different frequencies. (Although be warned that many bridges require any external oscillator to be floated from earth by a transformer.) Air-cored inductors are more easily measured at 20 kHz than at 1 kHz ($X_L = 2\pi fL$), whereas the primary inductance of an output transformer is traditionally measured at 20 Hz.

Historical (Wien) oscillators

All analogue oscillators require three blocks:

- A frequency-selective network.
- An amplifier having sufficient gain to overcome the losses in the frequency selective network, and positive feedback.
- An amplitude stabilisation circuit.

The amplifier could use valves, transistors, or integrated circuits. The frequency-selective network needs to be tunable, and since no network can be continuously tuned over the entire 20 Hz–20 kHz audio range, it is conventional to break the range into switched decades (20–200 Hz, 200 Hz–2 kHz, 2–20 kHz), and continuously tune over the resulting 10:1 ranges.

Traditionally, the most popular frequency-selective network was the Wien network. See Figure 4.40.

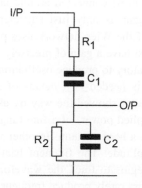

Figure 4.40
The Wien frequency-selective network.

The frequency of oscillation is given by:

$$f = \frac{1}{2\pi\sqrt{R_1 R_2 C_1 C_2}}$$

In order to change frequency, we only need to change the value of one component. The loss of the network at resonance is:

$$\frac{V_{\text{out}}}{V_{\text{in}}} = \frac{1}{1 + \frac{C_2}{C_1} + \frac{R_1}{R_2}}$$

The significance of the second equation is that for the oscillator to maintain constant amplitude as we tune from one end of the range to the other, we must change a pair of values simultaneously so that their ratio remains constant. Either the resistors must change together, or the capacitors must change together. Any tracking error between our ganged variable resistors or capacitors causes amplitude to change with frequency. This deviation from perfection is known as **range flatness**. In general, it is easier to achieve accurate tracking between variable air-spaced capacitors than variable resistors, so Wien oscillators usually use a dual-gang variable capacitor, but range flatness $< \pm 0.1$ dB is difficult to achieve. Variable capacitors only have 180° of rotation, so if you see an audio oscillator having a 180° scale (rather than the 270° achievable by a variable resistor) you can be pretty certain that you are looking at a Wien oscillator tuned by a variable capacitor.

Although the Wien network can be connected around any amplifier, sustained oscillation requires that the amplifier's gain is only **just** sufficient to overcome the losses in the frequency-selective network. If the Wien network uses paired components, it has a loss of one-third, so the amplifier must have a gain of precisely 3. Although amplifier gain can be critically adjusted in the laboratory to produce oscillation, such adjustment isn't stable and oscillation soon stops. What is needed is a means of stabilising the gain. The seminal Hewlett-Packard HP200 oscillator showed the way by using the change of resistance with temperature (and therefore applied power) of a fine tungsten filament isolated from external influences by placing it in a hard vacuum. In other words, it used a low-power incandescent lamp to stabilise amplitude. The filament heats but does not glow, and at low frequencies its temperature begins to track the waveform, increasing distortion, so thermally stabilised Wien oscillators rarely produce frequencies lower than 20 Hz.

Summing up, a traditional Wien oscillator is likely to produce sine waves from 20 Hz to 20 kHz in three ranges with a typical range flatness of ± 0.2 dB. Distortion is mainly dependent on the quality of the amplifier and can be made to be very low, with typical bench oscillators producing distortion ranging from 0.5% to 0.05% at 1 kHz (although 0.0003% has been claimed [1]), but rising at low frequencies if thermally stabilised. Ultimately, the low Q of the Wien network makes achieving low distortion difficult, so really low distortion analogue oscillators tend to be based on the higher Q of the state variable filter.

Analogue function generators

The Wien oscillator can produce a low-distortion sine wave, but an electronics laboratory often needs square waves or pulses. An alternative way of producing oscillations is to charge a capacitor from a constant current source (producing a rising ramp) until it reaches an upper threshold voltage, then immediately start discharging it with an equal and opposite constant current sink (producing a falling ramp). When the falling ramp reaches a lower threshold voltage, we repeat the cycle. The oscillator thus produces a triangular wave and is versatile because differentiation produces a square wave, and integration a crude sine wave.

Because the sine wave is derived from a triangular wave, it always contains significant distortion, and 1% THD is typical. The distortion can often be seen at the tip of the sine wave, where a pointed Gothic arch appears rather than a soft Norman arch. See Figure 4.41.

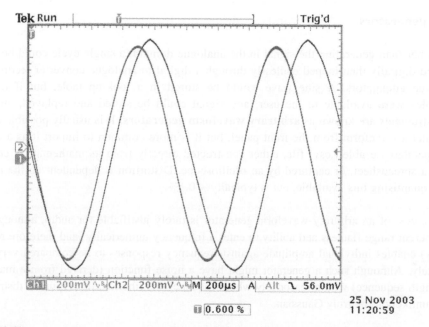

Figure 4.41
Comparison of low-distortion sine wave (smaller amplitude) with sine wave generated by function generator (higher amplitude). Note the sharpened Gothic arch produced by the function generator.

You might wonder why we should waste any time considering function generators for audio when they have such poor distortion, but even the cheapest function generators have two very valuable audio qualities:

• They produce good quality square waves (sharp edges and negligible ringing) that are good for testing amplifier stability.

- They do not require ganged controls to change frequency, so their range flatness is intrinsically perfect, making them excellent for critically measuring amplitude against frequency response of equalisation such as RIAA.

As an example of the second point, the author could not detect any range flatness error on his cheap function generator even though his dedicated audio test set can indicate 0.02 dB errors clearly.

Digital generators

Rather than generating the signal in the analogue domain, a single cycle could be generated digitally then looped endlessly through a digital to analogue converter feeding analogue attenuators. A sine wave would be stored in a look-up table, but if multiple tables were available to the user, any signal could be stored and replayed, and such instruments are known as **arbitrary waveform generators**. It is usually possible to construct a waveform from the front panel, but it is more common to import it as a comma separated variable (.csv) file, either constructed directly from its mathematical equation in a spreadsheet, or captured by an oscilloscope. Distortion is dependent on the number of quantising bits available, but is typically <0.1%.

The cost of an arbitrary waveform generator is rarely justifiable for audio, although their inherent range flatness and ability to enter a frequency numerically (and therefore repeatably) enables individual amplitude against frequency responses to be compared very accurately. Although such a generator might have a noise function (derived from a maximum length sequence) with a reasonably white frequency distribution, its amplitude distribution is unlikely to be truly Gaussian.

Dedicated analogue audio test sets

Neither a function generator nor a typical bench oscillator is ideal for audio testing. Fortunately, dedicated audio test sets are sometimes available second-hand.

An audio test set should at the very least contain:

- A fairly low distortion (<0.05%) sine wave oscillator (typically 20 Hz−20 kHz).

- A wide-band meter calibrated for reading sine waves.
- A simple form of total harmonic distortion (THD) measurement.

Better test sets may include noise and/or wow and flutter (W&F) measurements, but the audio test set's primary requirement is a wide-band meter and attenuators scaled directly in dB, plus some form of distortion measurement. (Some DVMs can express AC measurements in dB, but check their accuracy >60 Hz.)

When we looked at AC measurements, we saw that there were various options for specifying an AC waveform's amplitude:

- Rectify the waveform, apply the resulting DC to a mean reading meter, and calibrate it V_{RMS} of sine wave. This rectifier is ideal for driving an expanded scale having a range of only 1 dB because it tends to average noise to zero, improving accuracy, but is only suitable for measuring sine waves.
- Use a rectifier producing an output genuinely proportional to V_{RMS}, and apply it to the meter. This rectifier is appropriate for power and distortion measurements, and also for random noise.
- Use a rectifier that captures V_{peak} (both positive and negative), and uses a meter having **ballistics** that not only display short peaks accurately (fast **attack**), but allows them to be read (slow **decay**). This approach is exemplified by the PPM, which although primarily designed for programme is also useful for measuring interference, but typically over-reads random noise by ≈ 5 dB over a 22 Hz–22 kHz unweighted bandwidth.

Because different measurements require different rectifiers, dedicated audio test sets incorporate all three rectifier types, which are selected appropriately. Thus, to measure programme or man-made interference, a test set might have a moving coil meter having a sufficiently fast response time to meet the PPM attack specification (the slow decay is produced electronically) and driven by a peak-reading rectifier. To measure sine wave amplitude, a mean reading rectifier would be substituted, the meter ballistics would be slugged electronically, and the scale might be expanded electronically to ease measurement. To measure distortion, an RMS rectifier would be substituted because this sums harmonic powers correctly, and the scale might be calibrated as %. Finally, to measure random noise, the RMS rectifier would be retained, but the scale reverts to dB.

For simple amplitude measurements, the meter section is simply a meter/rectifier preceded by a calibrated amplifier. To measure harmonic distortion, a filter must be added to reject the fundamental. We want to measure the level of the harmonics, but must reject the fundamental without affecting the level of the second harmonic, which is one octave higher in frequency than the fundamental. This can be done in one of two ways:

- Use a high-pass filter. Since the second harmonic is an octave higher than the fundamental, if we want to measure THD to 0.1% (-60 dB), we need a filter having a slope of 60 dB/octave or more. This is achievable, but not easy, and such a filter is unlikely to be tunable for different frequencies, so this approach leads to a test set that can only measure distortion at one or two fixed frequencies.
- Use a notch filter. Active notch filters can easily achieve rejection of 80 dB or more at their notch frequency, but need precise tuning to achieve maximum rejection. Once auto-tuning has been added to maximise rejection, it is a small step to make the filter tunable over a wide range, and this approach leads to a test set that can measure distortion at any frequency, probably to better than 0.01% (-80 dB).

In the UK, various dedicated analogue audio test sets may be available, but all are likely to be at least 20 years old so reliability and serviceability could be questionable. Despite the previous caveats, examples worth considering include:

- BBC EP14/1: This was the first IC-based BBC test set, and includes a PPM. It was a true piece of laboratory equipment; using the expanded scale enables repeatable sinusoidal measurements to an accuracy of 0.05 dB. THD measurements down to $\approx 0.1\%$ can be made at 100 Hz and 1 kHz but a mean rectifier is used, rather than RMS, so summation of harmonics might be incorrect, reducing accuracy. The Wien oscillator is somewhat poorer distortion than the meter.
- BBC ME2/5: A "cooking" piece of test gear designed to replace the EP14/1 in less critical usage. The oscillator was digitally synthesised, and could sweep frequency (intended, but never used, for automated analogue music line testing). The meter section was only accurate to 0.1 dB but contains a useful if primitive digital frequency counter. The unit is newer, smaller, lighter, more expensive, less accurate, and less reliable than the EP14/1.
- Technical Projects MJS401D and Neutrik TT402 derivatives: A splendid piece of equipment with a wide bandwidth meter far better than the EP14/1, including THD measurement at **any** frequency, proper frequency counter, and comprehensive filters and rectifiers. Check for 1 kHz distortion at $+20$ dBu; with the 20 Hz$-$22 kHz filter selected, it should typically be better than 0.002%. Astonishingly, in these days of "flog it and forget it" manufacturing, Neutrik support their old products and were able

to provide a service manual for the author's 30-year-old MJS401D. Customer service like that is commendable and other manufacturers should take note of the brand loyalty that such a policy engenders. Options may add W&F and IMD. Later versions include a stereo phase meter (better than most oscilloscopes), so price may be variable; expect it to cost significantly more than an EP14/1, but perhaps need attention. Thoroughly recommended.

- Audio Precision System One and successors: Computer-controlled test equipment best used for production testing. Outstanding measurement ability optimised for sweeps. Only very lucky people find one cheap.

Professional audio equipment is inevitably balanced, so test equipment is often designed to interface with balanced audio. Traditionally, transformers were used, but electronic balancing is now common, which can cause problems when connected to unbalanced equipment. The problem usually occurs at the output of the oscillator. The balanced output is often provided by a pair of unbalanced amplifiers producing signals referenced to earth, one producing one polarity and the other inverted polarity. See Figure 4.42.

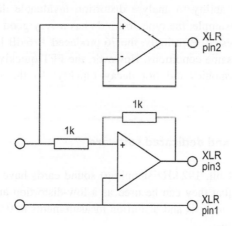

Figure 4.42
An electronically balanced output is often simply a pair of op-amps, one inverting, one non-inverting.

Because the output is referenced to earth, hum loops can occur. This is less of a problem in a balanced system, but when the unbalanced output is taken, loops cause problems. The scaling on an oscillator's attenuators refers to the voltage between the two balanced conductors. If we only use one output, the output is half the amplitude (-6 dB). It is easy to forget this and calculate gains incorrectly. Equipment using transformer balancing has neither of these problems — if we want an unbalanced source, we simply connect one leg to the earthy input of our amplifier under test.

Most audio test sets have a meter section monitoring output, typically called "oscilloscope" or "listen". This output is extremely useful when measuring noise or distortion. Listening to the character of the noise on headphones can often give clues as to its origin (using a loudspeaker is likely to provoke acoustic feedback).

When measuring distortion, the signal at the monitoring output is the distortion waveform and this can usefully be taken to an oscilloscope. Incorrect bias in a Class AB amplifier stage causes sharp crossover distortion spikes at 0 V transitions, so monitoring the distortion waveform can be a very quick method of checking bias.

Even better, if the monitoring output is taken to an oscilloscope or computer sound card that can perform an FFT, the spectrum of the distortion waveform can be investigated. The author finds the ability to analyse distortion invaluable during the early stages of audio design. As an example, the type 76 triode has a very good reputation for low distortion, so a batch was tested. Pleasingly, the 76 produced ≈ 6 dB less distortion than a typical 6J5GT under the same conditions. However, the FFT quickly revealed that, unlike the 6J5GT, the 76's harmonics did not decay quickly, so the type was set aside. See Figure 4.43.

Computer sound cards and dedicated software

Recording quality (24 bit, 192 kHz sampling) sound cards have sufficient dynamic range (typically >100 dB) that they can be used as a low-distortion audio generator and analyser, enabling detailed analysis and distortion measurements to 0.001%. However, there are some practical problems:

- Sound cards can only accept and deliver a very limited range of signal amplitudes without damage, so external interfacing that includes an amplifier/attenuator and overload protection is needed for both input and output. Fortunately, Pete Millett has very kindly done the hard work for us with his "Sound Card Interface/AC RMS Voltmeter" project (for which he can supply PCBs). Search out Pete Millett's website.

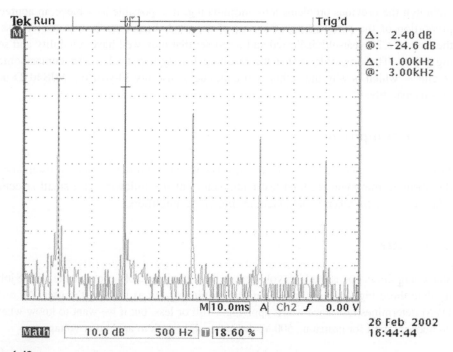

Figure 4.43

This FFT shows the distortion spectrum from second to sixth harmonic of a mu-follower using a 76 as the lower valve and triode-strapped D3a as the upper. Note that unlike most triodes, the levels of the second and third harmonics are comparable.

- The sound card and external interface deal with the hardware, but we still need to be able to generate and analyse audio signals. Fortunately, there is plenty of audio testing software available, such as "Audio Tester".
- Computers are electrically noisy and it can be difficult to eliminate computer hash from measurements. If located within the computer, the sound card should be positioned as far away as possible from all sources of interference, preferably at one end of the mother board near the case.
- External hum loops are possible, so audio transformers may be required, but bear in mind that although they may reduce interference, they will add distortion at low frequencies and loss at frequency extremes.

Although the previous problems look intimidating, it is possible to achieve an appreciable proportion of the performance of a state-of-the-art dedicated audio test set for less than the cost of a 20-year-old dedicated audio test set that may well have reliability and servicing issues. The downside is that the resulting test system tends to be cumbersome and fragile – which is why the author will continue to use his 30-year-old MJS401D until it dies irreparably.

Other test equipment

Most electronics engineers would regard a DVM, an oscilloscope, and an oscillator to be the absolute minimum required test equipment, but the following is a small selection of other items useful when building and testing valve amplifiers.

Insulation tester

Valve amplifiers operate at high voltages, so there are inevitably components whose job is to insulate those high voltages from sensitive terminals such as grids or the user. An ordinary DVM determines resistance whilst applying a volt or less, but if we want to know whether a component is fit for insulating 500 V, the only valid test is to apply that voltage.

A dedicated insulation tester comprises a high voltage generator in series with a sensitive ammeter calibrated either in leakage current or resistance. Early testers used a hand-cranked dynamo to apply a very approximate $500\,V_{DC}$, but modern testers replace this with a switching supply that applies a constant and selectable voltage. Thus, the author's fairly basic instrument offers a choice of applied voltages, and because it fundamentally measures current and has a minimum detectable current of 10 nA, its maximum measurable resistance increases with applied voltage. See Table 4.3.

Although primarily designed for testing insulation in power distribution systems and their transformers, insulation testers are ideal for detecting leaky coupling capacitors before expensive output valves reveal the problem by laying down their lives. Be warned that insulation testers tend to eat batteries because they are usually designed to be able to sink 1 mA at 1 kV – equating to 1 W of power.

Megger's newer insulation testers have an LCD display carrying an arced logarithmic scale and moving segments below it analogous to a moving coil meter's pointer. The

Table 4.3

Typical insulation testing voltages and their uses

Applied voltage (V)	Stated maximum measurable resistance (GΩ)	Typical audio use
50	5	Small-signal capacitors for equalisation, perhaps 63 V polystyrene
100	10	Input transformers
250	20	Most coupling capacitors
500	50	Higher voltage coupling capacitors
1000	100	Output and mains transformers

number of moving segments (one to four) signifies the rate of change of the measurement and a large digital display appears once the measurement has stabilised. This composite analogue/digital display is a triumph of ergonomics from which all other instrument makers could take lessons. See Figure 4.44.

Component bridge

A component or LCR bridge allows you to measure capacitors and inductors accurately. This is particularly useful when building filters or equalisation networks, and allows you to remove initial component value as a source of error. Component bridges can also measure AC resistance, and are only beaten when measuring resistances $<10\,\Omega$ by a DVM capable of a four-wire Kelvin measurement.

A Marconi TF2700 in good condition is potentially an excellent instrument, and you will still see it advertised second-hand (which says a great deal about an instrument that could be >40 years old). It uses a single PP9 battery, and current consumption is so low that it is not worth the bother of making a mains adapter, although replacing that expensive zinc carbon PP9 with a battery pack taking six AA alkaline cells would make a great deal of sense. The TF2700 can measure capacitance ($0.5\,\text{pF}{-}1100\,\mu\text{F}$), inductance ($0.2\,\mu\text{H}{-}110\,\text{H}$), and resistance ($10{-}11\,\text{M}\Omega$) to a basic accuracy of $\pm1\%$. It can indicate loss factor of capacitors (very useful), and can measure air-cored inductors — early digital bridges often can't. The circuit is very simple so it can be fixed easily if it develops a fault.

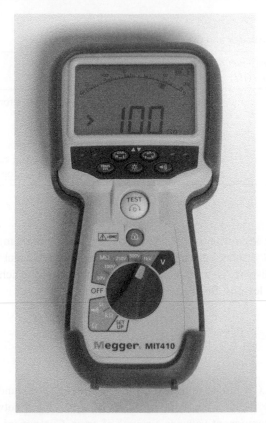

Figure 4.44
This excellent LCD display combines analogue and digital advantages.

Even better, the accuracy of the TF2700 can be improved quite easily. The main range switch uses 0.5% tolerance resistors, most of which can be replaced by $\pm 0.1\%$ metal film resistors, but the 10 Ω resistor must be replaced by a $\pm 0.1\%$ **non-inductive** wirewound resistor. The coarse balance control already uses $\pm 0.1\%$ resistors, so no changes are necessary.

The final error comes from the position of the dial on the fine balance variable resistor, and reducing this error takes a little more time. You need four 200 kΩ $\pm 0.1\%$ resistors and a 100 kΩ $\pm 0.1\%$ resistor. If these five resistors are wired in series, with the 100 kΩ at one end, you can pick off values from 100 to 900 kΩ in 100 kΩ steps. Use the bridge to measure the DC resistance of each of these values with the **main range switch set to 10 MΩ**. This forces the coarse balance control to be set to zero, and puts the onus of measurement on the fine balance control. If you now plot a graph of measured value against known value, and plot a straight line of best fit through it, you will quickly see how much rotation the dial requires on its shaft to minimise errors.

Assuming that the internal 100 nF ± 0.1% standard capacitor is not in error, the combination of replacing the main range switch resistors and fine dial adjustment usually reduces errors from ± 1% to ± 0.25%.

Despite its age and probable need for refurbishment (replace all the electrolytic capacitors and clean the switch contacts), the Marconi TF2700 is still a better choice than all but the better contemporary digital bridges, and it's worth noting that because it is a genuine bridge it needs no warm-up time, but instantly achieves its full accuracy. It took an expensive (by comparison) programmable impedance analyser having a basic accuracy of ±0.05% and 69 operating frequencies selectable between 20 Hz and 200 kHz to displace the author's TF2700 from his bench.

Impedance analyser

Cheaper impedance analysers are frequently described by their manufacturers as digital LCR bridges because this succinctly conveys the instrument's measurement ability to the user. Despite their appellation, they are not bridges because they do not adjust ratio arms in search of a null and thus compare the external impedance to their internal standard impedance (usually a 100 nF polystyrene capacitor). Instead, impedance analysers force a sine wave of known voltage and frequency across the component using a four-wire Kelvin connection and measure the resulting current together with its phase angle. They ultimately compare against a voltage reference, perhaps 4.096 V_{DC}.

Knowing the applied voltage, current, and phase, an internal Ohm's law calculation determines complex impedance, and once an appropriate component model has been chosen, further calculations incorporating the applied frequency derive a value of capacitance or inductance and associated resistance. Note that because an impedance analyser must measure amplitudes, the accuracy of its displayed value is critically dependent on components whose values drift from cold, so specified accuracy is only achieved after warm-up — perhaps 30 minutes.

The choice of component model is crucial when measuring any reactive component at a single frequency, but impedance analysers make the choice explicit and user-selectable, whereas it was hidden in analogue bridges such as the Marconi TF2700. It is assumed that any practical reactive component always has some associated resistance, and this may be in series or in parallel. Thus, a capacitor could be tested and its ESR might be explicitly included as a series resistance, but we also know that capacitors have leakage resistance, and this is in parallel, so we must select the most appropriate model for each and every measurement. The instrument might offer D or Q as an alternative to resistance, but such numbers are simply different algebraic interpretations of the basic two-component model selected by the user.

A single frequency measurement cannot distinguish between parallel or series resistances, but computer analysis of a sequence of swept frequency measurements allows for a much more sophisticated model (at least four components rather than two), and this is where the power of an impedance analyser lies. The computer analysis might be intrinsic to the instrument, or it might be obtained by exporting raw Z, Φ or $A + jB$ data to an external computer.

Getting the most out of an impedance analyser

Most impedance analysers are supplied with a pair of one metre leads that allow the four-wire Kelvin connection to be made, theoretically allowing accurate measurements to be made very conveniently. Don't use these leads for measuring capacitors $>1\,\mu F$ beyond 10 kHz as you will be measuring the leads rather than the capacitor. The problem is that the Kelvin connection has to return the voltage-sensing signal via a coaxial cable having shunt capacitance and this is manifested in the measurement results as 10−20 nH of additional series inductance (comparable with a small capacitor).

If we want to measure the component rather than the fixture, we need short leads (low series resistance and inductance), and if we want to minimise stray capacitance, those leads must either be unscreened or guarded. However, we know that measuring a small capacitance implies very small currents, so we need screening to prevent interference currents (such as from fluorescent lighting) from upsetting our measurement. The simplest and most repeatable solution is to enclose the component and its connections within an earthed screened box mounted directly on the instrument's measurement terminals. See Figure 4.45.

It appears to be an informal standard that instruments from different manufacturers (Agilent, Digimess, Hameg) separate the centres of their four BNC connectors by 22 mm. For ease of

Figure 4.45
This screened box enables precision component measurements with minimum interference and stray capacitance.

fitting to the instrument, only the box's two outermost BNCs retain their knurled grips. The centre two knurled grips can be removed by making two diametrically opposite cuts using a cutting wheel then cracking at the cut − take care not to accidentally apply force to the remaining part of the BNC as it is quite fragile. A die-cast box is better than folded because it has no corner gaps and because the lid has an internal lip, minimising the gap at this unavoidable join − the difference between the lid being screwed firmly closed and slightly loose is clearly measurable. The hinge needed a ≈3 mm spacer between it and the box to allow for the thickness of the hinge and to put the pivot point sufficiently far away from the box edge that it could swing the lid's lip into the box without binding. An explicit earth bond between the box and hinged lid maintains the lid's earth bond even if corrosion should develop over the years between the solid brass hinge and aluminium box. Hard foam lining of 6 mm thickness reduces stray capacitance from the component to chassis to <1 pF because:

- Capacitance is inversely proportional to separation, so a >6 mm gap enforces low capacitance.
- Capacitance is proportional to the relative permittivity (ε_r) of the intervening dielectric. For most plastics, $2 < \varepsilon_r < 3$, but foam is mostly air, making ε_r nearer to unity.

The test leads are deliberately short to minimise series inductance and resistance and stray shunt capacitance to chassis. The author hasn't yet found small true Kelvin clips, but if the instrument reports directly to a computer, a calibration sweep with the test leads shorted together enables residual series resistance and reactance to be stored and subtracted from later measurements (necessitating the data to be in $A + jB$ form at that point). If you do this, you will find that the leads not only possess inductive reactance that rises with frequency, but also rising resistance due to skin effect.

Even better, the computer can automatically calculate measurement uncertainties (using the uncertainty data provided by the manufacturer), add error bars to the resulting graphs, and thus warn the user when a measurement's accuracy becomes questionable. Be warned that correctly calculating associated uncertainties generally involves twice as much work as calculating any derived result. Fortunately, if you report the instrument's results directly to a computer, you only need to write its uncertainty calculation once.

The plot of an inductor's impedance magnitude against frequency is especially illuminating. See Figure 4.46.

The rising impedance from low frequencies is what we would expect from an inductor. As frequency increases, we see the resonant peak formed by the inductance and shunt winding capacitance. Finally, we see falling impedance caused by shunt winding capacitance C_{shunt}.

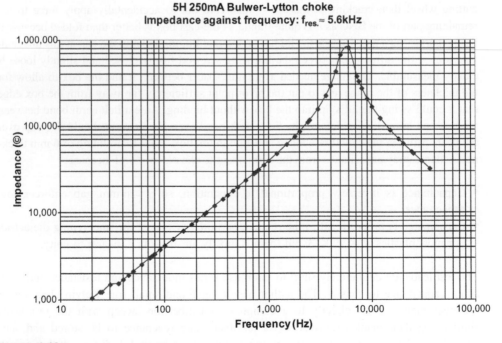

Figure 4.46
Practical inductors are only inductive up to their self-resonant frequency.

To a first approximation, C_{shunt} could be found by calculating and plotting reactance against frequency for a capacitor and adjusting its capacitance until the line of the theoretical capacitor overlaid the measured impedance above resonance. In principle, the same may be done for inductance below resonance, but beware that the inductance of iron-cored inductors is heavily dependent on both DC and AC excitation, so a small-signal measurement (although correct under measurement conditions) is unlikely to be representative of real use.

There is no reason to stop at individually overlaying and comparing C_{shunt} and series inductance. At resonance, chokes can be modelled quite accurately by choke inductance L in series with DC winding resistance R_{DC}, in parallel with C_{shunt} and shunt resistance R_{shunt}. Having chosen this four-component model, we can iteratively adjust model values until the measured and modelled impedances overlay perfectly, at which point the model's values may be read off. See Figure 4.47.

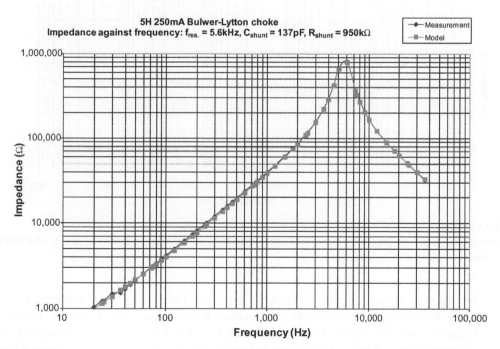

Figure 4.47

Overlaying measurement and model allows a choke's model values to be quickly determined.

Bearing in mind that the maximum attenuation of a practical LC power supply filter is set by the ratio of the choke's C_{shunt} to its following filtering capacitance, accurately determining C_{shunt} overrides all other considerations. R_{shunt} sets the depth of the null at the choke's high frequency resonance in a practical filter (which is rarely important), but needs to be approximated reasonably well in the model (so that the peak amplitudes match) to allow C_{shunt} to be found accurately.

Calculation and plotting of the proportionate error between model and measurement against frequency invariably reveals a discontinuity in the error trend at resonance, so fine-tuning of C_{shunt} and R_{shunt} that eliminates this discontinuity enables a reasonably confident estimate of C_{shunt} to be made. See Figure 4.48.

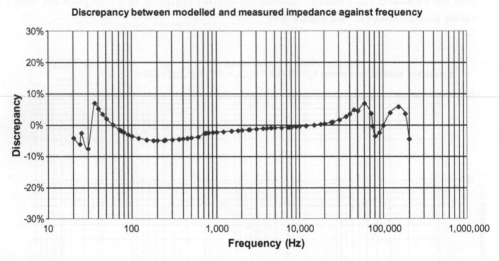

Discrepancy between modelled and measured impedance against frequency

Figure 4.48
Plotting proportionate impedance error and eliminating discontinuities around the ≈5.6 kHz resonance enables fine-tuning of C_{shunt} and R_{shunt}.

Having achieved a perfect fit at resonance in order to obtain an accurate value for C_{shunt}, the model invariably deviates from measurement at low frequencies. A practical core has hysteresis and eddy current losses that cause a loss of inductance at high frequencies, so optimising the fit at resonance requires reduced inductance, causing the low frequency discrepancy. Primarily considering small ferrite transformers, Wilson [2] proposed splitting the inductance into two series inductances, one having a parallel resistance, resulting in a six-component model that allows a much better fit.

For the four-component model, even $\pm 10\%$ discrepancy between modelled and measured impedance of an iron-cored inductance is a very good result, and $\pm 30\%$ is more likely. The six-component model can sometimes reduce discrepancy to $\pm 1\%$, but $\pm 5\%$ is more likely. LTspice's calculations are accurate, but they can only be as good as the component models they use, and modelling errors will be dominated by the inductors, so the four-component model is likely to cause ± 3 dB modelling errors, whereas the six-component model reduces them to ± 1 dB, which is negligible when analysing power supply filtering, but large if considering RIAA equalisation. Measurement shows that it is possible to manufacture signal inductors having inductance variation of $<1\%$ (<0.1 dB) over three decades, but it is difficult to minimise capacitance such that the self-resonant frequency does not fall into those three decades.

See the Appendix for equations shaped to allow any component model to be used easily in a spreadsheet. Alternatively, consider using a proper maths package for processing and displaying your data.

Transformer leakage inductance

Although the principles of iterative modelling were introduced using a smoothing choke, exactly the same model and swept frequency technique can be used for determining an output transformer's L_{leakage} and C_{shunt} by measuring across the primary with the secondary shorted. See Figure 4.49.

Output transformers invariably interleave primary and secondary sections to maximise magnetic coupling between primary and secondary windings and thereby reduce leakage inductance. Most transformers are wound in layers, so a proportion of primary is wound, followed by a proportion of secondary, and so on. This technique is known as horizontal sectioning because alternating primary and secondary layers are sectioned above and below horizontal lines. But a push—pull transformer having horizontal sectioning suffers asymmetry between the two halves of its primary unless it also has vertical sectioning. At its simplest, vertical sectioning splits the bobbin into two equal chambers, one for each half-primary, but better transformers reduce leakage inductance at the expense of shunt capacitance by swapping the positions of alternate layers between chambers so that each layer is tightly coupled to the other half-primary. One of the many advantages of the

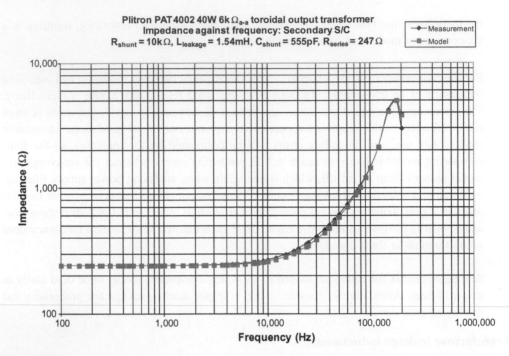

Figure 4.49
Modelling enables output transformer leakage inductance to be determined.

McIntosh output stage is that because both signal polarities are available on each valve, cross-coupling can be achieved without increasing shunt capacitance.

Asymmetry between half-primaries causes a pair of separated high frequency peaks rather than one overlaid peak. If leakage inductance is measured individually for each half-primary by not only shorting the secondary but also the other half-primary, the half-primary peaks may be investigated individually, whereupon the leakage inductances are likely to be similar but the shunt capacitances very different. Cheaper push–pull output transformers not only have half-primary asymmetry, but their high frequency resonances are barely out of the audio band. See Figure 4.50.

The ideal push–pull output transformer has both horizontal and vertical sectioning, allowing the twin peaks to overlay perfectly at >100 kHz (Radford claimed 300 kHz for his

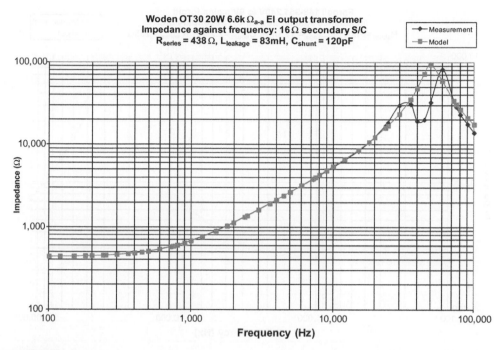

Figure 4.50

This terrible output transformer has high leakage inductance, and the twin peaks indicate asymmetry due to a lack of vertical sectioning.

transformers). Bifilar winding is a form of vertical sectioning because a pair of wires is laid down either side of a vertical line (albeit moving). Because dual voltage mains toroids commonly have their two 120 V primaries wound bifilar (cheaper to wind), this vertical sectioning means that the half-primaries are very similar, but not quite identical if used as a push–pull output transformer because the centre-tap is at opposite ends of each winding, slightly changing C_{ps} between halves. Note also that the bifilar winding causes crippling C_{shunt} (see later). See Figure 4.51.

Given that estimating $L_{leakage}$ required at least twenty accurate measurements plus careful interpretation of iterative computer calculation and subsequent graph plotting, the author was initially sceptical of data originally published in a 1957 GEC book [3]. See Table 4.4.

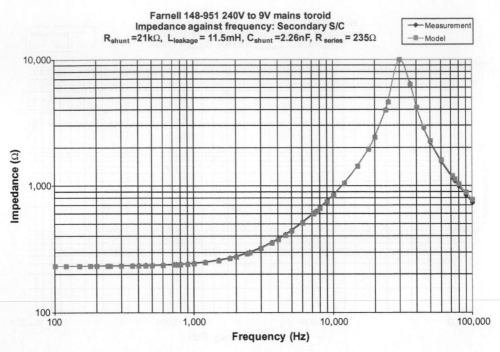

Figure 4.51

Because this mains toroid has bifilar primaries (vertical sectioning), it has a single resonant peak, albeit at a low frequency due to crippling bifilar shunt capacitance.

Table 4.4

GEC output transformer comparison

Relative age of transformer	Leakage inductance (mH)			Comments
	P-S	½P − ½P	½P − UL tap	
Early	20−25	20	50	a-g_2 capacitors required for stability
Medium	10−15	10	25	Stable without capacitors
Late	5−10	5	5	Stable with 30 dB of negative feedback

Making these leakage inductance measurements in a practical time using laboratory equipment available in 1957 would have required a hardware model of the transformer's leakage inductance using decade inductance, capacitance, and resistance boxes. The transformer and its model could have been fed via individual series resistors from a sweep frequency oscillator to make two potential dividers, and these outputs taken to the "X" and "Y" deflection plates of an oscilloscope. The decade boxes would have been iteratively adjusted for a straight line (rather than an ellipse) even in the face of changing frequency from the sweep oscillator, and transformer leakage inductance read off from the decade inductance box. GEC's Hurst Research Laboratories undoubtedly had the expertise and equipment to achieve the posited measurement but it would have been time-consuming and expensive, so the author doubts that transformer manufacturers of the time made such rigorous measurements.

Despite the GEC inference, $L_{leakage}$ alone is not an especially good indicator of output transformer quality because the high frequency loss caused by a given leakage inductance is related to transformer primary impedance (unstated in the GEC comparison). The author modelled some output transformers at high frequencies using the simple four-component model described earlier. See Table 4.5.

Starting from the bottom of Table 4.5, the full-size Woden OT30 is a wretched transformer not worth a fraction of the £20 the author paid at a radio fair. The guitar amplifier transformer has an excusably low resonant frequency because it does not need to cater for negative feedback, and there is a pleasing agreement between the manufacturer's data ($L_{leakage}$ = 80 mH, C_{shunt} = 700 pF) and tabulated measurements, even though the data sheet suggests a more optimistic primary inductance than the perfectly respectable measured value. The Rogers HG88 transformer has understandably low primary inductance because the HG88 was an integrated amplifier (restricting transformer size). The Parmeko (Partridge M.E. & Co.) P2925 more than doubles the size and primary inductance at the cost of a low resonant frequency, suggesting poor winding technique that limits the amount of negative feedback that can be applied safely, which is a surprise because Partridge senior certainly knew how to design good output transformers [4].

The Leak Stereo 20 is the first of the good transformers ($L_p/L_l \geq 10{,}000$), justifying its popularity as a donor amplifier in DIY designs (although note that this test broke the secondary into two halves and connected them in parallel to achieve 4 Ω with reduced L_l, rather than picking off a 4 Ω tap as in the original Leak design). The Haddon PE298 has far lower leakage inductance than the Leak, but resonates at a comparable frequency. The

Table 4.5

Output transformer comparison

Primary impedance (kΩ)	Rating (W)	Manufacturer	Type	L_p^* (H)	L_l (mH)	R_{series} (Ω)	C_{shunt} (pF)	R_{shunt} (kΩ)	f_{res} (kHz)	L_p/L_l
6	40	Plitron PAT4002	Audio toroid	202	1.53	248	555	10.5	168	132,000
1.75	100	Albion 93617 (ex-Westrex 100 W)	EI	75.3	1.3	49.5	950	3	143	57,900
6	15	Hammond 1615 (8 Ω secondary)	EI	111	7	337	500	50	94.5	15,900
5.5	–	Farnell 148-951 240 V:9 V/ 30 VA	Mains toroid	152	11.5	235	2,260	25	31.4	13,200
5	15	Hammond 1615 (4 Ω secondary)	EI	111	8.8	520	500	20	94.5	12,600
3.5	60	Haddon PE298 (ex-Westrex 60 W)	EI	39.5	3.5	139	630	7	108	11,300
8	10	Leak Stereo 20 EL84 (March 1959)	EI	136	13.5	535	210	38	88	10,000
8	10	Parmeko Atlantic series P2925	EI	116	30	1,445	220	60	54.5	3,870
8	10	Roger HG88 ECL86	EI	44.4	13.5	472	212	45	94.5	3,290
4	100	VDV-MAN100 guitar transformer	EI	177	76.5	500	742	69	21.6	2,269
6.6	20	Woden OT30 EL34	EI	32.6	83	438	120	100	56	393

$^*L_p$ measured at 20H$_3$ 1V$_{RMS}$ immediately following core demagnetisation.

diminutive Hammond 1615 is astonishingly good for its size, and is even better when all its secondary windings are used (8 Ω). The mains toroid seems surprisingly good (and passed an excellent low-amplitude 10 kHz square wave in a test mule 2A3 amplifier), but its core quickly saturates at low frequencies. Further, the steep (and elliptical) loadline of the mains toroid's 2.26 nF bifilar shunt capacitance inevitably increases high frequency distortion in the output valves, and its resonant frequency is low. The Albion 93617 is genuinely an excellent transformer, so the author is mortified that he has only one. The reason that the Albion 93617 is so good is that it came out of a Class B amplifier − Class B amplifiers require low leakage inductance to minimise switching spikes at crossover when one valve switches off. The toroidal Plitron PAT4002 appears to be in a league of its own, but note later comments about DC.

Given that the test methodology seemed sensitive enough to discern small changes in leakage inductance, the author tried the effect of adding a 0.001″ thick copper foil 40 mm wide to the VDV-MAN100 output transformer to see if an external Faraday shield would reduce leakage inductance by short-circuiting leakage flux. See Table 4.6.

Table 4.6

The measured effect of adding a Faraday shield to an output transformer

	$L_{leakage}$ (mH)	C_{shunt} (pF)	R_{shunt} (kΩ)
Before	76.5	742	69
After	74	745	57

As can be seen from the table, a 3% improvement in leakage inductance was detectable but scarcely worthwhile, and undoubtedly smaller than sample-to-sample variation.

Why go to the trouble of establishing model values?

The value of obtaining a power supply choke's six-component high frequency model is that it can be combined with a capacitor's ESR and leakage inductance in LTspice to quickly model and accurately predict a filter's performance, whereas measuring the >120 dB attenuation just above the choke's resonant frequency of a well-designed and implemented LC filter is quite difficult.

The value of determining an output transformer's high frequency model is that the L_p/L_l ratio unequivocally states transformer quality, allowing you to reject duffers **before** cutting metal. Sadly, deriving and using those model values in LTspice in order to determine the optimum feedback compensation of an entire amplifier is slower than intelligently twiddling all four compensation components whilst monitoring a 10 kHz square wave.

Provided that there isn't any sectioning, wound components can be modelled very well using either the four- or six-component models, so (usually unsectioned) toroidal transformers tend to model much better than sectioned output transformers.

Transformers and direct current

Output transformer design is ultimately about maximising the number of octaves between the low and high frequency limits, so skilled transformer designers sometimes state the L_p/L_1 ratio — and the essentially ungapped core of a toroidal output transformer produces an especially impressive number. Beware that a push—pull toroidal output transformer's primary inductance collapses at the merest hint of a direct current imbalance between the two halves, so it is difficult to maintain that impressive primary inductance in practice. See Figure 4.52.

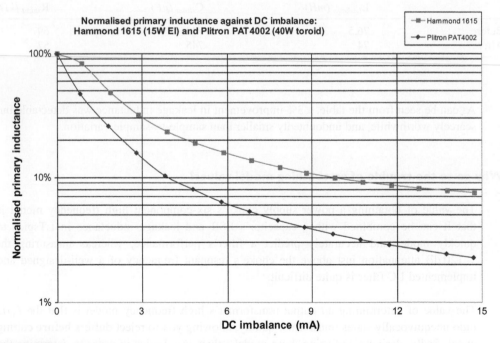

Figure 4.52
Note the severe dependence of output transformer primary inductance on imbalance current.

Note that primary inductance for both transformers falls catastrophically with even a small imbalance current, but that because interleaved EI cores still have an air gap (albeit very small); this causes the fall in inductance to be less severe than for the toroid. More importantly, when the imbalance was removed, the EI Hammond 1615 immediately recovered to 88% of its demagnetised primary inductance, whereas the toroidal Plitron PAT4002 only recovered to 20% (requiring explicit demagnetisation to restore full primary inductance). Thus, toroidal output transformers require a bias servo to ensure that there is **never** any imbalance of direct current, even during warm-up, so diligent toroidal output transformer manufacturers also supply bias servo modules. Conversely, a pre-set manual bias adjustment recovers most of an EI output transformer's primary inductance even if it suffers a warm-up imbalance.

When measuring primary inductance against imbalance current it is essential to confirm that the current source genuinely provides the expected currents **before** connecting it into circuit, and particularly that applied current actually is zero when set to zero. The author's example of the otherwise excellent Hameg 8118 has a constant offset of 0.8 mA in addition to the requested current the moment its current source is switched on (even if set to 0 mA). Worse, once a transformer's core **has** been magnetised by a direct current imbalance, its inductance remains low until the core has been deliberately demagnetised. Thus, when measuring primary inductance against direct current imbalance, we must start from zero current and work upwards.

Measuring resistance of iron-cored inductances

We often need to know a coil's resistance, and two common examples are:

- Choke DC resistance (to predict voltage drop or calculate Q).
- Power transformer primary and secondary resistance (to determine source resistance for modelling, perhaps PSUD2).

No problem. We set our DVM to its resistance range, connect, and wait for the reading to stabilise. At the instant we attempt to measure, Lenz's law causes the coil to produce an equal and opposite EMF, so no current flows. Current initially ramps from zero in accordance with $E = -Ldi/dt$, but becomes an exponential that tends towards a steady-state value determined by the coil's resistance. During that time, the core's magnetic dipoles lose their random orientation by progressively aligning themselves with the imposed magnetic field. Once the dipoles are aligned, the core can no longer resist current flow, and current is only limited by coil resistance, which we're trying to measure. A stable DC measurement of coil resistance only occurs after the direct current has aligned all the dipoles that it can, with dipole alignment being limited primarily by magnetic reluctance such as an air gap in the core, rather than current. Thus, DC coil resistance measurement of a toroid (essentially ungapped) is likely to align all the dipoles (known as saturation), whereas the deliberately gapped core of a power supply choke or single-ended output transformer retains a proportion of randomly aligned dipoles.

When we remove an imposed magnetic field, the core's magnetism falls back to a non-zero proportion known as the **remanence**. The ideal transformer core would have zero remanence, and this is approached by the more exotic core materials (mumetal, etc.), but they are expensive, limiting their use to small transformers. Larger transformers are iron-cored, and iron has significant remanence (which is why screwdrivers become magnetised). Thus, measuring the DC resistance of an iron-cored inductance magnetises the core, reducing inductance.

Demagnetising a transformer core

If a transformer core becomes magnetised, the most effective method of demagnetisation uses a function generator. First, connect a winding across a function generator set to its maximum sine wave output voltage at 0.1 Hz. The low frequency forces the inductive reactance of any practical transformer to zero, so current is set by:

$$I_{max} = \frac{V_{out}}{r_{out} + R_{DC}}$$

where:

V_{out} = function generator open-circuit output voltage
r_{out} = function generator output resistance (probably 50 Ω)
R_{DC} = transformer winding resistance.

The generator needs to be set to sweep (ideally logarithmically) from 0.1 to 100 Hz over a considerable time, certainly more than a minute, and its output should be switched off just before it repeats the sweep from the lowest frequency. If the generator voltage is observed on a DC coupled oscilloscope at the same time, dependent on the initial core magnetisation, the sine wave will be visibly distorted at the lowest frequency (as the generator struggles to drive the short circuit of the saturating transformer). Because the current drawn is inversely proportional to frequency, as frequency rises the transformer's core is taken through progressively smaller B/H loops, gradually demagnetising it. See Figure 4.53.

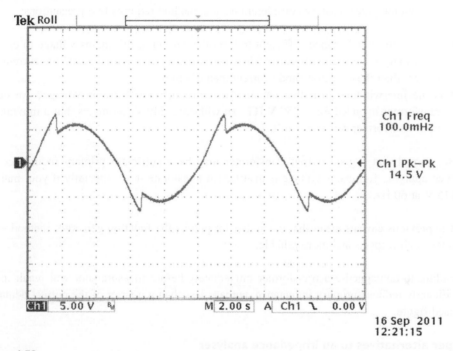

Figure 4.53
The symmetry of the distortion confirms that this output transformer core is being driven into saturation on both half-cycles. As frequency increases from 0.1 Hz, this distortion will disappear and the core will be demagnetised.

If you have a manual function generator, apply 0.1 Hz with the continuously variable frequency control set to minimum, slowly increase it to maximum (probably 1 Hz), then use the continuously variable amplitude control to gradually reduce amplitude to zero. It is essential that there is no DC offset at the output of the generator, or it will magnetise the transformer.

If you don't have an oscillator or function generator capable of producing 0.1 Hz, it may be possible to demagnetise a transformer using a higher frequency, but this requires brute force to drive sufficient current into the primary inductance's reactance to saturate the core, and you need to be certain that the coil and insulation can survive it. Thus, 50 or 60 Hz mains via a Variac (see later) may be used to demagnetise an output transformer when connected across its anode to anode terminals. Apply full mains voltage, then very slowly turn the Variac down to 0 V and this should leave the core demagnetised. Quite apart from the general safety considerations, this method requires two precautions:

- Is the primary inductance sufficient to limit the current at full mains voltage to one that the primary's wire can withstand? Measure the inductance and calculate the current − if it's less than the expected anode current then it's fine.
- Is the interwinding insulation capable of withstanding full mains voltage? An output transformer intended for >250 V HT can withstand 240 V_{RMS} mains, but a moving coil input transformer certainly can't.

Be warned that a good output transformer may have sufficient primary inductance that even applying full mains voltage is insufficient to saturate it − especially if your mains is 115 V at 60 Hz.

The previous caveats mean that the author far prefers the function generator method − it's safer in all respects and more reliable.

Failure to demagnetise a transformer immediately before measurement will result in significantly reduced primary inductance, perhaps less than a quarter of the correct demagnetised figure.

Cheaper alternatives to an impedance analyser

Although the author used a computer-controlled impedance analyser to determine the shunt capacitance of chokes and leakage inductance of output transformers, appropriate software plus a computer sound card can obtain impedance data just as quickly. However, a 192 kHz sample rate sound card can typically only measure to 85 kHz (not quite high enough to characterise output transformer resonance), and if we want to estimate a (<1 μF) capacitor's ESR and series inductance, we must measure to 2 MHz or beyond. There is rarely any need to determine a component's parasitic values to better than ±5% accuracy, so the following manual arrangement is perfectly adequate. See Figure 4.54.

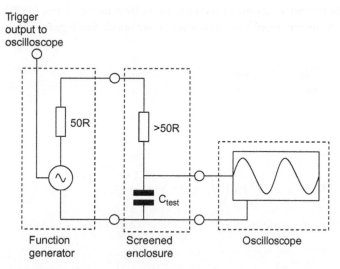

Figure 4.54

A function generator driving a simple potential divider with the output measured by an oscilloscope enables a capacitor's parasitic component values to be estimated.

A no-frills 20 MHz function generator applies its maximum amplitude sine wave to the component via a series resistance composed of its internal resistance (usually 50 Ω) plus a suitable external resistance ($> 50\,\Omega$ to limit the loading on the generator), thus forming a potential divider. The optimum value of external resistance for a given test is found by sweeping frequency to find the maximum amplitude (within the frequencies of interest), then adjusting resistance to give roughly half the open-circuit voltage. If a variable resistor is used, once the setting is found, its resistance can be measured using a DVM and the nearest standard value substituted. (A fixed resistor avoids the potential problem of wiper movement and fluctuating resistance.)

The amplitude at the output of the potential divider is best measured using a digital oscilloscope because it can cope with a large range of voltages, has automated amplitude and frequency measurements, and as a bonus displays the waveform. When the oscilloscope has insufficient gain, an external pre-amplifier (of known gain) can be inserted as necessary. Obviously, there will come a point when so much gain must be added that the signal becomes swamped by noise, but averaging reduces random noise by the square root of the averaging number, so the 256 averaging of an affordable oscilloscope such as the Rigol

DS1052E is perfectly capable of reducing noise by a factor of 16 (24 dB), allowing coherent signals to be recovered from below amplifier noise. See Figure 4.55.

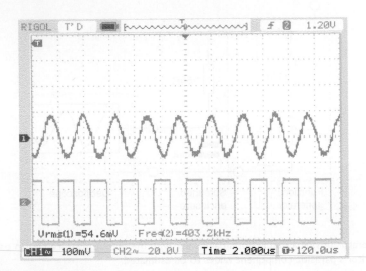

Figure 4.55
The combination of oscilloscope automated measurements and averaging enable measurements below the noise. Note that it is essential to centre the averaged signal vertically to avoid clipping the noise.

In the previous example, 60 dB ($\times$1000) of external gain was added, increasing the oscilloscope's stated sensitivity from 100 mV per division to 100 μV per division. Note that care had to be taken not to overload the oscilloscope's ADC with the signal plus noise prior to averaging – and this is why the smoothed sine wave occupies a smaller proportion of the screen than is usual.

With very little effort, the arrangement is good to 2 MHz, whereas the considerably more accurate Agilent E4980A 2 MHz impedance analyser costs £11 k.

Knowing the open-circuit output voltage of the function generator, its frequency, and the voltage across the component, a graph of impedance against frequency can be plotted. See Figure 4.56.

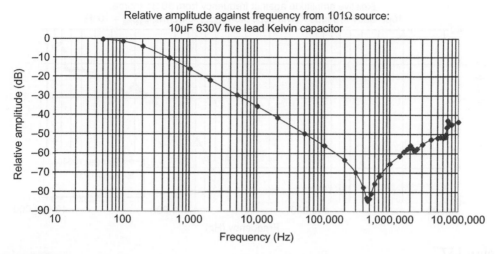

Figure 4.56
This plot of amplitude against frequency shows that the filtering effect of a practical capacitor is limited by foil and lead inductance and ESR.

As we should expect, the RC filter has attenuation that falls at 6 dB/octave until the capacitor's self-resonant frequency. If capacitance at the self-resonant frequency is known, series inductance can be calculated using:

$$L_{series} = \frac{1}{4\pi^2 f^2 C}$$

The capacitor's effective series resistance (ESR) can be found using:

$$ESR \approx \frac{V_{out}(min)}{V} R_{series}$$

Alternatively, if complex algebra holds no fears for you, you can use the equations in the Appendix to calculate the theoretical impedance against frequency of a capacitor having series resistance and inductance, and adjust its values until the model overlays the measurement at all frequencies. See Figure 4.57.

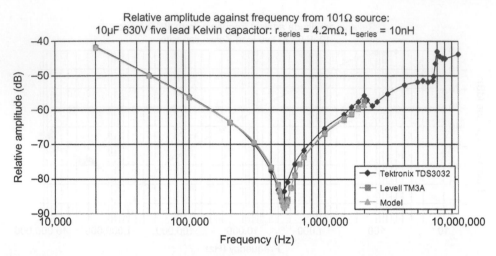

Figure 4.57
By modelling and overlaying, values of *L* and ESR may be determined.

Note that this graph contains two sets of measurements: one using the author's (mains powered, and therefore earthed) oscilloscope, and the other using a 40-year-old meter that was battery powered (and therefore floating). Note that the elimination of earth loops and consequent interference enabled the floating meter to measure a deeper null (−88 dB) at the resonant frequency rather than the −84 dB achieved by the oscilloscope. Note also that the oscilloscope measurement shows resonances above 2 MHz (probably a measurement artefact due to the multiple earth bonds and consequent loop), hence the previous caveat about LTspice modelling and frequencies above 2 MHz.

Other factors to bear in mind when determining values using the model and measurement technique are:

- To maximise measurement accuracy at high impedances, the external series resistor's value should be chosen so as to limit the maximum measured amplitude to about half the generator's open-circuit voltage. Function generators are designed to drive 50 Ω loads, so treat 50 Ω as the minimum external resistance. LTspice can be useful for quickly simulating the circuit to predict attenuation and help choose a suitable resistance.
- Final accuracy is determined by the cumulative errors of a number of voltage measurements, which are unlikely to individually better ±2% − so although model and

measurement comparison can yield a component's primary value, even a cheap component bridge gives a more accurate result.

- If capacitance falls noticeably with frequency (electrolytic capacitors), the resonance equation that implicitly assumes unchanging capacitance with frequency becomes inaccurate for estimating L_{series}, and an overlaid model including L_{series} becomes necessary. If required, a user-variable frequency-dependent capacitance term can be added to the model.

Magnetic shielding and electrostatic screening

We saw in Chapter 1 that penetration depth was inversely proportional to the square root of a magnetic shield's relative permeability:

$$\delta = \sqrt{\frac{2}{\omega \mu_0 \mu_r \sigma}}$$

where:

δ = penetration depth (m)
ω = angular frequency = $2\pi f$
μ_0 = permeability of free space = $4\pi \times 10^{-7}$ H/m
μ_r = relative permeability (1 for air, >5000 for steel)
σ = electrical conductivity (1/Ωm)

This equation is very significant because it implies that even thin tin plate becomes an excellent magnetic shield >200 kHz, and provided holes are minimised and the lid fits tightly, it is an electrostatic screen at all frequencies. In other words, tea and biscuit tins are excellent for enclosing sensitive measurements. All that is needed is to fit BNC connectors and interference-free high frequency measurements can be made. See Figure 4.58.

Note that the output connection to the meter (possibly isolated from earth) is via an insulated BNC and needs an explicit 0 V connection to the DUT, whereas the input connector from the function generator is an uninsulated type in order to earth the tin. The significance of this method of connection is that it permits correct measurement of four-wire Kelvin capacitors because it does not short their lower legs together.

Figure 4.58
When the (tight-fitting) lid is closed, a biscuit tin is an excellent low-interference measurement enclosure. Note that the input BNC is earthed to the tin whereas the output BNC is insulated.

Capacitor effective series resistance (ESR) meter

All capacitors have ESR, but the higher capacitance (and therefore lower reactance) of electrolytic capacitors makes ESR more significant and ESR is a good electrolyte health indicator. A seminal article by Bob Parker in the January 1996 edition of *Electronics Australia* introduced the world to an excellent instrument that could quickly and accurately measure ESR. The principle has since been developed into commercial products. See Figure 4.59.

Both instruments measure down to 0.01 Ω with a resolution of 0.01 Ω and are invaluable for quickly checking whether that electrolytic offered cheap on the junk stall should be bought or binned. The Bob Parker meter requires manual zeroing of lead resistance, but the commercial instrument has a pseudo-four-wire Kelvin connection to the crocodile clips that renders this unnecessary and has the bonus of also measuring capacitance.

The original Bob Parker meter kit sold by Dick Smith Electronics is no longer available, but the mkII version ("Silicon Chip", April 2004) is sold by Altronics. Given that the Altronics meter is a kit (and you therefore have the circuit diagram and PCB layout), there's no reason why the force and sense lines should not be broken on its PCB and taken

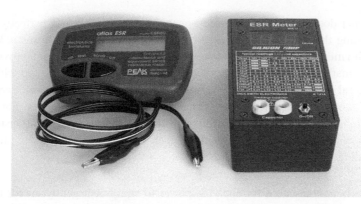

Figure 4.59
Either of these meters is useful for checking "bargain" electrolytic capacitors before purchase, and also when using PSUD2.

individually to the crocodile clips to make a four-wire Kelvin connection. Strictly, neither of the preceding examples makes a true Kelvin connection because a conventional crocodile clip doesn't make separate connections for force and sense and eliminate contact resistance, but it's good enough for the application.

The most significant use of an ESR meter is for obtaining ESR data for use in power supply modelling programs such as PSUD2. An ESR meter is invaluable for pinpointing otherwise inexplicable faults and particularly for comparing tantalum bead capacitors (which usually have a far higher ESR than aluminium electrolytics of the same capacitance).

"Voltstick"

This cheap indicator has various brand names and is incredibly useful. It usually looks like a fat pen with a white tip. If the tip is near to mains, an internal LED lights (runs off two AAA batteries for years). The indicator requires no contact and is potentially a life-saver, so always use one before cutting any cable that could carry mains. They're very useful for checking the fuse in a mains plug because they can just be waved near the cable. If the indicator detects mains, the fuse must be intact – and it has been checked without needing a screwdriver to open up the plug and check for continuity.

Continuously variable transformer (Variac)

A transformer does not necessarily need a secondary winding. It can perfectly well have a single winding that is tapped part way to produce the lower voltage; this is known as an **autotransformer**. Although an autotransformer is cheap, it does not provide isolation between primary and secondary. If an autotransformer were used to step 240 V to 12 V, but the neutral became disconnected between the mains outlet and the autotransformer, the full 240 V would appear on the 12 V circuit, with possibly fatal results. For this reason, the most common use of an autotransformer is the variable toroidal autotransformer, often referred to as a **Variac**, that has a rotating wiper that can move its tapping from one end of the winding to the other. See Figure 4.60.

Variacs are commonly used for applying power to equipment gently, or for testing tolerance to mains voltage changes. There are various reasons for gently applying power to a device under test (DUT):

- The DUT is newly built, and not known to work. Bringing power up gently minimises the smoke in the event of a wiring (or design) error.
- The DUT has not been powered for years, and although it worked once, there is a suspicion that it may have developed a fault.
- The DUT contains old electrolytic capacitors that if gently re-formed by ramping power up over the course of 30 minutes will subsequently be fine, but could fail with suddenly applied power.

Because mains voltage is not guaranteed to be exact, Variacs are also used to test that at the lowest expected voltage:

- Regulators do not drop out of regulation.
- Power amplifiers deliver the specified power.

And that at the highest expected voltage:

- Devices dissipating significant power remain at an acceptable temperature.
- Devices having voltage limits such as capacitors or output valves are still within their voltage limits.

Figure 4.60
A 10 A Variac is useful, if rather heavy.

In order to provide +10% output voltage, many Variacs have a 90% tapping to which the incoming mains may be connected. When the wiper reaches the 90% point, full mains is delivered, but as it sweeps past, the autotransformer steps the voltage up to a maximum of +10%. See Figure 4.61.

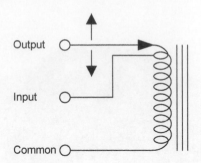

Figure 4.61
Applying the input to a tap part way down the winding allows a Variac to develop 0–110% output voltage.

A 50 VA transformer could provide 50 V at 1 A, 25 V at 2 A, or 10 V at 5 A, so secondary wire gauge is chosen to suit the expected current. Conversely, because practical considerations force a Variac to have a single winding of constant diameter wire, the maximum permissible load current must be constant, so Variacs are rated by load current, rather than VA.

Variacs are commonly available second-hand, but they are frequently naked, requiring a case to make them safe. New Variacs are available neatly cased with a mains outlet, complete with current and voltage monitoring meters, which can be very useful. If you have the choice, a 10 A Variac is more useful than a 2 A version, and although much heavier, only a little more expensive. Beware that Variacs necessarily have very little primary inductance so they tend to draw a large surge current when power is applied, often tripping breakers or perhaps blowing a fuse.

The ammeter on a commercial Variac must measure AC, which either means a moving-coil meter plus rectifier or a moving iron meter, but both are likely to have a cramped scale at low currents. Moving iron meters inherently have non-linear scales, but the moving coil meter acquires a non-linear scale because the fixed forward drop of the rectifier becomes a variable proportion of meter deflection. If the rectifier's forward drop could be minimised in comparison to meter deflection, a more linear scale would result. One way that this can be achieved is to use a mains toroid as a current transformer.

A 230 V:6 V toroid rated at 30 VA and having 12% regulation would probably have a turns ratio of $\approx$ 34:1 and a 5 A secondary rating. Inserting the 6 V winding in series with the load would step the current down by a factor of 34:1, so 1 A of load current would cause 29 mA of secondary current, and load resistance across the secondary would define the voltage developed by that 29 mA. If we tolerated 1 V across the 6 V winding, there would be 30−34 V of secondary voltage, requiring a load resistance including meter of ≈ 1 kΩ. More significantly, the 0.7 V diode drop becomes insignificant compared to the 30 V secondary voltage resulting in a linear scale.

A full-wave rectifier would measure using both half-cycles but a bridge rectifier doubles the diode drop (two diodes in series), whereas a centre-tapped rectifier (using two 0−115 V) windings halves the secondary voltage to 15−17 V, effectively increasing the proportion of the single diode drop to the same amount. Half-wave rectification risks saturating the toroid, so the simplest solution is to use a centre-tapped rectifier, and (approximately) halve the load resistance to 510 Ω.

We still need to scale the current through the meter correctly. A typical 1 mA movement might have 75 Ω coil resistance, so the shunt resistance required to pass the unwanted 28 mA would be ≈ 2.7 Ω and will need to be adjusted on test. Finally, we should remember that a rectified sine wave has a DC component (to which the meter responds) of $0.9I_{in}$. Thus, we need a fraction more current through the meter, but this will be taken care of when we fine-tune the ≈ 2.7 Ω shunt resistance on test. The shunt resistance is set by applying a 1 A load (as verified by a good DVM), and adjusting the resistance until the meter agrees with the DVM. It is essential that the 1 A load draws a sinusoidal current, so some form of (uncontrolled) heater such as a 250 W incandescent lamp is ideal. See Figure 4.62.

Testing mains transformer cores using a Variac

Variacs are invaluable for obtaining a plot of mains transformer's magnetising current against applied voltage, simply by measuring the current entering an unloaded mains transformer and its applied voltage. See Figure 4.63.

If a mains transformer's primary inductance was constant, we would expect magnetising current to rise linearly with applied alternating voltage. However, what we actually see is a slightly curved magnetising current. More usefully, since we know the frequency ($\pm 1\%$)

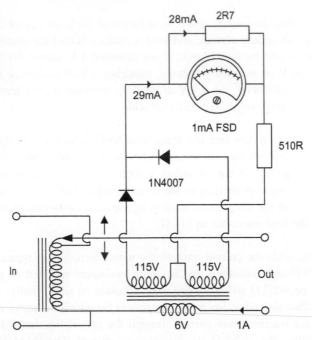

Figure 4.62
Using a mains toroid as a current transformer enables accurate load current measurement.

of our applied mains voltage, we can calculate the transformer's primary inductance at each applied voltage using:

$$L = \frac{V_{\text{applied}}}{2\pi f_{\text{mains}} I_{\text{magnetising}}}$$

The plot of primary inductance is most informative because it tells us about transformer core behaviour. Primary inductance rises from a low at low voltages to a maximum, and falls as saturation approaches. See Figure 4.64.

The significance of a transformer's primary inductance curve lies less in absolute values but more in its shape. See Figure 4.65.

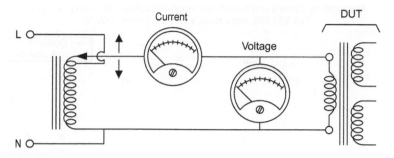

Figure 4.63
Measuring mains transformer magnetising current.

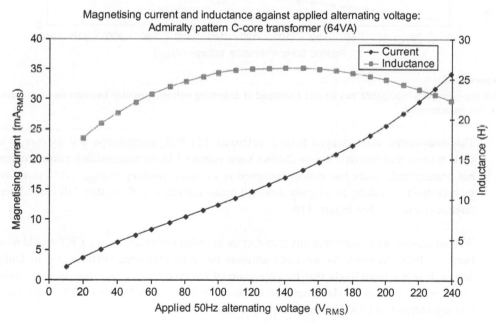

Figure 4.64
This curve is representative of a healthy mains transformer core.

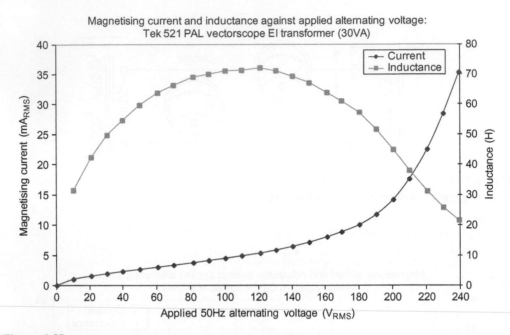

Figure 4.65
This transformer's inductance has almost collapsed at operating voltage, possibly because its core material has deteriorated.

This transformer was salvaged from a Tektronix 521 PAL vectorscope and its core is in such distress that not only has a distinct knee appeared in its magnetising current curve, but primary inductance has almost collapsed at its stated working voltage (70% down on its maximum), resulting in a highly distorted mains current waveform that will cause interference elsewhere. See Figure 4.66.

We can reasonably assume that any transformer intended to operate near a CRT would have been carefully designed, so we can't attribute the poor measured performance to faulty design. It seems more likely that the core material has deteriorated, and that the extra stresses of being tested on 50 Hz rather than 60 Hz ($X_L = 2\pi fL$, so 50 Hz forces 20% more magnetising current) and 240 V rather than 230 V (2×115 V) have tipped the balance.

By contrast, an instrument manufacturer might explicitly require their mains transformer to produce minimal leakage flux, and the manual for a Marconi sensitive valve voltmeter

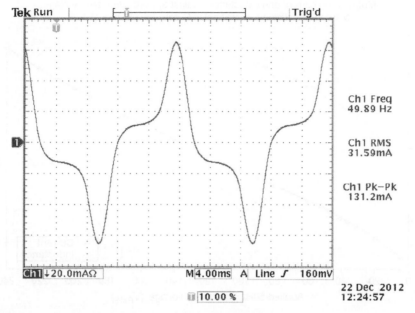

Tek Run | Trig'd

Ch1 Freq
49.89 Hz

Ch1 RMS
31.59mA

Ch1 Pk–Pk
131.2mA

Ch1 ↓20.0mAΩ   M 4.00ms A Line ʃ 160mV

10.00 %

22 Dec 2012
12:24:57

Figure 4.66
The saturating core indicated by collapsed primary inductance results in a high-amplitude distorted magnetising current.

specifically stated that its (C-core) mains transformer was operated at a lower flux density than usual for this very reason. See Figure 4.67.

Note that this final example has a beautifully flat primary inductance curve that barely falls with applied voltage, and is the only mains transformer you would dare position anywhere near your RIAA stage. The significance of the plot of primary inductance against applied voltage is that a flatter inductance curve implies reduced leakage flux and minimal magnetising current distortion. Once again, we test possible contenders for suitability and reject duffers **before** marking out and cutting metal.

Valve tester

Valve testers are useful if you have a large stock of valves or are a keen designer and want to plot your own valve curves. However, it should be borne in mind that a

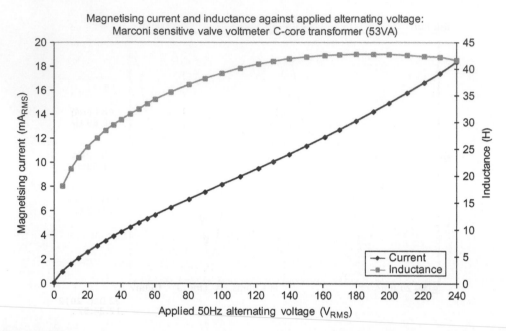

Magnetising current and inductance against applied alternating voltage:
Marconi sensitive valve voltmeter C-core transformer (53VA)

Figure 4.67

This mains transformer was designed to operate its core at a low flux density, resulting in an almost flat inductance curve.

universal valve tester inevitably exposes lethal voltages, so they are intrinsically danger-ous, even by the relaxed standards of their day, and should be used with caution. By definition, it is not possible to make a "safe" valve tester — the assumption was always made that they would be operated by people who were aware of the dangers and compe-tent to deal with them.

Typical receiving valves have heaters varying from 2.5 V at 2.5 A (2A3) to 40 V at 300 mA (PL519), with plenty of variation in between, so a valve tester needs a heater/fila-ment supply to cope with this. One possible solution might be a variable regulated DC supply, but a linear supply capable of providing 40 V would have to dissipate at least 94 W when supplying 2.5 V at 2.5 A, so the traditional solution was a tapped transformer. Inevitably, there is considerable wiring between the transformer and heater/filament, and this can cause a significant voltage drop at high currents, so checking the actual voltage on the valve's pins is well worthwhile.

Ideally, the supplies required by anodes and screen grids should not change their voltage under load. Typical valve testers provide up to 400 V at up to 100 mA, and this is just within sensible bounds of a linear regulator, but the cunning solution patented by AVO used a tapped transformer to apply AC directly to the valve electrodes, and used the valve's self-rectifying action. The reason for not rectifying the AC directly was that rectification and smoothing inevitably worsen the regulation of the supply, whilst simultaneously adding expense. The current through the valve therefore consists of half-wave rectified pulses. See Figure 4.68.

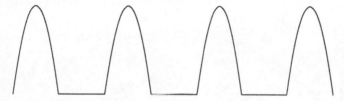

Figure 4.68
Valve anode current in scaled AC valve tester.

We saw earlier that the inertia of a moving coil meter causes it to respond only to the DC component of a current. A full-wave rectified sine wave is composed of:

$$v = V_{ac(RMS)}[0.90 + 0.6(2f) - 0.12(4f) + 0.05(6f) - 0.03(8f)...]$$

Thus, the DC component corresponds to $0.90\,V_{ac(RMS)}$ or, to put it another way, we require $V_{ac(RMS)} = 1.11\,V_{DC}$. This is not a problem – we simply scale our anode and screen grid tappings appropriately, so that when we select $V_a = 400$ V, the tester actually applies 444 V_{RMS} to the anode.

We can use almost the same trick at the grid, but we cannot allow the grid to go positive as the resulting grid current would almost certainly damage the valve, so we half-wave rectify the grid bias voltage. (Full-wave rectification is not necessary because the valve cannot conduct on the missing grid half-cycle when the anode and screen grid voltages are negative.) Half-wave rectification halves the mean voltage compared to full-wave rectification, so when measured by a mean reading meter, the control grid voltage **should** correspond to 0.555 of the claimed equivalent DC voltage. In practice, the calibration procedure for both the portable CT160 and the laboratory VCM163 specified a ratio of 0.52, whereas earlier instruments used 0.525, and the author needed 0.53 on his VCM163.

Grid voltage discrepancies aside, although the valve passes appropriate currents when conducting, it does so for only half the time, so the (mean reading) ammeter in the anode or screen grid circuit is calibrated to indicate double the current that would be indicated by an external mean reading ammeter. The ammeter may also have a scale for indicating R_{hk}. See Figure 4.69.

Figure 4.69
The AVO VCM163 (and earlier variants) also measures R_{hk}.

Summarising, don't go fiddling with calibration controls inside a scaled AC valve tester because its scales don't agree with a DVM unless you know exactly what you are doing. And even if you do, hesitate.

Despite the caveats, the scaled AC technique makes it possible to take reasonably accurate measurements over a wide range at far less cost than a true laboratory test rig using pure DC supplies.

Although one use of a valve tester is to take a sufficient number of measurements to be able to plot curves, the more common requirement is to investigate performance at a single point, perhaps to match valves in push—pull output stages. When matching valves, we should not only match anode currents (to avoid output transformer core saturation), but ideally also match mutual conductance (to maximise cancellation of even harmonic distortion). Mutual conductance is a moving target because it changes with anode current, so it is a small-signal parameter that should be measured by a low-amplitude alternating voltage applied to the valve's control grid. A laboratory DC test rig could use any convenient frequency, but a scaled AC tester applies mains frequency plus a spray of harmonics to the grid, so a higher frequency is required, but not so high that Miller capacitance can cause a problem. In their VCM163, AVO applied 15 kHz to the control grid at different amplitudes depending on the g_m range selected, sensed the anode (or screen grid) current with a 10 Ω resistor, then amplified the resulting (rather small) voltage via a high-pass filter in order to reject the (much larger) mains related signals. Because this enabled accurate determination of mutual conductance at any operating point, it became known as a **dynamic** measurement, and required a separate meter. See Figure 4.70.

Figure 4.70
Adding the dynamic g_m measurement requires a second meter.

By contrast, earlier AVO testers made a **static** measurement of mutual conductance. They biased the control grid to −0.5 V, measured the resulting anode current, and the meter current was then manually nulled using a "backing off" control. Finally, grid voltage was increased to +0.5 V and the increase in anode current could be directly read as mutual conductance in mA/V. Interestingly, the complete patent specification [5] mentions that using −0.5 V to +0.5 V doesn't give the right answer, but that if the voltage change is doubled from −1 V to +1 V, *"the mutual conductance figure is substantially correct"*.

A scaled AC valve tester produces test voltages that are directly proportional to mains voltage, so not only does the mains transformer primary need coarse tappings to set the tester to the nominal mains voltage, it also needs fine tappings that can be adjusted from the front panel to compensate for short-term fluctuation. We can now draw a simplified diagram of a final generation scaled AC valve tester. See Figure 4.71.

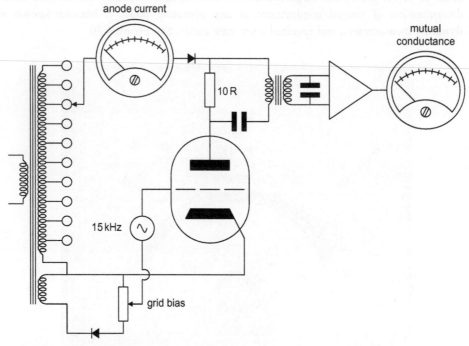

Figure 4.71
Simplified diagram of a final generation scaled AC valve tester.

In practice, although the anode and screen grid are able to self-rectify, AVO found part way through production of the mkIV tester that adding a series silicon diode prevented parasitic oscillations, so later mkIV instruments and the VCM163 (but not the CT160) already have half-wave rectified voltages at these electrodes.

When scaled AC testers were designed, mains demand was far less variable than it is now, and sudden current demands caused by millions of 2.4 kW electric kettles being switched on at the instant of the adverts part way through a popular television programme just didn't occur. In short, scaled AC testers need their mains to be regulated if they are to be used for generating sets of curves.

Synthesised mains supplies having excellent regulation are available, but they are expensive (as much, if not more, than the valve tester). Constant voltage transformers can't be used because although controlled saturation of their iron core holds V_{RMS} constant, it distorts the waveform and therefore invalidates the AC/DC relationships upon which scaled AC testers are based. The cheapest solution is a second-hand traditional AC voltage regulator based on a servo motor-driven Variac.

Incidentally, if you have an AVO tester with a dirty electrode selection switch, they can be removed, stripped down, thoroughly cleaned, reassembled, and refitted. The job takes an entire afternoon, but if you opt to do it, take the opportunity to check and, if necessary, correct an early design flaw in the rotor's phosphor bronze leaf springs, which should sit flush in their moulded cavity, but some did not have cropped corners, so they sat proud and wore the adjacent rotor. As you reassemble the rotors onto the shaft, gently check that each one rotates freely in both directions and that the leaf springs don't catch.

AVO testers have double-pole mains switches that are likely to be worn or have dirty contacts, possibly both, which can add significant (and variable) resistance in series with the tester, and significantly distort results. It's far cheaper and simpler to replace it immediately than have to diagnose it later as being the fault.

A modern problem unique to the AVO VCM163 can arise due to its 15 kHz g_m measurement relying on high-pass rather than notch filtering to attenuate mains fundamental and its harmonics. Domestic photovoltaic arrays are becoming popular but their associated inverters occasionally create high frequency mains interference that upsets the sensitive VCM163 g_m measurement.

Since the modern use of valves is primarily for audio, the voltages required are rather more restricted, and the scaled AC technique is no longer essential. Rather than having many sockets wired in parallel with electrode connections selected by a large (and expensive) multi-pole switch, modern testers tend to have dedicated plug-in boards for each socket and removable wire links to select electrode connections.

Curve tracer

One of the uses of the valve tester was to generate data from which curves could be plotted, and another was to match valves at a particular operating point. If we had an instrument that could plot curves directly, we would not only save time, but we could match valves over an entire set of curves rather than at a single point. A curve tracer is effectively a low-bandwidth oscilloscope having a fixed time base and step generator instead of the trigger block, so comparison between one set of curves and another either requires a curve tracer with digital memory, or an analogue tracer and a camera. (Traditionally, a Polaroid film camera would have been used, but a digital camera does the job far more conveniently — just disable the flash.)

The most useful curves for a valve are the mutual characteristics (I_a against V_g) or the anode characteristics (I_a against V_a). In each case, anode current is monitored and applied to the "Y" deflection plates, and a swept voltage is applied to the "X" deflection plates to produce a curve. See Figure 4.72.

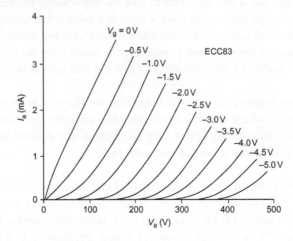

Figure 4.72
Valve anode characteristics.

Since the swept voltage is applied to the "X" deflection plates and the valve simultaneously, it can have any waveform, and the cheapest waveform is a half-wave rectified sine wave derived from a mains transformer. Unfortunately, the half-wave rectified sine wave slows towards the end of the sweep, causing uneven brightness when displayed on a cathode ray tube, and other errors can cause the retrace (as the voltage returns to 0 V) not to overlay perfectly. A ramp waveform solves both these problems but requires a generator and power amplifier to feed the valve.

Ideally, we would like to produce a family of curves simultaneously, perhaps a complete set of anode characteristics for a 6080 power triode. To do this, we would need to sweep V_a from 0 V to perhaps 200 V, supplying a current of up to 65 mA (to limit P_a to the maximum allowable dissipation of 13 W), requiring a power amplifier based on an HT regulator. In addition, we would need a step generator and high voltage amplifier for V_g that quickly and repetitively sequences from 0 to -80 V in 10 V steps to generate the individual grid curves. See Figure 4.73.

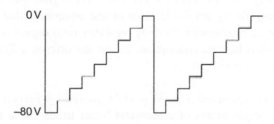

Figure 4.73
A repetitive stepped grid waveform enables a family of anode curves.

The previous arrangement would display triode anode characteristics, but displaying pentode anode characteristics is somewhat harder. Measuring pentode anode current forces the current sense resistor to be in the anode circuit, but as the absolute voltage of this resistor is being swept from 0 V to perhaps 400 V, a differential amplifier having very good common mode rejection is required, whereas we could declare that $I_a = I_k$ for the triode, and measure cathode current without worrying about a superimposed sweep voltage. The other problem with measuring a pentode is that when $V_a = 0$ V, $I_a = 0$, so the screen grid becomes an anode and passes the entire cathode current, yet its maximum dissipation is strictly limited. As a consequence, we cannot simply apply a constant voltage to the screen grid; it must be switched on only as the anode voltage is swept.

Another problem is that as the anode voltage approaches its maximum, so does anode current, causing anode dissipation to approach its maximum, but a hot anode changes the valve's characteristics slightly. (This is one cause of retrace error in the half-wave rectified sine wave sweep.) To avoid this error, the sweep can be modified to consist of a series of short pulses having amplitudes that follow the path of the original ramp waveform. As an example, if the anode voltage pulses caused anode current to be switched on for only 10% of the time, they would reduce anode dissipation to one-tenth of the continuous ramp.

The preceding thoughts only consider an analogue curve tracer, yet they show that an ideal instrument must be quite complex. Very few laboratories needed a curve tracer, so development cost had to be recovered over a limited number of sales, and prices were stratospheric. Worse, it appears that development stopped when a design was "good enough", which is presumably why the author's Tektronix 571 digital transistor curve tracer was quite the most staggeringly unreliable piece of test equipment that he has ever endured, releasing its analogue smoke with alarming regularity until digital death and EPROM corruption finally took it beyond redemption. If you are offered a Tektronix 571, accept it only as a no-strings-attached gift.

More recently, a microcontroller-based module offered by French eBay seller "rimlok" provides the core requirements of a computer-based triode curve tracer. Anode voltage sweep is provided in the traditional manner by half-wave rectifying the output of an external high voltage transformer, and anode current is assumed to be identical to cathode current, which is sensed by a 10 Ω resistor whose voltage is applied to an ADC. Grid voltage is controlled in 1 V steps by a DAC. See Figure 4.74.

Because the results of each measurement are stored, each need only be made once and errors due to anode heating are reduced. The module has some limitations:

- Op-amp voltage limits within the module mean that V_{gk} can only be adjusted between 0 and −31 V, limiting testing of low-μ high voltage valves.
- V_{gk} can only be adjusted in 1 V steps.

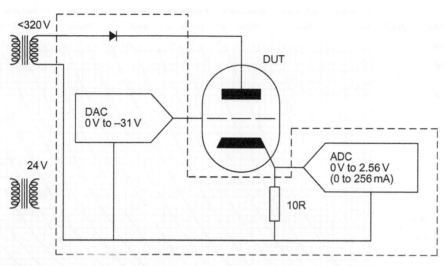

Figure 4.74
This simple modular curve tracer enables quick comparisons between valves.

- Because the module has only one ADC, it cannot measure anode voltage and cathode current simultaneously. Instead, it switches its ADC between anode and cathode terminals, measuring swept anode voltages on its first mains half-cycle with grid voltage set to −31 V, then switching to the cathode terminal, and assumes that mains half-cycle voltages are unchanged as grid voltage is stepped, but fluctuating mains voltage and transformer regulation mean that this is not necessarily true, limiting absolute accuracy.
- Because it measures cathode rather than anode current, the module cannot produce true pentode curves.
- The measurement of valve parameters is made as the curves are captured rather than on stored data, so changing the measurement point requires the time-consuming generation of a new set of curves.

The module's limitations are entirely forgivable at its price, and the addition of a few junk box transformers enables the construction of a triode curve tracer of minimal cost and complexity yet capable of generating useful data that can be saved, relegating classic valve-based analogue curve tracers to the museum. See Figure 4.75.

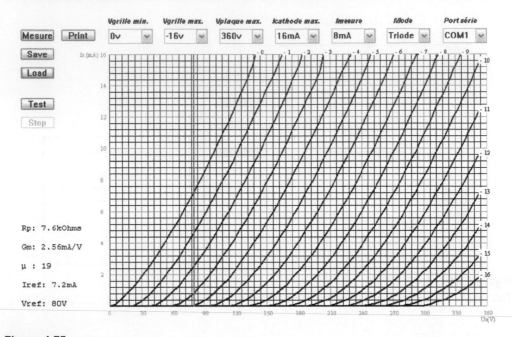

Figure 4.75

This set of Pinnacle 6J5GT curves was produced using the curve tracer board offered by French eBay seller "rimlok".

Although not mentioned in its accompanying documentation, the best way to provide the anode voltage is from a transformer having a 320 V secondary, but with its mains supplied via a Variac to allow maximum anode voltage to be continuously varied between 0 and 480 V. A potentially better pulse-based valve curve tracer is available from www.dos4ever.com.

Semiconductor component analyser

This is a tiny handheld device having three leads and an LCD display that scrolls to give comprehensive information. It is incredibly useful for quickly determining the pin-out of a known component, identifying an unknown one, and can also be used for faultfinding if you don't mind removing components from their surrounding circuitry. See Figure 4.76.

The more recent DCA75 variant has some useful improvements over the DCA55, one of which is a very limited curve tracing ability, but capable of matching input JFETs or BJTs.

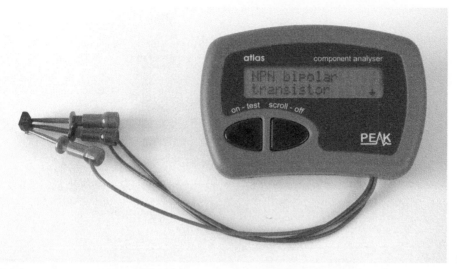

Figure 4.76
Semiconductor component analyser. Although useful, this analyser and its sister ESR meter switch on and flatten their batteries when stored loose, but a cotton bud box is just the right size to prevent this problem.

A handy constant current regulator for DC measurements

The author often needs a constant current regulator during testing and prototyping. The device is simply placed in series with any power supply, and turns it into a programmable constant current source, so it only needs two terminals. See Figure 4.77.

The circuit is a modification of standard constant current heater regulator circuits, but with the addition of a multi-position switch and variable resistor it can very conveniently set any current between ≈3 mA and 1.5 A, so it can be used for anything from plotting diode curves to powering "P" series valve heaters. See Figure 4.78.

The BZY79C22 diode protects the 317 from over voltage and stored charge in the load. The 317 must be bolted to a heatsink, perhaps the aluminium chassis. The logarithmic range switch doubles current at each click, whilst the 25 kΩ variable resistor and associated 43 kΩ resistor ensure that each range just overlaps the next. The entire circuitry can

375

Figure 4.77
This device converts any voltage supply into a regulated constant current supply.

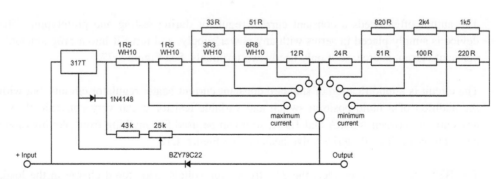

Figure 4.78
Circuit diagram of 3 mA to 1.5 A adjustable constant current sink.

be very conveniently hard-wired on the lid of a $6'' \times 4'' \times 2''$ die-cast aluminium box, making it very easy to make or repair. See Figure 4.79.

Figure 4.79
The components are hard-wired onto the lid of a die-cast box.

A handy constant current regulator for AC measurements

Although the preceding circuit's DC performance was good, its AC performance was poor. Conversely, the following circuit's DC stability is quite poor and its current range limited, but its AC performance is quite reasonable (and could be improved if really necessary by adding a second FET). See Figure 4.80.

The author finds the circuit particularly useful when measuring choke inductance as a function of applied current, so it was worth making a neat version. See Figure 4.81.

As can be seen, the circuit is hard-wired on a small heatsink and uses a wirewound variable resistor to set current. The optional conductive plastic rheostat in series with the

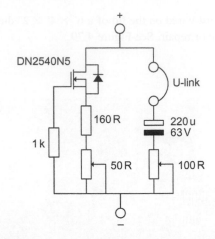

Figure 4.80
Circuit diagram of 20–200 mA adjustable constant current sink.

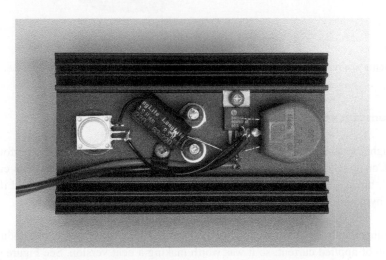

Figure 4.81
The components are hard-wired onto a heatsink.

220 µF capacitor (engaged by inserting the U-link) allows harmonic resistor values to be found quickly without disturbing DC conditions. Once the rheostat has been set, unplugging the U-link allows its resistance to 0 V to be measured. If a harmonic resistor value >100 Ω is needed, the author substitutes a decade resistance box for the U-link, adjusts to minimise third harmonic amplitude, measures the rheostat resistance, then adds it to the decade box resistance.

References

1. Linear Technology application note AN43-33.
2. P.R. Wilson, *Effective Modelling of Leakage Inductance for Use in Circuit Simulation.* Available from Southampton University UK website.
3. GEC, *An Approach to Audio Frequency Amplifier Design.* GEC (1957). Reprinted by Audio Amateur Press, Peterborough, NH (1994).
4. N. Partridge, Distortion in transformer cores. *Wireless World* (1939). A series of four articles: 22 June, pp. 572−574; 29 June, pp. 597−600; 6 July, pp. 8−11; 13 July, pp. 30−32.
5. Sydney Rutherford Wilkins & The Automatic Coil Winder & Electrical Equipment Company Ltd. An Improved Method and Apparatus for Testing Radio Valves. UK Patent 480752 (1938).

Recommended further reading

R. Cordell, *Designing Audio Power Amplifiers.* McGraw-Hill (2011), ISBN 978-0-07-164024-4.
This excellent book devotes two chapters to introducing LTspice modelling with an audio bias, whilst the main material (solid state amplifiers) is peppered with useful practical observations.
R. Metzler, *Audio Measurement Handbook.* Audio Precision (1993).
As should be expected from a manufacturer of audio test equipment, this little book is excellent on purely analogue techniques but its age precludes recent digital advances.
Keithley Instruments, *Low Level Measurements*, 4th edn. Keithley Instruments (1992).
Now updated and available as a free download, this book is written by a manufacturer that specialises in making low current (perhaps fA) measurements, so although not explicitly written for audio, it contains many useful principles.
Agilent Technologies, *Agilent Impedance Measurement Handbook*, 4th edn.

Available as a free download, this document is clearly the result of engineers wearying of answering the same questions over and over again. It's a mine of useful information, particularly on the problems of test fixtures.

F.E. Terman and J.M. Pettit, *Electronic Measurements*, 2nd edn. McGraw-Hill (1952).

As would be expected from the publication date, this book covers fundamental laboratory measurement principles with the emphasis on elegant configurations rather than brute force expensive instruments, but at the expense of measurement time.

CHAPTER 5

FAULTFINDING TO FETTLING

In the previous chapter, we looked inside test equipment to see how it worked, to explore its design limitations, and to understand how best to use it. We are now in a position to move on, so this chapter assumes familiarity with test equipment and is more concerned with bringing a built circuit to a usable state.

There are three stages of testing:

- Safety testing: Will the device under test (DUT) endanger the user?
- Functionality testing: Does the DUT work as it should?
- Faultfinding (hopefully optional): Having established that the DUT doesn't work correctly, what needs to be fixed?

Rather than scribbling results on scraps of paper and subsequently losing them, you will find it extremely useful to keep a hardback logbook of your tests — large desk diaries make good logbooks and they're cheap if bought in February. The ideal logbook entry includes:

- Circuit diagram of the circuit tested.
- Diagram or description of how the test was undertaken.
- Results of the test.
- Subjective comments on the test. (Were the results expected?)
- Date of the test, and test equipment used.

Building Valve Amplifiers. DOI: http://dx.doi.org/10.1016/B978-0-08-096638-0.00005-9

- If test data has been saved to computer (spreadsheet, photograph, etc.), include full file names adjacent to the test diagram/description.

The purpose of the logbook is to enable you to look back and **know** the results of a test, rather than scratching your head and saying, "I'm sure I've tested this before."

Safety

Before we go any further, we should understand that there is no such thing as perfect safety. Everything we do, from crossing the road to eating a chicken sandwich carries an element of risk. Lawyers currently enjoy a climate where courts award ludicrous damages to plaintiffs who deny responsibility for their own actions – of course a cup of hot coffee scalds if you spill it over yourself! The dangers of hot liquids are well known, but the dangers of valve electronics are not as well known as they used to be, and even professional engineers can become complacent about putting their hands inside live equipment.

The aim of this short section is to point out some of the more common dangers and show simple ways by which risk may be reduced. Nevertheless, it is impossible to cover all possible situations, and this is why electronics workshops forbid work on live equipment unless two people are present. In that way, one person is ready to rescue the other in the event of an accident. If you have a more experienced friend, it would be a good idea to emulate commercial practice by asking them to attend when live working is necessary and to check your work beforehand. Ensure that they know how to switch off the power quickly.

An understanding of the physiological effects of electric shock, whilst somewhat ghoulish, serves to underline why electrical safety is so important.

How an electric shock can be received

The best way of avoiding electric shock is to understand how it can be received.

The electricity supply leaving the wall socket in your home is a very good approximation to a pure Thévenin source of zero resistance, with one side of the source connected to earth. In this instance, we do not mean earth in its purely technical sense, the supply really is connected to the planet Earth. You and I perform most of our activities on the surface of the earth, and we are therefore electrically connected to it, albeit usually via skin resistance of $\approx 2 \text{ k}\Omega$. We can reduce our earth resistance by standing barefoot on a damp floor, or by firmly gripping something that is electrically bonded to earth.

Manual activities like hobbies are precisely that, **manual**. They involve our hands going inside objects and touching them. Humans generally possess two hands, and so what is more natural than to put **both** hands inside a piece of equipment? The scene has now been set for one hand to be holding the (earthed) chassis of a piece of equipment, whilst the other is moving around and accidentally comes into contact with live mains.

Look at your hands. They are on the ends of your arms, and your arms are joined to your torso. Trace a line from the fingers of one hand to the fingers of your other hand, without the line leaving your body. Note the path.

The easiest path for an electric current to flow from one hand to the other crosses near the heart, and once a current has passed through the skin, blood conducts almost as well as a wire, so it will pass through the heart. Similarly, a shock from one hand to earth passes through the heart.

The effects of electric shock

The main electric shock danger is fibrillation of the heart. The heart normally beats at a slow pace regulated by electrical impulses from the brain. If we apply 230 V/50 Hz to the heart, it pulses quickly, the flow regulating valves do not operate correctly, and no blood is pumped, leaving the brain to die of oxygen starvation in about 10 minutes. A sustained current of 20 mA through the heart is sufficient to cause fibrillation, resulting in the adages "Twenty mills kills" and "It's the volts that jolts, but the mills that kills."

A sustained current might not kill, but could cause irreversible injury due to the internal heating effect of the current; radio frequency burns are notorious for this, and can result in limbs having to be amputated to prevent the spread of gangrene. A lower sustained current may cause injury from which the victim does recover. Eventually. Skin grafts may be necessary.

A current of 20 mA results from a 230 V supply connected across a resistance of 11.5 kΩ. But the IEC [1] accepted value for skin resistance (at 220 V/50 Hz) is:

< 1000 Ω for 5% of the population
< 1350 Ω for 50% of the population
< 2125 Ω for 95% of the population.

In other words, skin resistance is insufficient to reliably reduce shock current from 230 V mains to below the lethal value, and very few power supplies have a source resistance as high as 10 kΩ, so shock current is determined primarily by skin resistance, making a shock from a puny transformer delivering 230 V just as dangerous as the shock received from a 100 A mains feeder. Skin resistance is reduced by damp hands, and standing in the rain in a puddle of water reduces earth resistance. Conversely, standing on an insulating rubber mat increases earth resistance, increasing the chance of surviving contact with the mains.

All of the previous considerations refer to the direct consequences of electric shock, but do not consider secondary effects. A minor shock that causes the victim to lose their balance and fall could be fatal if they happen to be standing on a ladder 30 ft above concrete. Another possibility is that the reflex muscle jerk in reaction to the shock could cause the victim to throw themselves onto glass and subsequently bleed to death.

Even after a minor electric shock, the victim will be confused and disoriented, and anaphylactic shock, in its full medical sense, is a possibility. Shock kills.

Burns

Although the primary electrical hazard is shock from high voltages, it should be realised that low voltages can also be hazardous. A low voltage/high-current DC supply will have a large low ESR reservoir capacitor capable of delivering many amps of current into a short circuit.

Rechargeable batteries are even more dangerous because they are capable of sourcing substantial current for a significant time. The lead—acid batteries in cars are rated by their cold cranking amperes (CCA), and the (low compression) original Mini required 380 A. Even small rechargeable nickel metal hydride (NiMH) cells can source significant currents, and electric flight enthusiasts cheerfully and repeatedly draw 100 A from AA cells to fly their model aircraft. Rechargeable batteries are not merely capable of burning; they can vaporise rings, metal bracelets, and tools.

Do not wear jewellery when working near batteries.

Avoiding shock and burns

It should now be obvious that electric shock is potentially lethal, burns can be serious, and that both must be avoided.

Provided that you have made, or modified, your equipment carefully, there will be no exposed voltages, and all metalwork will be earthed, resulting in a very low risk of shock. The danger arises when you **deliberately** remove the safety covers, and start testing the equipment with power applied.

Some authorities suggest that you should only work on live equipment with your left arm behind your back, so that any shock received cannot pass from arm to arm across the heart. Whilst it is true that this would reduce a shock's severity, it tends to increase the risk of receiving that shock.

The best way of improving safety is to think about safety, and to **think about what you are doing**. It might seem obvious to think about what you are doing, but for most of our lives we think about many things at once. For instance, when driving, are you thinking **only** about driving, or are you actually thinking about what you are going to say to your boss when you arrive late for work, and when is that idiot in front of you going to turn into the junction, and isn't that a rather attractive male/female/alien over there by the bus stop?

Thinking about what you are doing means not working late. Do not attempt to test a newly completed project at 11.30 at night; you will not be alert and could damage the project and/or yourself. Intriguingly, the engineering test that precipitated the Chernobyl nuclear disaster began at 01:00 local time. What **were** they thinking of?

Electrical safety testing

Consumer electronics falls into two safety categories. Class I has an earthed conductive barrier enclosing the high voltages, whereas Class II has two independent insulating barriers enclosing the high voltages. Many appliances are actually a combination of the two categories because although the appliance could be Class I, the mains lead is invariably Class II. Class I relies on **quickly** interrupting the supply (blowing a fuse or tripping a breaker), whereas Class II relies on undamaged barriers. Commercial safety testing therefore consists of a careful inspection by a competent person to check barrier integrity, plug wiring, and correct fusing. These visual tests are backed up by electrical tests.

A safe piece of Class I equipment has an earthed conductive barrier with no conductive path to any of the enclosed high voltages and a sufficiently low-resistance path to earth that any contact to the mains line causes such a large current to flow that the mains fuse ruptures quickly. In addition to informed inspection, electrical safety can be quantified by measuring leakage current between the barrier and line, and by measuring the resistance from the barrier to the earth pin of the mains plug. Since many pieces of equipment have to be tested, it makes sense to have a dedicated instrument to make the tests.

Portable appliance testers (PAT)

PAT testers are simply specialised resistance and leakage testers having appropriate connectors for **quickly** testing mains portable appliances. The emphasis on the word "quickly" is important because even a small business could have hundreds of appliances needing annual safety testing. In order to cause minimum disruption each appliance must be tested quickly and unambiguously, and a record made of test results. Older PAT testers simply made the required pass/fail electrical tests, whereas newer testers guide the operator through the inspection, make the necessary measurements, give pass/fail status, and log the data together with the appliance's identification (often a bar code) for later download to a central database. Because the newer testers make it easier to produce evidence demonstrating compliance with safety legislation, older manual testers are available in good working order, and you might want to acquire one. See Figure 5.1.

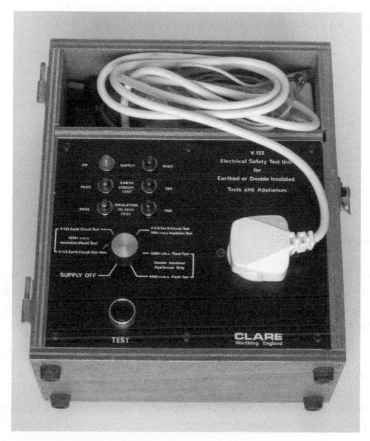

Figure 5.1
Older PAT testers are still perfectly capable of safety testing.

As part of a Class I test, a PAT tester measures earth loop resistance (resistance from the chassis to the earth pin of the plug) by sinking a considerable current, usually >15 A. The purpose of such a large test current is to detect frayed earth connections. A frayed connection down to one strand would still have low resistance because it would be very short, even though its cross-sectional area would be small, so a simple resistance measurement could not detect the problem. However, the PAT tester deliberately seeks to rupture such frayed connections, which subsequently fail the resistance test. Simultaneously, the tester applies a high voltage (>500 V_{RMS}) simultaneously to the line and neutral terminals and monitors leakage current returning through the earth circuit. Do not touch the appliance whilst the PAT tester applies this dangerous voltage.

Be aware that PAT testers are not infallible, and can develop faults just like any other piece of equipment. If you rely on a tester to determine safety of other equipment you have to be certain that the tester works correctly and that you are operating it correctly. Most testers include operating instructions and a selection of "faulty" test jigs that should be regularly tested to verify correct operation of the tester. You might question the validity of using a second-hand tester, particularly since it will almost certainly be sold without any warranty of fitness for purpose. Provided that you are **not** using it to assist in selling goods or services, the question is not whether the tester perfectly meets all the latest safety legislation, but whether it reduces risk compared to not being able to test at all.

Functionality testing

The word "testing" implies a degree of ambiguity about the results of the test. If we **knew** that our new amplifier was going to work perfectly from the moment that it was completed, we would not need to test it. However, we know that wiring mistakes can be made, and that components could be faulty, so we test our work. Cautiously.

Second-hand equipment versus freshly constructed new equipment

Both types of equipment should be treated with a great deal of suspicion and apprehension. The only sensible state of mind when first switching on is controlled fear.

Quite clearly, a piece of newly constructed equipment **will** be switched on at some point, but some old equipment may be so dangerous, or riddled with faults, that it should never be energised, other than applying kinetic energy to throw it into a skip. The state of old equipment can easily be determined by looking at the components.

Things to avoid are:

- Wire insulated with rubber and covered with cotton.
- Enormous resistors marked with tip, body, and spot colour codes.
- Electrolytic capacitors with bulges in the rubber surface supporting the tags.

- Previous evidence of fire.
- Insulating tape anywhere!

Vintage radios may incorporate any or all of these features yet still have value, so try to check with someone else before destroying them.

For a professional, the worst possible sign is previous modification by an amateur. The professional then has to decide whether or not the amateur knew what they were doing. What is the effect of their modification, and was it done competently and safely? For this reason, modified equipment is usually worth **less** than unmodified equipment, so bear this in mind before you embark on modifications.

Second-hand equipment can often be dated by the date on the electrolytic capacitors. If it is over 40 years old, it is likely to need major refurbishment just to make it work, so this should be taken into account if you are considering purchase.

The first application of power

The author strives for dullness, predictability, and a complete lack of excitement during testing. Before applying power for the first time, the following "Ten Commandments" should be observed:

- Inspect the earth bonding. Does the chassis appear to be properly earthed? (Does an undamaged earth wire make a good connection to the chassis? If a tag is used, is it tightly bonded with a serrated washer **between** it and the chassis?)
- When measured, is the resistance from the earth pin of the mains plug to the chassis of the equipment significantly less than 0.5 Ω? (Preferably use a PAT tester.)
- With the power switch on and the equipment switch (if fitted) switched on, is the resistance from the other pins of the mains plug to the earth pin infinite? (Preferably use a PAT tester or insulation tester.)
- Does the mains cable look safe? It should not be frayed, perished, cut, or melted by a soldering iron.
- Is the mains plug wired correctly?
- Does the plug grip the sheath of the cable correctly?
- Does the plug look safe? It should not have cracks, chips, dirty pins, etc.

- Is a fuse of appropriate rating fitted? (It is unlikely that the fuse rating should be greater than 3 A.)
- Does all the internal wiring of the chassis look secure?
- Is the chassis clear of swarf, wire off-cuts, and loose strands? Turn it so that bits can fall out, and give it a really good shake, whilst blowing vigorously into the chassis to free small parts. Alternatively, if it is too heavy to lift and shake, use a ½″ paintbrush and a powerful vacuum cleaner to remove debris.

If all appears to be well, you can move on to the next stage.

The author **always** assumes that when power is applied, the amplifier will explode or, at the very least, catch fire. It does not make sense to stand with your face directly over the amplifier, or to place it in the middle of a pile of inflammable debris. Power amplifiers should have dummy loads or cheap loudspeakers connected across their outputs.

The safest way to test a valve amplifier is in stages. Many smaller amplifiers use a valve rectifier, so if this is removed, the mains transformer and heaters can be tested before energising the high voltage supply. Silicon rectifiers make disabling the high voltage supply a little harder, requiring the AC to the rectifiers to be removed. If this is a new amplifier, you will have planned ahead by testing the heaters **before** installing any other wiring (that way, the heater wiring is easily accessible should a fault surface).

The heaters in indirectly heated valves take a moment before they begin to glow, so connect a meter across the heater supply to give an instant indication of whether the heater supply is present, and leave the meter in a clearly visible position. If you have a Variac, you can apply power gently, and if the meter monitoring the heater supply doesn't immediately respond when you advance the Variac from zero, back it off and investigate. The advantage of using a Variac is that even a short circuit across the heater supply would be unlikely to cause damage because you would spot it before applying significant voltage.

If you don't have a Variac, retire to a safe distance, and switch power on in silence. This way, you will hear any unusual noises, such as the crackles or pops that presage destruction. If the meter monitoring the heater supply responds appropriately and nothing untoward happens, move a little closer, and sniff the air. Can you smell burning? Are there any little wisps of smoke leaving the chassis? If all still seems to be well, look closely at the heaters — they should be glowing, but should not have hotspots. See Figure 5.2.

Figure 5.2
This 13E1 had a nasty cathode hotspot, so it was rejected without power being applied to its anode.

Leave the valves to warm for a while, perhaps put the kettle on and make a cup of tea. Ideally, at first switch-on, unused valves should warm their heaters for 30 minutes before anode voltage is applied. However, even 10 minutes gives time for your nerves to calm down and for you to consider your next step. Having left the heaters glowing for a while, check the temperature of the mains transformer, which should be cool. Switch off power, and **unplug** the mains lead, placing it in plain sight. If the amplifier has heater regulators, carefully inspect them for signs of damage, but be careful near the valves, which will still be hot.

If nothing has been damaged by warming the heaters, it is time to test the other supplies. Depending on the complexity or output power of the amplifier, this might be done in a number of ways:

- Classic amplifiers simply need their rectifier valve to be inserted and the amplifier to be switched on.
- Modern amplifiers usually use silicon rectifiers. Suddenly applying full voltage to an electrolytic that has sat quietly on a shelf for several months is unkind. Use a Variac to increase voltage gently over 20 seconds or so.
- If the amplifier has a separate heater transformer, it is worth applying full heater power, but powering other transformers via a Variac and gently testing their wiring before finally connecting them to the amplifier's internal mains distribution.

Individual circumstances will determine how you test with full voltages for the first time, but if it's possible to do it gently, then do so — it minimises the amount of smoke. Test using the cheapest, most horrible loudspeakers you can find — you wouldn't want to destroy an irreplaceable LS3/5a.

Having chosen your method of testing, apply power. Listen. Are the loudspeakers making any unusual noises? In this instance, silence really is golden. Is the main HT voltage correct? If this voltage is correct, then it is highly likely that the circuit is working as it should, and you may breathe a quiet sigh of relief.

If you are testing a newly built power amplifier, there is a 50/50 chance that global negative feedback taken from the loudspeaker output will turn out to be **positive** feedback, and the amplifier will become a power oscillator. Often, before oscillation starts, a quickly increasing hum will be heard from the loudspeaker, if the amplifier is switched off at this point, only a brief shriek of oscillation will be suffered.

If, at any point, something untoward happens, switch off immediately at the mains outlet, and unplug the mains plug.

Usually, if anything is wrong in an electronic circuit, heat is generated and components are burnt, so look for charred resistors or wires. Once the burnt parts are found, the fault is usually blindingly obvious.

If you switched off hurriedly because of a burning smell, what sort of a smell was it? Old equipment will often be dusty, so a slight burnt dust smell is normal. Bacon smells are sometimes produced by burning mains transformers, whereas scorched PCBs often smell like underground railway stations with a hint of charcoal, but burning wiring gives off an acrid smell.

If the amplifier appeared to be satisfactory, leave it switched on for a minute or two longer, whilst keeping an eagle eye on everything, particularly output valves, which should not develop glowing anodes, or purple and white flashes. Switch off, and sniff the internals closely for unusual smells. Some engineers go one step further, and touch components with their finger to check temperature, but this is not recommended as high voltages may still be present. If all seems well, the amplifier can be switched on again, and all the voltages carefully checked; if it still looks good, then it probably **is** good.

The EU and mains voltage

European mains voltage is presently **specified** as being 230 V + 10%/ − 6%, but this is simply a paperwork ruse intended to enable electrical goods to be sold freely within the European Union. There has never been any intention to reconfigure national electricity supply networks and change the voltage at your wall socket, so actual mains voltage is the same as it always was (240 V in the UK, 220 V in France). Thus, 230 V + 10% = 253 V ≈ 240 V + 6% (the old UK upper limit), and 230 V − 6% = 217 V, which only allows for a 1.4% fall in the French nominal voltage, and this is why the specification will broaden to 230 V ± 10%, requiring electrical goods to operate correctly on a supply anywhere between 207 and 253 V.

Electronics using linear regulators wastes more energy as the required input voltage range broadens and valves require heater voltages to be within ± 5%, so traditional mains transformers had tapped primaries to accommodate regional mains voltage variations. But the EU directive deliberately promotes ignorance as to the actual mains voltage, rendering the tapped primary technique less viable, thereby forcing domestic electronics either to be unnecessarily wasteful of energy or to require more specialised design.

If you have a Variac, now is the time to check how well the amplifier responds to mains voltage variations. Use the Variac to check that capacitor voltage limits are not exceeded when 253 V (230 V + 10%) is applied, and that regulators don't become excessively hot. Similarly, drop the mains voltage to 217 V (230 V − 6%) and check that regulators do not drop out. Combined, these two tests are quite severe (and a ± 10% test would be very severe), so if you **know** that your mains is more stable, or centred upon a different voltage, you might decide to adopt a less stringent test.

For the first few weeks of service, a new amplifier should be watched like a hawk for signs of incipient self-immolation, and should not be left unattended whilst switched on.

Faultfinding

Unfortunately, we all have to do some faultfinding at some time or another, either because we made a design or construction mistake, or because we need to repair an amplifier that has failed due to old age. Before we dive into details, we should ask one very important question, "Did it work once?"

If it worked once, you are looking for a faulty joint or component, and the resistance range of your DVM will prove invaluable for finding carbon resistors that have "gone high" in value from age, or capacitors that have become leaky. If the circuit is freshly built, then you are probably looking for a wiring or design error.

Individually testing each component

One way of faultfinding an amplifier might be to remove each component individually and test it for faults:

- Resistors: Check claimed value against measured value using the resistance range of a DVM.
- Non-polarised capacitors: Use an insulation tester to check leakage at the working voltage, then check claimed value of capacitance against measured value on the capacitance range of a component bridge, and question "d".
- Electrolytic capacitors: Measure capacitance and ESR using an ESR meter.
- Inductors: Use a DVM set to its resistance range to confirm continuity, then measure inductance on a component bridge. Use an insulation tester to check leakage from the winding to the chassis.
- Valves: Use a valve tester or curve tracer and compare with a manufacturer's data sheet.
- Semiconductor diodes: Use the diode check range on a DVM to check for open circuit when reverse biased, and correct forward drop when forward biased.
- Transistors: Use a semiconductor analyser to check functionality and perhaps compare measured h_{FE} with manufacturer's data. Alternatively, use a curve tracer to plot full characteristics. Bipolar junction transistors can be tested using a DVM's diode check range to check the base to emitter junction and base to collector junction.
- Transformers: Connect to an oscillator and check each output with an oscilloscope to verify functionality and turns ratio. Faster, check each winding for continuity using the resistance range of a DVM, then measure primary inductance using a component bridge — a shorted turn anywhere will cause primary inductance (H) to collapse to leakage inductance (mH). Inter-winding and winding to chassis insulation is best checked with an insulation tester.

You will notice that the previous tests presume a good deal of expensive test equipment and full manufacturer's data and/or copious experience for assessing measurement results.

We only use the shotgun technique of removing and testing each component when:

- We know no better.
- All else has failed.

The very best piece of test gear is your brain. It's readily available, so it seems a shame not to use it.

DC conditions

Most faults can be found very quickly by measuring the **DC conditions** of the circuit. For a well-designed circuit to be observably faulty, its DC conditions usually need to be very wrong. Consequently, checking the measured voltages against the design voltages quickly pinpoints the fault — perhaps a confused resistor multiplier band colour. Mark the measured voltages in pencil on the circuit diagram (not the manufacturer's original, but a photocopy or print from file). This usually has the effect of making the fault appear blindingly obvious.

Sometimes you will not have the circuit diagram of the amplifier, let alone its design voltages. No matter, there were very few variations in classic valve circuitry, and their circuits were so simple that it is not difficult to produce a block diagram of the amplifier. At this point, consider how you would design an amplifier using those valves, and look for similarities in the actual amplifier. It should now be possible to obtain a rough idea of what sensible voltages might be, and these can be checked against the faulty circuit. Modern amplifiers are likely to be more complex and include silicon, so they may require careful circuit tracing. Fortunately, there are now so many websites carrying information on valve audio that it is highly likely that the information you need is somewhere on the web — you just have to find it (and check that it's correct). If you can't find a complete circuit diagram, valve data sheets are an excellent second best because they give maximum ratings, typical applications, pin connections, and sometimes application notes.

A calculator is invaluable for calculating currents through resistors, and generally deciding whether measured voltages make sense.

Unless you are very unlucky, you probably have one working channel, so you can compare voltages measured at the same point between channels. It is very tempting to swap valves to see if the fault moves with the valve, but this might generate more faults in a modern amplifier that includes silicon, so think carefully before using this technique.

Don't unthinkingly believe a measurement. Even the standard 10 MΩ input impedance of a DVM loads some circuits, particularly the grid circuit of cathode followers or circuits with grid battery bias. The author was once convinced that audible distortion was due to the DC conditions within a valve active crossover, and a 10 MΩ DVM appeared to confirm the theory, but a valve voltmeter having 90 MΩ input resistance measured a more correct value, and the distortion finally turned out to be due to an intermittently scraping loudspeaker voice coil.

When televisions had CRTs, service technicians commonly had EHT probes for checking that the (typically 10−15 kV) final anode voltage was present and correct. These oversized bright red probes having multiple protective hilts before their handle look like props from a science fiction film, and their 1000:1 voltage division and voltage rating not only permits measurement to 40 kV, but increases the accompanying DVM's input resistance from 10 MΩ to 1 GΩ (yes, only 1 GΩ, not the 10 GΩ you would expect from 1000 × 10 MΩ). Nevertheless, 1 GΩ is perfectly adequate for testing most high-impedance circuitry even if the dagger-sized probe is a little clumsy. These probes frequently appear at radio fairs for quite reasonable prices.

Whilst on the topic of DVM accessories, almost all DVMs are capable of measuring temperature with the addition of a probe, usually a (cheap) Type K thermocouple. The significance of using a thermocouple and DVM is that the sensing element is very small (smaller than a match head), giving it a very low thermal mass, making it ideal for pressing against components and measuring their temperature reasonably accurately ($\pm 2°C$).

Turning very briefly to AC, some circuit nodes are such high impedance that even the 0.8 pF of an active probe can be sufficient to completely change operation. It is not uncommon for the 8 pF of a passive probe to suppress radio frequency oscillation at a sensitive node, so it's worth leaving one channel of an oscilloscope monitoring oscillation while the other is used to track the cause − that way, if the roving probe suppresses oscillation, the effect will be seen.

Blocks and attitudes

Imagine that you have just been told that the Hi-Fi isn't working. If you have simultaneously been plunged into darkness, you assume that mains power has failed. Alternatively, if you look at the digital source, and see that its display isn't counting, despite you having pressed "play", you conclude that either the player or its source data are faulty, and you try another disc or file. In each instance you are breaking the system down into blocks, and checking each block. Exactly the same technique can be applied to internal electronic faultfinding.

A power amplifier example

Imagine that you switch the amplifier on to play some music, yet nothing comes out of the loudspeakers. You look at the amplifier and observe that the heaters are glowing. You have just eliminated the mains lead and associated fuse. Next, it could be that the amplifier simply isn't receiving any signal, or that the loudspeakers have become disconnected. Of course, the loudspeakers could be faulty, but most loudspeakers contain two drive units, and possibly more, so for both loudspeakers to be faulty, you require four simultaneous failures. That just isn't likely, even if you had a **very** good party the night before. Similarly, although the leads to both loudspeakers could both be faulty, it's unlikely. It is far more likely that we are looking for a single fault that affects both channels.

We ought to check that a signal is reaching the amplifier. We could select a different source on the pre-amplifier, or if a source has an integral volume control (and some do), we could plug it directly into the power amplifier, and increase source volume gently from zero.

Assuming that there is still no sound, we know that we are looking for something that is common to both channels, and that's usually the power supply. We know that the heaters work, so that narrows the investigation to the high voltage supply. There haven't been any nasty smells, explosions, or fizzing noises, so something has died quietly and completely. If a component was merely poorly, the voltage would be low, but probably not zero, and the amplifier would have produced some sound, although distorted.

All power rectification is full-wave, requiring either two or four diodes, so it is unlikely that two or more diodes have failed simultaneously, unless it's a valve rectifier, where the common cause for two diode failures would be the heater, but we already checked that all the heaters were glowing. We can lose voltage either because something in series has failed open circuit, or because something in parallel has failed short circuit. If a reservoir or smoothing capacitor had failed short circuit, it would probably announce its failure with some noise (explosion, fizzing, distorted sound plus hum from the loudspeakers). It is far more likely that a series resistor or choke has failed open circuit.

Now that we know what we are looking for, we can switch off, unplug the amplifier from the mains (leaving the plug in plain view), take the covers off and (carefully) investigate. There could still be charged capacitors, so it's worth using the voltage range of your DVM to check that they are discharged, but be careful when probing inside even a notionally unpowered amplifier, as more than one piece of equipment has been destroyed by the slip of a probe. If capacitors retain charge, a 10 kΩ wirewound resistor is a handy way to discharge them safely. See Figure 5.3.

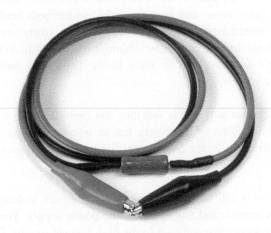

Figure 5.3
A 10 k 6 W resistor with leads and insulated crocodile clips is very handy for safely discharging capacitors.

Quite apart from the fact that a capacitor's residual charge could give you a nasty surprise, it would interfere with any attempt at resistance measurement by your meter. Bear in mind that electrolytic capacitors suffer badly from dielectric absorption, so leave your meter monitoring their voltage whilst discharging. You must pull the capacitor's voltage to well below 1 V, and this can easily take 20 seconds. Assuming that all the capacitors are discharged, we can use the resistance range to check the choke and series resistors in the high voltage supply, and find the problem quickly.

A pre-amplifier example

When the RIAA pre-amplifier is selected, one channel is very low level and distorted. Bear in mind that the cartridge or associated pick-up arm wiring could have failed. One obvious test is to set your DVM to its resistance range and check continuity from tip to sleeve of each phono plug, but momentarily passing DC through a moving magnet cartridge just might magnetise its core and increase distortion, so a risk-free alternative is simply to swap the phono plugs over and see if the fault swaps channels. If it does, the fault is in the arm or cartridge; if it doesn't, the fault is in the pre-amplifier.

If, like the author, you have chosen to use a DIN plug as your connection from pick-up arm to pre-amplifier, you can't easily do this swap (although you could make up a short adapter lead for swapping channels). There are other ways of testing each channel of the pre-amplifier:

- Tap the input valve of each channel gently and listen to each loudspeaker. Since valves are invariably slightly microphonic, the thump/ting should be equally loud from each loudspeaker.
- Turn up the volume fully, and listen for hiss on each channel. If one channel is significantly quieter than the other, it suggests that the signal from the first stage is not being amplified. The noise should be a clean hiss. Uneven noise suggests a faulty connection.
- With the volume turned fairly well down, put your finger on each input of the amplifier. This should cause a loud hum. (If you do this test at the cartridge pins, beware that the loud hum doesn't make you jump and hit the stylus.)

Having established that the pre-amplifier is genuinely at fault, we know that it is unlikely to be power supplies (unless built as dual mono) because one channel works, so it's time to look at a circuit diagram. See Figure 5.4.

Looking at the circuit, we see that there is a lot of fragile silicon (is there any other kind?). We ought to first work out what the circuit is doing. The first stage is pretty conventional, although the LED cathode bias is a useful indicator that shows whether the stage is passing cathode current. The first stage is followed by conventional 75 μs passive equalisation, although the 12 k resistor in series with the 270 pF capacitor indicates that 3.18 μs has also been implemented. The second stage also has LED cathode bias, and has a cascode constant current load (to minimise distortion), and this is direct coupled to a cathode follower. The cathode follower has a simple constant current load and provides a

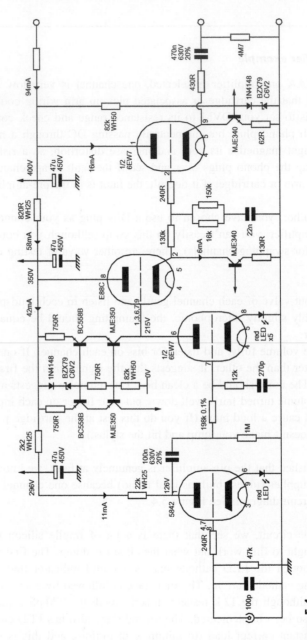

Figure 5.4
RIAA stage with active loads.

low (and unchanging) output resistance to drive the 3180 μs/318 μs passive equalisation that is direct coupled to the output cathode follower.

The quickest way to find the fault in this circuit would be to start at the input valve, and measure the DC at the output of each valve with the circuit fully powered. We should be careful not to slip with the probes and create extra faults!

The input valve is a conventional common cathode, so even if we didn't know that the design voltage is 126 V, we could check to see that it is somewhere between ⅓HT and ⅔HT. When faultfinding, we really don't quibble about precise voltages, we look for things to be "roughly right" or "really wrong".

Although the anode circuit of the second stage looks complex, we can ignore the silicon for the moment, and just check the anode voltage using the same ⅓HT to ⅔HT criterion. Because it's so very easy to slip with a DVM probe, it's safer to measure the anode voltage on the valve socket rather than on the collector of the MJE350 (where you could slip and short to either of the adjacent pins). Because the third stage is a cathode follower, its output is on its cathode, and for any cathode follower DC coupled from the preceding anode, its cathode should be a few volts higher than the preceding anode voltage. The fourth stage is also a DC coupled cathode follower, so its voltage should be a few volts higher than the preceding cathode follower's cathode voltage.

Perhaps when we measure, we find that the anode of the second valve is at 30 V. This is too low, and indicates a fault. We also notice that the LEDs in the cathode circuit aren't glowing, and this suggests that the valve is not passing any significant current. It looks as though we need to investigate the silicon. If the valve is not passing current, the fault is far more likely to be in the constant current load, so we need to check that the constant-current load is being told to do the right thing.

When working with transistors, we generally make the sweeping assumption that the base draws zero current. This assumption hugely simplifies faultfinding because it means that we can predict the voltages looking down the bias chain formed by the 1N4148 diode, BZX79 C6V2 Zener diode, 750 Ω resistor, and 82 kΩ wirewound resistor. Because the wirewound resistor is nice and big, it is easy to touch with the DVM's probe. Togther, we would expect the two diodes to drop 6.9 V, and in comparison with the 82 kΩ, the 750 Ω won't drop much, so we should expect the voltage across the 82 kΩ resistor to be about 10 V less than the HT.

Having measured the drop across the 82 kΩ resistor, and found that it looks roughly correct, that suggests that the bias chain is correct. We now need to check the transistors. The easiest way to check a transistor is to check the voltage drop across its base–emitter junction, which should be ≈ 0.7 V. It's very tricky to connect two probes directly to a transistor safely, so we find other, larger, points that are connected to the base and emitter to check the transistors. They turn out to be correct, so we deduce that the 750 Ω current programming resistor has failed open circuit. Just to be certain, we switch the power off, leaving a meter monitoring the HT, and using a DVM that is guaranteed not to switch diodes on, measure the 750 Ω resistor before removing and replacing it. Unfortunately, when we measure, we find that the 750 Ω resistor is innocent.

Despite having confidently predicted that the silicon circuitry would be at fault, it seems blameless. If it isn't faulty, then it must be sourcing a current down to 0 V, and if the current isn't going through the valve, it must be going somewhere else. The next check we could make is to measure the voltage across the 750 Ω current programming resistor. Between them, the Zener and 1N4148 diodes drop 6.9 V, and this voltage is across the base–emitter junction of the transistor plus 750 Ω resistor. The base–emitter junction drops 0.7 V, so we should expect to see 6.2 V across the 750 Ω resistor, exactly the same as the Zener voltage. We measure the drop, and it is correct. There's no longer any doubt about it, the constant current load is working correctly, and is delivering 6.2 V/750 Ω = 8.27 mA; it's just not going where it should.

We now need to look very carefully, perhaps with a magnifying glass and bright single-chip LED torch, to see if there are any whiskers of wire lurking on the second valve's socket, and because it is DC coupled to the third valve, we need to check that too. Cleaning the bases using a hog's hair brush whilst vacuuming it is often a good idea. Unfortunately, even this doesn't clear the fault.

Power transistors have their collectors connected to their case, so when we bolt them to a heatsink, we have to use an insulating kit. The MJE350 dissipates 1.1 W, which is more than it can comfortably dissipate without a heatsink, so it was screwed to the chassis. (This is not ideal because it adds $\approx 6-8$ pF to the output capacitance of the constant current source.) We unscrew the transistor, lift it clear, and apply power for just long enough to see the five LEDs light up and the anode voltage of the second stage rise to 180 V. The insulating washer was faulty!

Although it is satisfying to find the fault, we must think a little further. What caused the fault? Did it fall, or was it pushed? There's no point in replacing a component only to have it fail again three months later. We need to know what caused the failure. The insulating washer **might** have been of faulty manufacture, but the far more likely reason is that it has been pulled against a hole that was not deburred fully. Check the hole on the chassis and the transistor very carefully. Do not reuse the fixing screw — it might have had a sharp burr that damaged the washer.

Unusually, because of the capacitance problem, the best repair would be to fit a small heatsink to the transistor and lose the heat to the air, rather than to the chassis.

These two examples demonstrated that a little thought at the scene of the crime can take you to the guilty component, and that no special test equipment is needed other than a willingness to observe clues and think about what they are telling you. The second example, in particular, demonstrates that we need to think about currents and where they flow. Current doesn't just flow into somewhere and disappear — if it did, there would be untidy heaps of electrons everywhere.

Hum

Mains hum is one of the more common problems to afflict a new project, and because there are so many causes, it can be awkward to diagnose. There are various ways that hum can find its way into an amplifier:

- Directly injected from a power supply: Poor smoothing or excessive current draw.
- Poor heater/cathode insulation within a valve: Not always detected by a valve tester's hot R_{hk} test, so best detected by valve substitution. Replacing the faulty valve is the only cure, although operating the heater at 150% voltage for 10 s might burn off lint bridging heater to cathode and you have nothing to lose.
- Electrostatic pick-up: Unwanted capacitance between signal and power wiring requiring increased separation or an interposed earthed conductive screen.

- Electromagnetic pick-up: Unwanted mutual inductance between signal and power wiring that can be reduced by increasing separation between the two coils (one of which may be a simple wire) or rotating one coil.
- Hum loop: Check for excess signal connections to the chassis.
- Poor earth contact to the chassis: Undo the wanted contact to the chassis, clean the contact surfaces, reassemble, tighten with spanner. Check the soldered joint.

Assuming that hum has been discovered, there are some quick checks that can be made before dragging out the oscilloscope. Does the hum disappear the instant that you switch the amplifier off, or does it gently fade away? If it disappears instantly, it is from the amplifier's own power supply, either electromagnetic pick-up (invariably 50/60 Hz) or ripple due to rectification/smoothing (invariably 100/120 Hz). If it gently fades away, the hum is at the input of the amplifier, and could be electrostatic pick-up, a hum loop, or it could already be present on the signal entering the amplifier.

Does the volume control affect how loud the hum is? If it does, then the hum is probably being injected before the volume control — although as all volume controls contain a shunt element, electrostatic pick-up on the volume control's output wiring remains possible. Is the hum only present when one particular input is selected? If so, unplug that input and if the hum goes away, you are looking for hum on that particular source equipment. Does the hum change when you touch the chassis or a lead? If so, you have electrostatic pick-up due to a failed earth connection. Does the hum change when you move a lead? Leads might be screened, but if you have a moving coil cartridge, and trail pick-up arm leads across a mains transformer, you can expect electromagnetic pick-up. Is the hum dependent on which pieces of equipment are plugged into the mains? If so, this suggests a hum loop. Unplug all the audio and mains leads, and for each piece of equipment, check continuity ($<0.5\,\Omega$) between the earth pin of the mains plug and the body of its phono sockets. In the unlikely event that two pieces of equipment have continuity between mains earth and signal earth, you have a hum loop, and it may be necessary to break the bond between the chassis and 0 V signal earth within one piece of equipment.

Assuming that you have pinned the hum problem down to one piece of equipment, it is now time to switch on the oscilloscope. Trigger the oscilloscope from "line" so that it is always triggered as you poke around looking for hum, then touch the probe tip with your finger to check that the oscilloscope genuinely is triggered, and that it is ready and able to detect hum. See Figure 5.5.

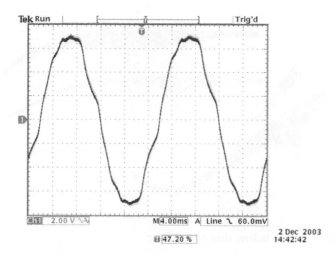

Figure 5.5
Typical hum waveform obtained by touching probe tip with finger.

A power amplifier example

Since the patient is a power amplifier, the oscilloscope may be able to show the hum at the output, so it's worth looking, because the shape of the hum gives a clue as to its origin. Bear in mind whilst investigating that you are unlikely to find squeaky clean waveforms, they will always be messy — either covered in noise, or distorted. If you find something approximating to a sawtooth waveform at the output of the amplifier, you have power supply ripple from somewhere near a reservoir capacitor, and it's likely that the hum is being injected directly into the output stage. The next step is to look for hum on the HT supply, but this needs to be done very carefully.

Switching the input coupling of the oscilloscope to AC will reject the DC, allowing the sensitivity of the oscilloscope to be increased until the hum is clearly visible, but the full HT is still being applied to the probe. Some ×10 probes can only withstand 200 V, and the HT at a power amplifier output stage is likely to be >300 V. In addition to your ×10 probes, you also need a ×100 probe that is rated for high voltages. Bought new, high voltage oscilloscope probes are expensive, but you may well be able to find a second-hand probe for a more reasonable price. If you're rummaging through a box full of old probes, bear in mind that high voltage probes tend to be bulkier than normal probes. See Figure 5.6.

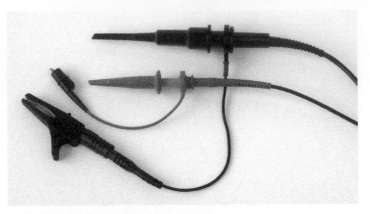

Figure 5.6
High voltage probes tend to be bulkier than normal ones.

Assuming that you have a safe means of looking at the HT feeding the output transformer, $<1\,V_{pk-pk}$ is an acceptable ripple amplitude for a push–pull amplifier, but a single-ended output stage is unable to reject power supply ripple, so $<30\,mV_{pk-pk}$ ripple is preferable. These rough guides are appropriate for conventional loudspeakers, but high-efficiency loudspeakers such as horns are less tolerant, so $<100\,mV_{pk-pk}$ (push–pull) and $<3\,mV_{pk-pk}$ (single-ended) might be more appropriate. The Quad II has a trap for the unwary in that it has substantial ripple (typically $>60\,V_{pk-pk}$) at the HT feeding the output transformer, but smooths the HT to the screen grids and relies on tetrode action to reject ripple at the anodes.

Assuming that there is excessive ripple on the HT feeding the output stage, something must be done about it. If it is a new fault on an old amplifier, then an electrolytic capacitor has probably dried out, and it must be replaced by another of similar value and the same, or higher, voltage rating. If it is a new amplifier, a component fault is unlikely, and the design needs to be changed, perhaps by adding a stage of LC smoothing.

Beware that more powerful or sophisticated amplifiers are likely to grid bias the output stage, so hum could be injected directly into output valve grids due to excess ripple on the bias supply.

A valve microphone example

It's not unusual for hum to come from a variety of sources. Condenser microphones require an amplifier behind the capsule that not only has high input impedance, but amplifies a very small signal, so eliminating hum is quite a problem. An early 1960s valve microphone had hum, and although replacing connectors/cables and attending to earth bonds substantially reduced the hum, it did not eliminate it, so attention turned to the HT supply. See Figure 5.7.

The supply is a conventional bridge rectifier feeding a reservoir capacitor followed by resistor/capacitor smoothing and a neon regulator valve. When the hum at the reservoir capacitor was investigated, instead of it being an even sawtooth, it had alternating large and small teeth, suggesting that one path of the bridge rectifier was less able to charge the capacitor than the other. See Figure 5.8.

Replacing the bridge rectifier would restore an even sawtooth, but only make a very minor difference to the hum, so the hum on the next capacitor down the chain was investigated. The hum on this capacitor should have been an almost pure 100 Hz sine wave, yet noise and spikes were present. This capacitor was simply not doing its job. See Figure 5.9.

Given that the power supply was 40 years old at the time, it seemed likely that if one electrolytic capacitor was faulty, then they all were, and if they weren't, then they soon would be. Replacing the bridge rectifier and **all** the capacitors made sense.

However, the nastiest fault came to light as the power supply was being gutted of its (chassis-mounting) capacitors. Instead of having positive and negative solder tags, each capacitor had a single positive solder tag, and the negative connection was made simply by pressing the aluminium capacitor can onto the aluminium chassis. Because there wasn't a star washer between the can and the chassis, the mechanical joint was not gas-tight and developing aluminium oxide gradually increased the resistance over the years, so much of the hum was due to this increased resistance. This is a very poor construction technique, and should be replaced on sight. See Figure 5.10.

Modern capacitors are so much smaller that the entire replacement circuit was built on a small piece of strip board, and this cured the hum. Moral: Sometimes there are so many small faults that nothing less than a comprehensive rebuild can effect a full cure.

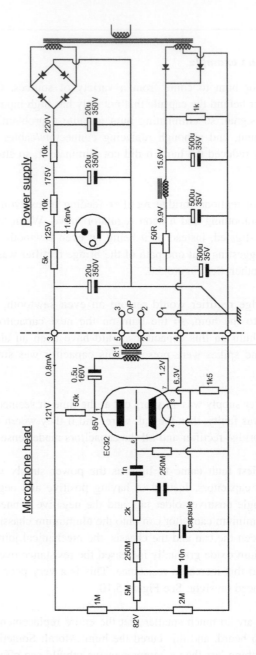

Figure 5.7
The power supply for this valve microphone employs extensive filtering.

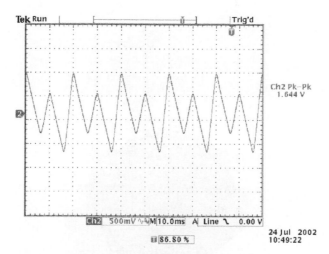

Figure 5.8
The unequal amplitude teeth on this reservoir capacitor ripple waveform were caused by faulty rectification.

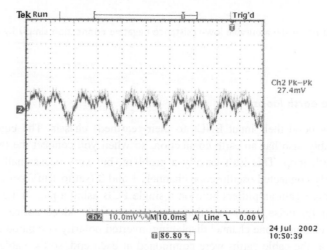

Figure 5.9
The hum on this capacitor **ought** to be almost pure 100 Hz, without noise spikes.

Figure 5.10
This capacitor could not make a durable low-resistance negative connection simply by contacting the chassis!.

Oscilloscope earth loops

Oscilloscopes bond their input BNCs to their (earthed) chassis. The equipment you are testing probably also has an earth connection, so when you connect the two together, you create an earth loop. This isn't usually a problem, but very occasionally it can be. The author recently connected oscilloscope channels 1 and 2 (set to 1 mV per division) directly to a pair of high-gain amplifiers built on a single PCB using a pair of 1.2 m BNC coaxial cables to test for noise, but found that moving one cable added noticeable hum to both oscilloscope channels, one channel displaying inverted polarity compared to the other. The cause was that the cable earths were commoned at each end, so the cables formed a hum loop and moving one cable changed the loop area, thereby changing the induced hum current. The current circulated round the loop, so from the point of view of the oscilloscope, it entered at one channel and exited at the other, causing polarity inversion of the resulting voltage drop along the cables. The hum loop couldn't be eliminated, so the cure was to minimise its loop area by passing both cables down nylon braid, thereby maintaining minimum loop area even when the cables were moved. This is the first time the author has eliminated hum by the addition of **nylon** braid! Perhaps more significantly, it suggests that there is a good case for multi-channel phono-to-phono connections (which invariably create hum loops) from a source to its destination to have their coaxial cables tightly bound together rather than leaving individual cables hanging loose.

Because an oscilloscope invariably adds earth loops, it's tempting to leave its earth clips dangling. Quite apart from the fact that this inevitably means that a clip will brush across (and destroy) circuitry, it always causes interference. Accept that you will create an earth loop and minimise its effect by ensuring that the oscilloscope's earth clips have a good (low-resistance) contact to the chassis or 0 V signal earth under test and close to the signal being probed.

Oscillation

There are various causes of oscillation, and each produces a characteristic frequency:

- Incorrect polarity of global feedback in power amplifiers tends to produce oscillation between 1 and 20 kHz.
- Excessive global feedback around a power amplifier with poor compensation and/or poor output transformers is most likely to occur between 30 and 300 kHz, but 1 MHz is possible.
- Oscillation in individual stages tends to be at radio frequencies, and could be anywhere between 1 and 100 MHz.
- Motorboating is due to feedback around a number of stages via the common power supply, and tends to occur at 1 Hz.

Global feedback and power oscillators

The simplest fault is that a new amplifier has been built, but when the global negative feedback loop is connected, the amplifier turns into an oscillator. If, as is usual, the negative loudspeaker terminal of the amplifier is connected to 0 V, then an amplifier having series applied negative feedback forces the other terminal to become the non-inverted output. This means that in order to maintain correct polarity, oscillation in a push–pull amplifier is best cured by swapping the signals to the output grids from the driver stage. See Figure 5.11.

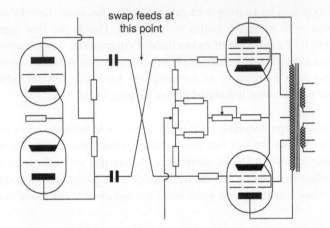

Figure 5.11
Correcting polarity in a push–pull amplifier is most easily done by swapping over feeds at the output of the driver.

Single-ended output transformers identify their anode and HT tails because incorrect connection increases shunt capacitance, so the only cure for oscillation is to transpose the secondary tails. See Figure 5.12.

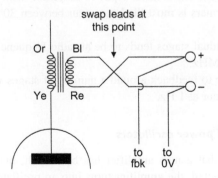

Figure 5.12
Correcting polarity in a single-ended amplifier can only be done by swapping over the output transformer secondary.

In theory, we always know the transformer phasing, and we know which way they should be connected, so the previous problems never occur. In practice, the author finds it quicker to wire one channel of a stereo amplifier to one configuration and the other channel to the opposite polarity. This ensures that one channel is wrong and one right, so the incorrect channel is easily identified and corrected even if there are other stability issues.

Some amplifiers apply local negative feedback by including an output transformer secondary in the cathode circuit of their output valves. Practice shows that it is quicker simply to wire the cathode feedback winding one way round, apply a 1 kHz signal from an oscillator, and monitor the output amplitude on an oscilloscope. Then, without changing signal settings, swap the connection of the cathode feedback winding. The connection that produces the lowest output voltage (and no oscillation) is the correct one.

Global feedback and compensation

The amount of global feedback that an amplifier will tolerate is governed primarily by its output transformer, so a new transformer requires new compensation. For various reasons [2] it is the interaction between the input stage and output transformer that determines the stability of an amplifier when global feedback is applied, so most amplifiers conform to a pattern:

- The input stage's anode load R_L is shunted by a capacitor C_1, which may be in series with a resistor R_1.
- The global feedback resistor (R_{fbk}) is bypassed with a capacitor C_2, which may be in series with a resistor R_2.

We can therefore draw a generic diagram that shows possibilities for adjusting compensation once R_{fbk} has been set to give the required gain. See Figure 5.13.

The way to determine the optimum values is to fit variable resistors and capacitors in the appropriate positions, apply a 10 kHz square wave to the input of the amplifier, monitor the output across a dummy load made from power thick-film resistors, and adjust the compensation components until the cleanest, sharpest square wave results. Two variable capacitors are needed, and these are best obtained from radio fairs. Any dual-gang variable capacitor that could tune a medium wave radio will do,

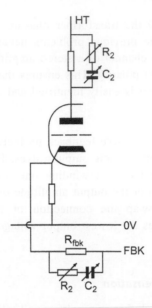

Figure 5.13
Global negative feedback usually requires additional compensation components.

because they are typically 50−500 pF (best) or 30−365 pF (smaller, and slightly newer). Mica compression trimmers are often fitted across the main capacitor, so remove them to reduce the minimum capacitance and maximise the range. The author has needed this test jig sufficiently frequently that it became worthwhile to make a dedicated version in a plastic lunchbox (thus insulating all the high voltages). See Figure 5.14.

Note that the variable capacitors have scales calibrated in E24 values so that the effect of using the nearest standard capacitance value can be easily seen. The resistance controls are uncalibrated as it is assumed that any required resistance is readily available.

- The value of C_2 is critical. Too little causes the amplifier to oscillate, too much doesn't significantly reduce ringing amplitude, but rounds the leading edge about which the ringing occurs.

Figure 5.14
This dedicated measurement jig enables fast determination of global feedback compensation values.

- R_2 may need to be zero to prevent oscillation. A good starting point is $R_{fbk}/10$.
- The value of R_1 is critical for damping the ringing – too little causes overshoot. A good starting point is $R_L/10$.
- The value of C_1 is not critical, but too high a value rounds the square wave, and too little causes overshoot at the leading edge. Don't touch the body or shaft of the variable capacitor while the amplifier is powered – note that the author's box has fully insulated knobs.

The quickest way to find the optimum settings is to start without C_1 and R_1 connected. Adjust C_2 to give minimum ringing concomitant with the ringing decaying exponentially and being superimposed on a sharp leading edge, rather than a curving edge. (You will be relieved to learn that this condition takes less time to identify than to describe!)

Switch off the amplifier and unplug it from the mains, leaving the plug in clear sight. Connect C_1 and R_1. Apply power to the amplifier, and adjust C_1 and R_1 simultaneously for minimum exponentially decaying ringing on a sharp leading edge. A minor adjustment of C_2 may be necessary. Strive for an optimum setting of all components with a minimum setting of C_2.

When the input stage is single-ended, symmetry between the positive and negative edges can never be achieved because the input valve's non-linear anode characteristics mean that its output resistance is slightly higher swinging positive than negative. However, if the input stage is a differential pair, symmetry **is** possible, provided that the compensation network is connected between the anodes. See Figure 5.15.

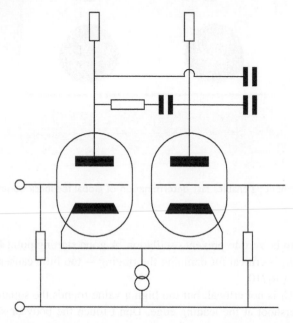

Figure 5.15
A differential pair input stage should connect the compensation components between the anodes.

For a differential network, start with $R_1 = R_L/5$, and use only one section of the variable capacitor.

Having carefully set the optimum values, dab a 220 nF (or similar) capacitor across the output of the amplifier whilst observing the 10 kHz square wave. Note that adding the capacitor ruins your carefully optimised square wave response — this implies two things:

• There's no point in struggling to fit precise values of resistors and capacitors in the compensation networks. They are a compromise. This is a **good** thing, because it means that the nearest standard values in your stock will probably do.

- It would be a very good idea if you knew precisely what sort of a load the amplifier was actually going to drive. And it would be even better if the load could be made resistive at high frequencies; we will investigate how this can be done in Chapter 6.

Once you have determined the optimum settings, switch the amplifier off, and having checked that all the capacitors are discharged, use a DVM to measure the value of the resistors and a component bridge to measure the capacitors. Substitute fixed components and confirm that the amplifier still works as expected. Optimising the compensation components doesn't take long, and the results are well worthwhile. See Figures 5.16 and 5.17.

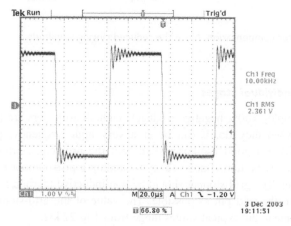

Figure 5.16
10 kHz square wave before compensation.

As an unusual commercial example, the Rogers Cadet III doesn't directly shunt its anode load but shunts V_{gk} of the second stage, which amounts to the same thing, but avoids injecting interference from the power supply. It bypasses its global feedback resistor with a capacitor, but sets both compensating resistors to 0 Ω. In addition, a small amount of neutralisation (positive feedback) has been applied from one output valve's anode to the grid of its counterpart. If the output transformer were perfectly balanced (split bobbin winding), and the phase splitter were perfectly balanced, we would expect to see a similar capacitor from the other anode. See Figure 5.18.

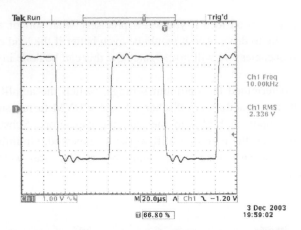

Figure 5.17
10 kHz square wave after compensation. Note the reduced ringing amplitude.

Oscillation in individual stages

Because they employ 100% negative feedback, cathode followers are particularly likely to oscillate, and when they do, it can be at a very high frequency, possibly as high as 100 MHz, requiring a good oscilloscope to spot the problem. Fortunately, cathode followers can generally be tamed quite easily by adding a grid-stopper resistor, and possibly a cathode-stopper. The grid-stopper should be non-inductive (carbon film is ideal) and soldered as close to the grid pin as possible. The value of the grid-stopper resistor must be found by experiment, but typical values range from 1 to 22 kΩ.

If needed, the purpose of a cathode-stopper resistor is to buffer a capacitive load from the (slightly inductive) output impedance of the cathode follower. One of the cathode follower's virtues is its low output impedance, so it seems a shame to raise it by adding series resistance. Fortunately, cathode-stopper resistors can usually be quite low value, typically 47–470 Ω.

Another cause of local oscillation is poor HT decoupling. Adding a 100 nF capacitor directly between the top of the anode load (if present, otherwise, the anode pin) and the bottom of the cathode resistor may help. See Figure 5.19.

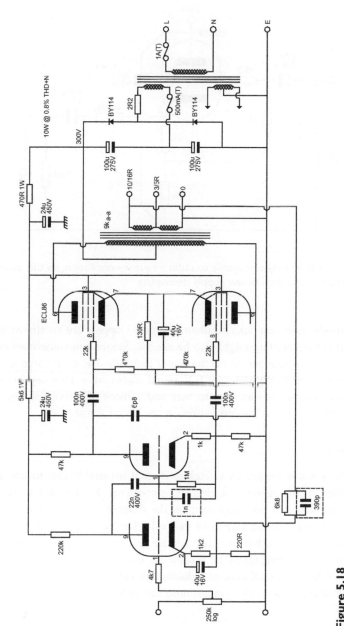

Figure 5.18

Rogers Cadet III power amplifier.

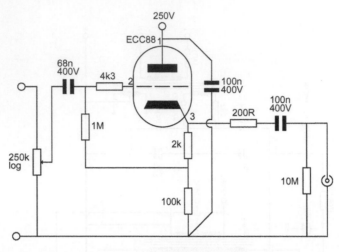

Figure 5.19
Cathode followers are particularly susceptible to radio frequency oscillation, so this example has a grid-stopper, cathode-stopper, and local power supply decoupling.

Common cathode stages can also oscillate, and grid-stoppers are an almost universal panacea, but local HT decoupling might also be needed. Stages can sometimes couple together and oscillate at radio frequencies via the heater path. Ideally, each valve should have each heater pin decoupled to **chassis** (not 0 V signal earth) by a 10 n capacitor having short leads, but this tends to be awkward to wire and a more effective solution can be to use individual heater wiring back to decoupling at the heater transformer.

Motorboating

Motorboating is a low frequency (1 Hz) oscillation invariably due to unwanted coupling within a power supply using cascaded RC smoothing. Experiment to see if increasing smoothing capacitance at one stage changes the frequency. If you can change something, then you must be near to the source of the fault. The best cure is to reduce source resistance. A regulator is ideal, but reducing a series resistor might work. Alternatively, reducing smoothing capacitance might be acceptable, depending on hum. If all else fails, you could resort to the traditional cure of reducing the value of audio coupling capacitors, but this is not recommended — why compromise the audio design because of a power supply problem?

Noise and crackles

Noise and crackles are due to an intermittent conduction path, and the most likely cause is either a dry soldered joint, or a dirty contact in a connector or switch.

Because the fault is caused by intermittent contact, tapping soldered joints with a plastic (insulating) pen can be effective, but an even better alternative is to poke the joint with a sharp (insulated) probe in the manner of an unsympathetic dentist. See Figure 5.20.

Figure 5.20
A sharp (insulated) probe can be useful for finding dry joints.

If the amplifier is old, the fingers in the valve socket may be making poor contact with the valve pins, so try moving each valve gently in its socket. The real cure is to replace the socket, but removing the valve and squirting contact cleaner into the socket (with power off) may effect a temporary cure. Alternatively, using a small probe to tighten the fingers may be a longer-lasting solution.

Although DIN and XLR connectors have a wiping action when the plug is inserted (unlike phonos), the silver plating on the better quality connectors can oxidise when a plug or socket is left unmated. The pins in a plug can easily be recovered by sliding the body back to expose the pins and dipping them in Goddard's Silver Dip until they are bright and shiny. Sockets are harder to clean, and it may be easier simply to replace them.

Luck also plays a part. The author spotted a bulge on the base of an electrolytic HT capacitor, and immediately knew that it should be replaced. Unfortunately, one of the wires just wouldn't desolder, and closer inspection revealed that the joint had **never** been soldered! When bought at a market, the amplifier was fraudulently described as "works well" — but with a production fault like that it could never have worked well in its life, and the unsoldered joint explained the odd crackles heard during preliminary testing. See Figure 5.21.

Figure 5.21
Not only is this electrolytic faulty (note the bulge and cracks adjacent to the tag) but its earth wire was never soldered!.

Electrolytic capacitors fail when their electrolyte evaporates, causing poor and possibly intermittent contact to one of the capacitor plates. Old electrolytic capacitors are best replaced en masse, rather than searching to find the noisy offender from a bunch of equally likely suspects.

Intermittent faults

These are the worst to diagnose. They are usually mechanical, so poking around with a sharp insulated probe can be useful. A traditional test for **semiconductor** electronics was to heat it with a hairdryer to produce the fault, then selectively cool parts with aerosol freezer. This is expensive in an environmentally unfriendly freezer, so this technique should only be used as a last resort after probing or thumping with a screwdriver handle. Squirting freezer spray near a hot valve is likely to crack the envelope.

Test mules

It can be seen that faultfinding takes time and the implicit assumption is that the electronic design was correct, but this may not always be the case. Semiconductor designers wallow in the luxury of SPICE simulation, confident in the knowledge that semiconductors rely on fundamental physics that can be reliably simulated, but valves rely primarily on physical dimensions that are subject to production tolerances and are not so reliably simulated — a genuine test is needed.

Rather than construct a work of art and discover that its electronic design is flawed, it is useful to test a prototype of a particularly adventurous design beforehand to prove the principle and determine component sensitivities. Such prototypes are often known as test mules, and whilst they are rarely things of beauty, they save tears later on. See Figure 5.22.

Figure 5.22
This test mule is pig-ugly and the flimsy folded chassis sags under the weight of its transformers but it enables quick testing.

The author's "Plug and Pray" output stage test mule includes Crystal Palace driver circuitry because it can accommodate any output stage, plus a regulated (and therefore adjustable) HT supply. Various output sockets were fitted to enable different valves to be quickly plugged in but there is no protection of any sort (hence the mule's name). No permanent output transformer was fitted as this changes with test. Output stage current and balance are provided as front-panel controls to allow easy adjustment (you wouldn't dream of providing such dangerous controls on a finished amplifier), and output stage current monitoring is also available on the front panel as the voltage dropped across individual 1 Ω resistors. The input is balanced to allow easier interfacing to the author's audio test set.

This formal test mule took considerable time to make, but was deemed worthwhile because of the expected number of uses. Conversely, you might want to quickly try something a little less formal. See Figure 5.23.

Figure 5.23
A quick rat's nest can yield valuable data.

This circuit is a differential pair having a low output capacitance cascode CCS and was built primarily to test distortion against quiescent current, hence the adjustable programming resistor in the CCS. The circuit took longer to test than it took to construct (yes, it shows) and adjustment of programming current and HT voltage yielded valuable data that allowed confident construction of the later project. Obviously, high voltage rat's nest construction requires a certain degree of caution.

Classic amplifiers: Comments

The following remarks relate to the author's personal experience of only a few samples of each amplifier, but the comments are included because some guidance is better than none. Various amplifiers, such as Radford and Dynaco, are not included, not because the author has any bias against them, but simply because he has not been fortunate enough to own one.

As a very broad generalisation, classic amplifiers using more expensive output valves are likely to be better simply because their designer had to justify the increased cost. Amplifiers using KT66 may be better than EL34, which will be more powerful than EL84, and ECL82 or ECL86 are at the bottom of the heap. Curiously, amplifiers using KT88 **may** be worse than any of these, because they may have been designed as public address amplifiers, purely for their high output power.

The quality of output transformers is crucial. Poor output transformers are likely to be small for their rating, although C-core transformers may be an exception to this rule, and generally imply that good quality was intended.

Quad II

There are still an awful lot of these about, and they generally require very little work to restore them to their original performance. Typically, the $180\,\Omega$ 3W cathode resistor and its associated $25\,\mu$F capacitor need to be replaced. Quad IIs are popular with tweakers, so there are various modifications. Mostly the modifications replace the GZ34 with silicon to increase output power, and others replace the GEC KT66 with EL34 because NOS GEC KT66 valves are fearfully expensive, although current production KT66 is said to be satisfactory.

Williamson

These were most frequently made by amateurs, so finding a matching pair is tricky, and build quality may be less than wonderful. However, at the very least, the output transformers are well worth salvaging. In the UK, Williamsons used KT66, but some American and Australian variants used the somewhat less linear (but much cheaper) 807.

Leak TL12 and BBC LSM/8 derivative

This amplifier has an output stage very similar to the Williamson (triode-strapped KT66), and has become very fashionable (read expensive), but they are likely to be in quite poor condition because of their age.

The BBC LSM/8 lived in a compartment at the bottom of the LSU/10 loudspeaker (where it became very hot). Because it was intended for studio monitoring, the amplifier has a transformer balanced input and a volume control. In common with many BBC loudspeaker amplifiers, some versions included a bass equaliser. Once these input modifications are removed, the two amplifiers are identical.

BBC amplifiers

Oddly, many BBC amplifiers were designed for 25 Ω loudspeakers, and cannot be modified for 8 Ω without replacing the output transformer. It is also well worth asking why the BBC disposed of the amplifier, particularly if it is thought to have come from an impoverished local radio station. Under the latter circumstances, it is most unlikely that it spent its time cherished in a protective box in a dry cupboard.

Leak TL10

This is similar in design to the Mullard 5-20, but uses a 6SN7 phase splitter. It is quite a nice conservatively rated amplifier designed for KT61 (tricky to find), but can be modified for EL34 or 6L6. Although many TL10s were made, few seem to have survived.

Leak TL12+

Despite its name, this is a completely different beast from the TL12, and is very similar to a Mullard 5-10 using EL84 output valves. They are comparatively recent (perhaps only 45 years old), but by modern standards they are noisy, owing to high sensitivity and the EF86 input pentode.

Leak Stereo 20

Far more common than the TL12+, this is almost a pair of TL12+ on one chassis but sharing a slightly under-rated mains transformer, and with a few other corners cut. Input valve is the dual triode ECC83, but they are still noisy because of their excessive sensitivity. Thanks to the continuous demand by musicians (guitar amplifiers), modern EL84s are cheap and plentiful, but NOS Mullard and Siemens EL84s are rare and therefore expensive.

All Leak amplifiers are ridiculously over-sensitive; 110 mV for full power (Stereo 20) is far too sensitive, and causes all sorts of hum and noise problems.

Rogers Cadet III

Rated at 10 W using ECL86 output valves. There were two versions, an integrated version and a separate chassis version. These are ideal for beginners to cut their teeth on, but not of great intrinsic value, being let down by their tiny output transformers. Beware that further rather nasty cost-cutting means that cathode bias resistors and decoupling capacitors were shared between the stereo channels, so poor stereo crosstalk is probably due to a failed electrolytic capacitor! Bizarrely, the power supply uses a voltage doubler, and because this imposes high ripple currents, the two doubler capacitors are highly likely to need replacement. Oddly, the disc input stage was very good for its time, but the pre-amplifier section needs three ECC807s ($\mu = 140$, and irreplaceable), so if you are considering buying one, you **must** check that the amplifier has its full complement of undamaged ECC807s.

Westrex 2331-A

Westrex was the UK subsidiary of Western Electric and their (ex-cinema) amplifiers turn up from time to time. The (mono) 2331-A was rated at 60 W using two parallel pairs of cathode biased push−pull EL34 and there was a 100 W fixed bias variant employing a bigger power supply and output transformer. The Class B 100 W variant needed an excellent output transformer having very low leakage inductance to minimise switching spikes, whereas the Class AB 60 W variant could tolerate a lesser transformer that was merely very good. The driver circuitry is fairly dreadful, although the push−pull feedback from output valve anodes is interesting, and the motorboating cure that wastes current in an HT potential divider laughable. See Figure 5.24.

Arguably, the worst driver design feature is following the transformer phase splitter (excellent balance) with a ganged volume control that variably destroys push−pull balance dependent on volume setting. If you insist on restoring and using these amplifiers, remove

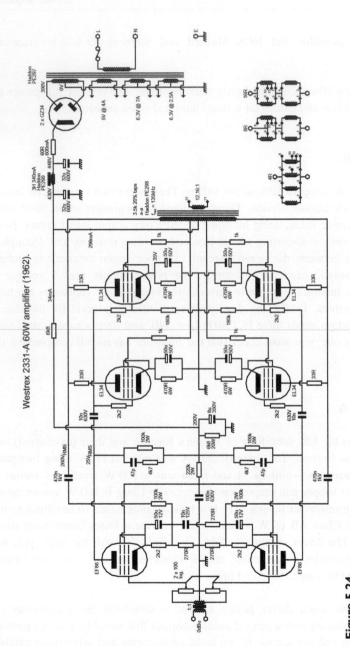

Figure 5.24

Westrex 2331-A 60 W power amplifier.

that volume control, but maintain the input transformer loading by substituting 100 kΩ 1% resistors.

Both amplifiers are built on a pig-ugly steel chassis that weighs almost as much as the transformers and chokes. If you find a pair of 2331-As, salvage the output transformers and power iron but rebuild it with driver circuitry worthy of that output transformer. If you find only one of the 100 W variant, contact this author.

References

1. IEC publication 497, Geneva (1987).
2. M. Jones, *Valve Amplifiers*, 4th edn. Newnes (2012), ISBN 978-0-08-096640-3.

CHAPTER 6

PERFORMANCE TESTING

The previous chapter enabled us to bring a piece of audio equipment to a state where it could be used safely. In this chapter, we will assume that we have a working piece of equipment, but that we want to measure precisely how well it works — we might do this for a variety of reasons:

- We hope to verify that it works as designed.
- We suspect that it can be improved, so we measure with a view to correction.
- We **know** that it works well, and want some numbers to boast about.

Linear distortions

Linear distortions do not change with amplitude and do not introduce frequencies that were not present at the input, but they cause a parameter to vary with frequency, so the most common linear distortion measurement is amplitude against frequency. By contrast, non-linear distortion always introduces new frequencies and is almost invariably level-dependent, so the most common non-linear distortion measurement is total harmonic distortion against level.

We should always measure linear distortions first because an unexpected result could force a minor circuit change affecting non-linear distortion. By contrast, is it unusual for minor

Building Valve Amplifiers. DOI: http://dx.doi.org/10.1016/B978-0-08-096638-0.00006-0
Copyright © 2014 Elsevier Ltd. All rights reserved.

changes of bias required as a result of measuring non-linear distortion to significantly change a linear distortion such as amplitude against frequency response.

Gain and attenuation

The first measurement is invariably gain measurement. We apply a signal of known amplitude to the input of the amplifier and measure the amplitude at the output. The ratio of the two is the **gain** (G):

$$\text{Gain} = \frac{V_{\text{out}}}{V_{\text{in}}}$$

The audio band is traditionally considered to range from 20 Hz to 20 kHz, so to avoid errors caused by high frequency and low frequency roll-offs, gain is measured near to the geometric mean of this band using a sine wave having a frequency of 1 kHz.

You will occasionally find that the inverse of gain, **attenuation** (A), is measured, in which case:

$$\text{Attenuation} = \frac{V_{\text{in}}}{V_{\text{out}}}$$

If signal amplitude is excessive, the amplifier's output signal will become distorted, making the measurement inaccurate. To avoid distortion, and maximise accuracy, we must apply a sufficiently small signal to the input of the amplifier to ensure that the output is undistorted. For this reason, gain is known as a **small-signal** measurement.

The whole point of making technical measurements is to provide a prediction of how well the equipment will work when used for its intended purpose, so our test level is important. Professional audio equipment has tightly defined levels, but even domestic audio now has reasonably well-defined levels. For example, the CD standard defined that a player reproducing an undistorted sine wave at the maximum level permitted by the digital code (known as 0 dBFS, or 0 dB full scale) should deliver an analogue amplitude of 2 V_{RMS}. If a player is capable of cleanly delivering 2 V_{RMS}, it is not unreasonable to expect a pre-amplifier to be able to cope with this amplitude without distortion, so we could use this as our test level.

Although it is traditional to specify sine waves in terms of V_{RMS}, and meters are calibrated in terms of V_{RMS}, gain should not change with amplitude, so there is no reason why we should not modify our test level slightly to better suit particular test equipment. Thus, if we were using an analogue oscilloscope to measure gain, we might find it more convenient to use a sine wave having an amplitude of 6 $V_{pk\text{-}pk}$ (equivalent to 2.12 V_{RMS}) because this would occupy exactly six vertical divisions at 1 V/div. (Although most oscilloscopes have eight vertical divisions, analogue display tube linearity tends to deteriorate at extreme deflections.) Conversely, if we were using a digital oscilloscope, we might use an amplitude of 4 $V_{pk\text{-}pk}$ (equivalent to 1.414 V_{RMS}) because this would occupy exactly eight vertical divisions at 0.5 V/div. (We need to fill the screen vertically to minimise ADC errors.)

It doesn't matter which form of measurement we use, just so long as we measure the input and output in the same way:

$$\frac{V_{out(RMS)}}{V_{(RMS)}} = \frac{V_{out(pk\text{-}pk)}}{V_{in(pk\text{-}pk)}} = \text{Gain}$$

Dedicated audio test sets

Dedicated audio test sets invariably express their results in terms of **decibels** or **dB**:

$$\text{Gain}_{(dB)} = 20\log\left(\frac{V_{out(RMS)}}{V_{in(RMS)}}\right)$$

It's probably some time since you covered logarithms at school, so at the risk of offending those with better memories, the equation is used by first calculating the voltage ratio, then taking the common logarithm of the result (the "log" button on your calculator), and finally multiplying the result by 20. Remember: Terms inside brackets first, then functions (logs, trig, etc.), then powers, then multiplication and division, and finally addition or subtraction.

One justification for this apparent complication is that if we have a chain of amplifiers with their gains specified in dB, we can find the total gain simply by adding all the gains (in dB) together. If we want to find the total gain when individual gains are expressed as ratios, we have to multiply all the gains together, which is somewhat harder.

Note that gain expressed in dB is still a **ratio**, albeit a logarithmic ratio. If we want to use dB to express an absolute voltage, we must choose a reference voltage, and compare our measured voltage to that reference. Many years ago, telecommunications companies sent analogue audio over 600 Ω telephone lines, and in order to express levels in terms of dB, they defined a reference level of 1 mW dissipated in 600 Ω, and called this level **0 dBm** (m for milliwatt). Power is quite difficult to measure, so their meters actually measured the voltage across a 600 Ω resistor and the meter scales were calibrated in terms of the power dissipated in that resistor.

We can find the voltage required to dissipate 1 mW in 600 Ω by rearranging $P = V^2/R$ to give:

$$V = \sqrt{PR} = \sqrt{0.001 \times 600} = \sqrt{0.6} \approx 0.775 \ V_{RMS}$$

Because this voltage is derived from power, it must be specified in terms of V_{RMS}, and this is one reason why V_{RMS} is popular for audio.

Nobody has used 600 Ω for decades, yet nonsense is solemnly talked about the need to drive 600 Ω loads, and designers of analogue audio test sets and DVMs seem unable to shed their designs of the dBm legacy, despite measuring voltage across entirely incorrect impedances. The correct modern parlance is **dBu**, which uses the common reference level of 0.775 V_{RMS}, but explicitly neglects the 600 Ω resistance requirement of dBm.

20 V_{RMS} in dBu:

$$dBu = 20\log\left(\frac{V_{RMS}}{0.775 \ V_{RMS}}\right) = 20\log\left(\frac{20}{0.775 \ V}\right) = +28.2 \ dBu$$

Note that when using dBu, it is conventional to state explicitly when the number is positive. Smaller voltages can easily produce negative values.

3.54 mV_{RMS} in dBu:

$$dBu = 20\log\left(\frac{V_{RMS}}{0.775 \ V_{RMS}}\right) = 20\log\left(\frac{0.00354}{0.775 \ V}\right) = -46.8 \ dBu$$

It is also conventional to specify AC measurements in dB to only one decimal place unless you have astonishingly accurate test equipment, because it is quite difficult to measure AC more accurately than 1%, and this corresponds to ≈ 0.1 dB.

Conversely, you might have made a measurement in dB but need to convert it back into a voltage ratio. In which case:

$$\frac{V_1}{V_2} = 10^{\left(\frac{dB}{20}\right)}$$

16 dB expressed as a voltage ratio:

$$\frac{V_1}{V_2} = 10^{\left(\frac{dB}{20}\right)} = 10^{\left(\frac{16}{20}\right)} = 6.31$$

-16 dB expressed as a voltage ratio:

$$\frac{V_1}{V_2} = 10^{\left(\frac{dB}{20}\right)} = 10^{\left(\frac{-16}{20}\right)} = 0.158$$

$+16$ dBu expressed as an absolute voltage:

$$V_{RMS} = 0.775 \; V_{RMS} \times 10^{\left(\frac{dB}{20}\right)} = 0.775 \; V_{RMS} \times 10^{\left(\frac{16}{20}\right)} = 4.89 \; V_{RMS}$$

-16 dBu expressed as an absolute voltage:

$$V_{RMS} = 0.775 \; V_{RMS} \times 10^{\left(\frac{dB}{20}\right)} = 0.775 \; V_{RMS} \times 10^{\left(\frac{-16}{20}\right)} = 123 \; mV_{RMS}$$

Note that the sign is absolutely critical, and that it is very important to be clear about whether you are using dB as a ratio or dBu as an absolute level. In addition to dBu, other audio references are in common use. See Table 6.1.

Table 6.1

Common audio reference levels

Nomenclature	Reference level
dBV	1 V_{RMS}
dBW	1 W (usually into 8 Ω, but check)
dBFS	Maximum undistorted sine wave amplitude permitted by digital code (full scale)

Beware that radio frequency engineers (and their instruments) also use dBm but refer their 1 mW to a 50 Ω load, making their voltages entirely different. Because they genuinely must terminate transmission lines correctly, their use of dBm is entirely correct.

Analogue peak programme meter (PPM) scales

Analogue PPMs were mentioned in Chapter 4, so we will briefly investigate the scale to understand the logic behind it. The analogue PPM was originated by the BBC for measuring the level of live programme material and there were various requirements:

- The programme material would subsequently be presented to transmitters employing amplitude modulation, which can be damaged by overload, or to tape machines, which distort on overload.
- Transmission was generally live, so there was no opportunity for a second take. Levels had to be right first time, every time.
- The meter would be read and interpreted by operators who were primarily concerned with making artistic adjustments to achieve the best sound balance.
- The operators would be working under pressure, possibly under subdued lighting.

Taken together, these requirements meant that the meter needed a fast response, and a clear uncluttered display. An equal increment logarithmic scale of 4 dB/division meant that meter deflection was broadly proportional to loudness. See Figure 6.1.

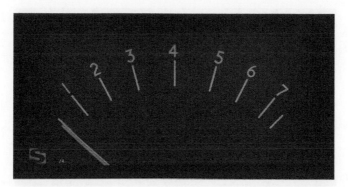

Figure 6.1
The PPM scale is 4 dB per division, with 0 dBu = PPM4. (This is a stereo PPM having a pair of front-to-front movements, hence the two pointers).

White lettering on a black background made the scale easy to read. 0 dBu = PPM4, and with 4 dB per division, PPM1 = −12 dBu and PPM7 = +12 dBu. Normal practice used 0 dBu as line-up level, and programme was mixed so that peaks did not exceed PPM6, or +8 dBu. Although UK broadcasters used the 1−7 scale, European broadcasters marked their scale −12 to +12. The meters were identical in all other respects [1].

Some early valve-driven PPMs such as the BBC MNA/3 couldn't achieve 4 dB/division over the entire range and were 6 dB between PPM1 and PPM2, but this failing was considered inconsequential because "Programme shouldn't be down there".

When a PPM is used for engineering, all controls are referenced to PPM4. Thus, the level of the incoming signal can be read directly from the controls provided that they have been adjusted to set the PPM to read PPM4. PPMs intended for engineering use generally also include an expanded scale in addition to the PPM scale, and may include other scales, making them more cluttered, but they're not being used to mix live programmes. See Figure 6.2.

The unfortunate consequence of the 600 Ω legacy

The reason that the telecommunications companies used 600 Ω is that their equipment drove cables that were so long that they genuinely were transmission lines at audio

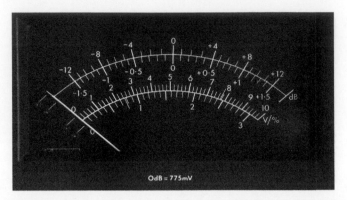

Figure 6.2
Engineering PPMs always include an expanded scale in addition to the PPM scale, and may include other scales.

frequencies. Because the cables were transmission lines, impedance matching was important, so the cables' **characteristic impedance** of 600 Ω had to be driven from a 600 Ω source and be **terminated** by a 600 Ω load. As a consequence, exchange (telecomms) and studio (broadcast) plant (electronic equipment) was all designed for 600 Ω impedance matching. The significance of this is that the input resistance of the destination formed a potential divider in conjunction with the source resistance driving it. See Figure 6.3.

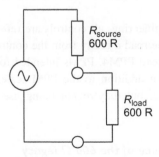

Figure 6.3
The legacy of 600 Ω analogue telecommunications is a 6 dB potential divider to trap the unwary.

Using the potential divider equation:

$$\frac{V_{\text{out}}}{V_{\text{in}}} = \frac{R_{\text{lower}}}{R_{\text{upper}} + R_{\text{lower}}} = \frac{600 \ \Omega}{600 \ \Omega + 600 \ \Omega} = 0.5 = -6 \ \text{dB}$$

The modern technique is to make all devices have high input impedance (often known as **bridging** impedance) and low output impedance. Thus:

$$\frac{V_{\text{out}}}{V_{\text{in}}} = \frac{R_{\text{lower}}}{R_{\text{upper}} + R_{\text{lower}}} = \frac{\infty \ \Omega}{0 \ \Omega + \infty \ \Omega} = 1 = 0 \ \text{dB}$$

We can now see that the modern technique avoids a wasteful signal loss of 6 dB. The practical consequence is that if you set the attenuators on a 600 Ω audio oscillator to produce a specific output level, its output will be 6 dB high unless it is loaded or **terminated** by 600 Ω. Unless you know your oscillator, it is wisest to **measure** the signal at the input of the amplifier, rather than rely on the attenuators.

Source and load impedances for the device under test (DUT)

Having ensured that the oscillator is correctly terminated, we must also ensure that we terminate our DUT correctly.

If we were measuring a very simple unbuffered valve pre-amplifier, it might have an output impedance of $\approx 6 \ \text{k}\Omega$ and expect to see a 1 MΩ load. A typical audio test set has an input impedance of 100 kΩ on its "high" setting, so this would cause an additional loss of 0.53 dB compared to a measurement with the correct 1 MΩ loading. (Beware that BBC equipment is more likely to be 60 kΩ, not 100 kΩ.) Conversely, measuring with a 10:1 oscilloscope probe (10 MΩ input resistance), would theoretically give a reading 0.054 dB high, but since this corresponds to +0.6%, it is comparable with oscilloscope error, so we would probably never notice it.

If we measure a power amplifier, we should terminate it with a load resistance appropriate to the secondary setting. Thus, if we have set the secondary to match a 4 Ω load, we need a 4 Ω resistor, often known as a **dummy load**. Since we will almost certainly attempt to determine maximum output power, the resistor must be capable of withstanding this power without damage and without changing its value due to heating.

Unfortunately, wirewound resistors are not ideal as dummy loads because values $<100\ \Omega$ have significant inductance compared to their resistance. Fortunately, 50 W non-inductive thick-film resistors are available, and multiples of these can be fastened to a large heatsink to make a non-inductive dummy load. (Although miniature high-value thick-film resistors produce measurable distortion, it is inversely proportional both to the resistor's physical size and value, so they are innocent in this application.) See Figure 6.4.

Figure 6.4
This 0.5—19.5 Ω 4.5 A dummy load has thick-film resistors screwed to a large heatsink.

Because of the way it is designed/connected, the author's dummy load has a 4.5 A current rating irrespective of setting, but a variable power rating. This occurs because the three main sections are each made of a pair of 10 Ω 50 W resistors in parallel, resulting in three 100 W 5 Ω resistors each having a 4.5 A rating. The four 1 Ω sections only need 20 W rated resistors to maintain the 4.5 A current rating. The final 0.5 Ω section is overrated because it was made from a pair of 1 Ω 20 W resistors in parallel.

Rather than add switches (and fret about contact resistance and current rating), load resistance is selected by plugging on the 4 mm sockets. Values in 0.5 Ω increments are obtained simply by appropriate plugging of the two load leads, and the addition of a single shorting lead allows selection of any resistance from 0.5 to 19.5 Ω in 0.5 Ω steps. See Figure 6.5.

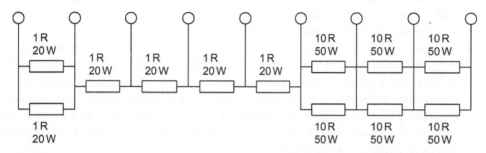

Figure 6.5
Circuit diagram of the 0.5–19.5 Ω 4.5 A dummy load.

Whether the DUT is a power amplifier or a pre-amplifier, it should be driven from an appropriate source resistance. Modern line-level sources are generally based on op-amps and typically have an output resistance of <600 Ω. Some audio oscillators have selectable output resistance, whereas others are fixed, but the highest output resistance is generally 600 Ω, so any of these settings is appropriate. However, if you know that you will drive your DUT from a higher source resistance, perhaps directly from a 100 kΩ logarithmic volume control, a 25 kΩ series resistor should be added to emulate the maximum expected source resistance.

Summarising, test conditions should replicate expected conditions of use.

Measuring gain at different frequencies

Once we have taken the trouble to measure the gain correctly at 1 kHz, it is easy to leave all other controls alone, change frequency, and take additional measurements to produce a graph of amplitude against frequency response, which is habitually abbreviated to "frequency response".

When we measured gain, it was equally valid to present the result as a pure voltage ratio or in dB, but the purpose of measuring the amplitude against frequency response of an amplifier is to see if the amplitude **deviates** significantly from the 1 kHz level. This observation has two important implications:

- When we are worried about the audibility of amplitude deviations, we should use a logarithmic measurement because the ear/brain responds logarithmically. We should either measure directly in dB or convert our measurement into dB.
- Absolute gain (or loss) is irrelevant; only deviations from the 1 kHz reference level matter. If we measured 1 kHz gain on a cathode follower buffer, we might have applied 0 dBu to the input of the DUT and measured -0.6 dBu at the output. Rather than do unnecessary arithmetic, it is far better to adjust the oscillator output level to produce precisely 0 dBu at 1 kHz at the **output** of the DUT, then (leaving oscillator level unchanged) measure output level at different frequencies and obtain the amplitude against frequency response directly.

Which frequencies to use?

Because the ear/brain responds logarithmically, audio amplitude against frequency graphs are plotted with a linear vertical scale (remember that dB are already logarithmic) and a logarithmic horizontal scale. The most useful graph has equally spaced points, rather than a desert punctuated by oases of closely clustered points. Thus, we should choose measurement frequencies that produce equally spaced points on a logarithmic scale. We could use the oscilloscope attenuator logarithmic sequence of 1, 2, 5 to give equally spaced points, resulting in test frequencies of: 20 Hz, 50 Hz, 100 Hz, 200 Hz, 500 Hz, 1 kHz, 2 kHz, 5 kHz, 10 kHz, and 20 kHz. Unfortunately, these points are rather sparse, so a better choice is to use International Standards Organisation (ISO) Recommendation 266 ⅓ octave frequencies:

20 Hz, 25 Hz, 31.5 Hz, 40 Hz, 50 Hz, 63 Hz, 80 Hz, 100 Hz, 125 Hz, 160 Hz, 200 Hz, 250 Hz, 315 Hz, 400 Hz, 500 Hz, 1 kHz, 1.25 kHz, 1.6 kHz, 2 kHz, 2.5 kHz, 3.15 kHz, 4 kHz, 5 kHz, 6.3 kHz, 8 kHz, 10 kHz, 12.5 kHz, 16 kHz, 20 kHz.

Twenty-nine points are thus needed to cover the range from 20 Hz to 20 kHz, and these frequencies would be entirely necessary if we were measuring an RIAA pre-amplifier because they potentially have deviations at almost any frequency across the audio band. However, if we were manually measuring a typical power amplifier, using all the ISO one-third octave frequencies would be wasted effort because it is designed to have a low-frequency roll-off, a high frequency roll-off, and be flat in between. See Figure 6.6.

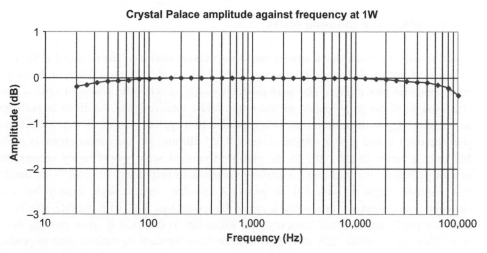

Figure 6.6

Crystal Palace amplifier amplitude against frequency response taken at an output power of 1 W.

ISO one-third octave frequencies (and their multiples beyond 20 kHz) were used, resulting in a graph with evenly spaced points. As can be seen, the amplitude response against frequency is very nearly flat, so taking the full set of measurements was both tedious and unnecessary. When measuring a valve power amplifier, it is better to sweep low frequencies slowly from 2 Hz up to 50 Hz, to find any low frequency bumps, then sweep from 10 to 500 kHz to spot undamped high frequency resonances due to the output transformer, and if no unpleasant resonances are found, simply determine the low and high frequencies for which the response is either 1 dB or 3 dB down compared to the 1 kHz reference level.

An even faster method than sweeping is to apply a 10 kHz square wave and check for negligible ringing at the leading edge, then change to 100 Hz square wave and select DC coupling at the input of the oscilloscope before checking that the top and bottom horizontal lines are truly **horizontal**, rather than sagging (sag indicates low frequency loss). Assuming that the square wave response **is** satisfactory, the −3 dB points can be determined using sine waves. If it isn't, there's not a lot of point in plotting a pretty graph of a high or low frequency aberration; the problem needs to be corrected!

Plotting the graph and choosing scales

Traditionally, we would have dashed out to a specialist stationer, then winced at the price of a pad of four-cycle logarithmic paper. These days, results are more likely to be plotted by a computer plus printer. You were probably taught at school to choose a vertical scale that used all of the graph paper, so you might find a total range of ±2 dB to be sufficient to encompass your measurements. When printed to a typical size of 60 mm by 100 mm, this approach would give a vertical scale of 0.067 dB/mm, yet the measurement is probably only accurate to ±0.1 dB, so the expanded vertical scale would imply unwarranted accuracy. Worse, it could make a perfectly reasonable response look poor − an audibly flat response ought to **look** flat on paper. A sensible vertical scale usually has 2 dB between major markers, producing a curve that correlates well with audible effects and does not imply unwarranted accuracy. The exception to this rule is when plotting deviations from equalisation such as RIAA, because time constant deviations tend to produce shelf errors occurring over a broad range of frequencies that only become inaudible <0.2 dB.

Measuring RIAA equalisation via an inverse RIAA network

It is easy to make a practical RIAA stage having an imperfect frequency response due to:

- Incorrect calculation of component values.
- Deviation from calculated value of actual value (component tolerance).
- Inappropriate capacitor dielectrics (polyester, paper) cause capacitance to fall by ≈3% at 20 kHz compared to 20 Hz.
- Valves vary from sample to sample and output resistance rises as they wear.
- Failing to damp transformer resonances correctly (particularly moving coil input transformers).
- Failing to consider the effect of loading the output with 10 kΩ resistance rather than 100 kΩ.

It is very convenient to have a reference inverse RIAA network that can be connected between an oscillator and the input of an RIAA stage so that the combination theoretically yields a flat amplitude response against frequency. Provided that the inverse RIAA network is perfect and presents the expected source resistance to the RIAA stage, any deviation from flatness is due to the RIAA stage, so corrections may be made until acceptable flatness is achieved.

Unfortunately, designing a "perfect" inverse RIAA network is not trivial. Factors that must be taken into account are:

- The network is sensitive to source resistance, so it needs to be matched to oscillator output resistance.
- Any passive network has some sensitivity to load resistance and capacitance, so possible errors due to loading need to be checked and minimised.
- Capacitors are typically only available in 1% tolerance, but resistors are available in 0.1%, so errors should be limited to the capacitors. Further, it would be convenient if standard capacitor values could be used.

With these considerations in mind, the author began with the fundamental RIAA replay equation:

$$G_s = \frac{(1 + 318 \times 10^{-6} \times s)}{(1 + 3180 \times 10^{-6} \times s)(1 + 75 \times 10^{-6} \times s)}$$

where:

G_s = gain as a function of s
$s = j\omega$
$\omega = 2\pi f$.

However, practical measurement of equalisation produced by this equation requires a steadily rising response that is not achievable from a passive inverse RIAA network — we are forced to include an additional time constant, so it might as well be the 3.18 µs time constant that had to be implemented at the time of cutting to protect the (probably Neumann) cutting head. The RIAA replay equation therefore becomes:

$$G_s = \frac{(1 + 318 \times 10^{-6} \times s)(1 + 3.18 \times 10^{-6} \times s)}{(1 + 3180 \times 10^{-6} \times s)(1 + 75 \times 10^{-6} \times s)}$$

Rather than reinvent the wheel, an existing inverse RIAA network developed by Jim Hagerman [2] from Stanley Lipshitz's [3] seminal AES paper was investigated, and the equation for its attenuation was written from first principles and multiplied by the fundamental RIAA replay equation to generate a graph that should ideally show zero error. See Figure 6.7.

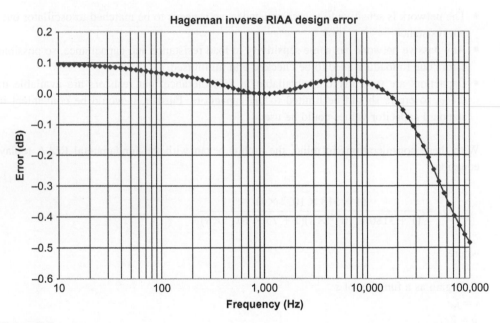

Figure 6.7

The design error curve for this commercial RIAA inverse network is quite respectable despite being restricted to E24 component values.

Despite being restricted by E24 component values, the response is surprisingly good over the audio band. Nevertheless, every BBC audio test set from the (valve) ATM/1 onwards is capable of measuring to better than 0.1 dB, so it seems worthwhile to improve matters.

The error worsens towards 100 kHz because the original Hagerman article assumed 3.5 μs rather than 3.18 μs. Although the less popular Ortofon cutting heads used 3.5 μs rather than Neumann's 3.18 μs, it is the author's experience that it is always preferable to use too little replay equalisation than too much. Changing component values to set 3.18 μs and making the fine adjustments possible by choosing 0.1% tolerance resistors available in E96 values significantly reduces design errors. See Figure 6.8.

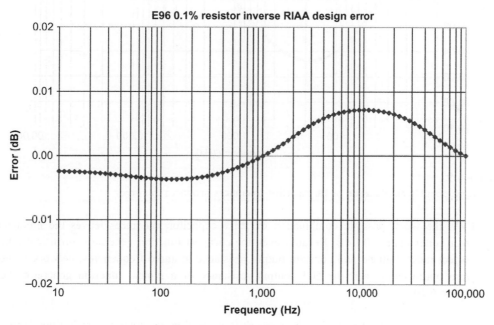

Figure 6.8
Using a mix of E96 and E24 resistor values allows the design error to be reduced.

The design errors using single E24 capacitors and combinations of E96 and E24 resistors are now comfortably within ±0.01 dB, so the effect of practical components can be investigated. Since capacitors are available in ±1% tolerance at best, the worst possible combination of capacitor tolerance extremes was investigated. See Figure 6.9.

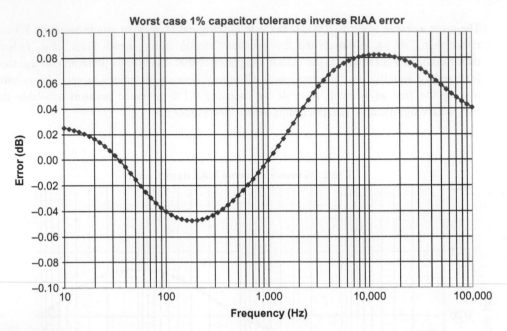

Figure 6.9

The worst possible combination of 1% capacitor tolerances produces this error.

Even the worst possible combination of ±1% capacitor tolerances leaves the predicted error better than ±0.1 dB. Because even the cheapest function generator intrinsically has perfect range flatness and constant output resistance at audio frequencies, network values were optimised to suit the 50 Ω output resistance of a typical function generator. See Figure 6.10.

Another advantage of matching the network to a function generator is that it allows square wave testing at the flick of a switch. The network was built into a tobacco tin and hangs off the front of the function generator in order to minimise incorrectly terminated lead length that would otherwise cause overshoot on the leading edge of square waves. See Figure 6.11.

Normal audio leads can be used to connect the output of either version to the pre-amplifier under test. The author selected from ±1% tolerance polystyrene capacitors using his Marconi TF2700 component bridge, but used unselected ±0.1% and ±1% resistors. See Figure 6.12.

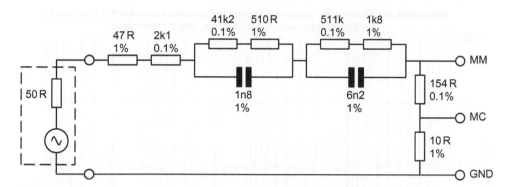

Figure 6.10
Final circuit of inverse RIAA network.

Figure 6.11
Practical implementation of the inverse RIAA network.

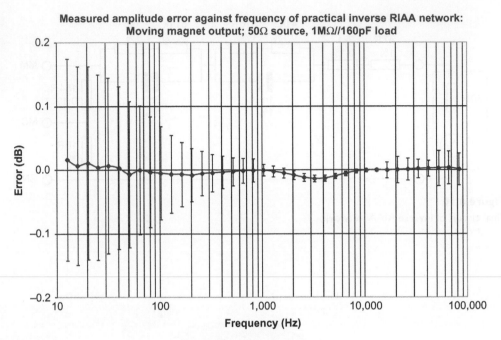

Figure 6.12
Measured practical inverse RIAA error.

As can be seen, the network's measured errors are well below their uncertainties, and certainly good enough for the intended use. Load capacitance forms an additional low-pass filter in conjunction with the network's output resistance, but modelling with LTspice suggests that 1.5 nF can be tolerated across the moving magnet output ($r_{out} < 164\ \Omega$) before 0.01 dB error results at 20 kHz, whereas any load resistance $\geq 10\ k\Omega$ causes a pure attenuation error, independent of frequency. Because it is derived from a resistive potential divider, the moving coil output is even less sensitive to loading errors.

Output resistance of the moving coil output is 10 Ω, and if the intended cartridge's output resistance is higher (40 Ω for the Denon DL103), an additional series resistor should be added between the output of the pre-RIAA network and RIAA stage to ensure that the RIAA stage is being tested under the actual conditions of use. This precaution is particularly important when a moving coil step-up transformer is to be used as incorrect source resistance changes damping of its ultrasonic resonance, causing in-band response errors or perhaps compromising overload performance that makes clicks and scratches more noticeable.

In the unlikely event of proposed cartridge source resistance being lower than $10\,\Omega$, it is perfectly permissible to substitute a reduced resistance for the network's $10\,\Omega$ resistor, but note that it further attenuates output voltage.

In summary, provided that $\pm0.1\%$ tolerance resistors and selected $\pm1\%$ tolerance capacitors are used, the function generator's source resistance is $50\,\Omega$, and the network's output resistance matched to proposed cartridge resistance, this inverse RIAA network may be considered to be suitable for testing any unbalanced MM or MC RIAA stage with negligible error.

There is no longer any excuse for RIAA errors, yet the author has recently seen errors on commercial products from companies who ought to know better.

Impedance against frequency

It can be extremely useful to measure **impedance** against frequency. Power amplifiers are carefully designed by choosing loadlines that cunningly maximise output power and minimise distortion, yet we connect them to real loudspeakers or headphones having wildly varying impedance. Thus, we might wish to determine the actual impedance of our chosen transducer in order to optimise its matching to the amplifier, or to see if it has any unusual features that might cause problems. (A peak in impedance at high frequencies could cause instability in some amplifiers.) Alternatively, we might want to determine the output impedance of a power amplifier to verify that it is sufficiently low not to disturb the Thiele−Small [4] parameters of an associated loudspeaker, causing an unwanted bump at low frequencies.

Measuring the impedance of a loudspeaker

A single moving coil loudspeaker has a non-flat impedance curve for two reasons:

- The moving mass and its suspension combine to produce a low frequency mechanical resonance that is transformed through the motor into a peak in electrical impedance. A typical moving coil bass driver has a resonant frequency between 20 and 50 Hz in free air. Dome tweeters resonate between 500 Hz and 1.5 kHz, but their resonance may be so heavily damped (perhaps by Ferrofluid®) as to be unobservable.

- As implied by the name, moving coil drivers have a coil that possesses inductance, causing impedance to rise with frequency. Although the moving element of a ribbon tweeter is very nearly resistive, the necessary impedance-matching transformer has leakage inductance, producing an impedance curve very similar to a moving coil tweeter.

Although it is theoretically possible to compensate the impedance curve for the low-frequency mechanical resonance, it usually requires inconveniently large component values and the mechanical parameters being compensated are not stable with temperature. Fortunately, compensating for voice coil inductance **is** practical, requiring a simple CR Zobel network across the loudspeaker terminals. See Figure 6.13.

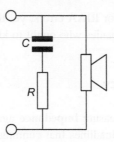

Figure 6.13
A simple Zobel network can compensate for a loudspeaker's voice coil inductance.

In theory, the required resistance is equal to voice coil DC resistance (in practice $1.2R_{\mathrm{DC}}$ is more likely), and the required capacitance is likely to be:

$$C_{\mathrm{Zobel}} = \frac{L_{\mathrm{voice\ coil}}}{R_{\mathrm{DC}}^2}$$

Many manufacturers specify voice coil inductance (rarely accurately), so it is better to connect a rough approximation to the calculated Zobel network across the driver, then trim its values until a flat impedance results. A decade capacitance box makes life much easier, and they are surprisingly cheap second-hand. See Figure 6.14.

Figure 6.14
Capacitance (and resistance) decade boxes are surprisingly cheaply available second-hand, and are very useful.

The easiest way to measure loudspeaker impedance is to make a potential divider from your oscillator's output resistance and the loudspeaker's impedance, then measure its attenuation. See Figure 6.15.

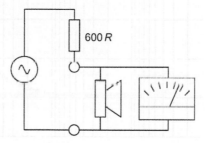

Figure 6.15
Measuring loudspeaker impedance.

We make the assumption that the oscillator has a constant output resistance with frequency, then use the potential divider equation in reverse to find the lower resistance:

$$\frac{V_{\text{out}}}{V_{\text{in}}} = \frac{R_{\text{lower}}}{R_{\text{upper}} + R_{\text{lower}}}$$

Rearranging:

$$Z_{\text{loudspeaker}} = \frac{R_{\text{oscillator}}}{\left(\frac{V_{\text{loudspeaker}}}{V_{\text{oscillator (open circuit)}}} - 1 \right)}$$

Note that the attenuation term has been inverted, and that we must measure the output voltage of the oscillator without any loading, **before** connecting the loudspeaker across it. Since an awkward calculation is required to derive the impedance, and it is quite likely that V_{in} and V_{out} were actually measured in dBu but must be converted back into absolute voltages, a spreadsheet is ideal both for the calculations and for plotting the subsequent graph. As an example, a pair of Sennheiser HD650 headphones were measured before and after a Zobel network was added. See Figure 6.16.

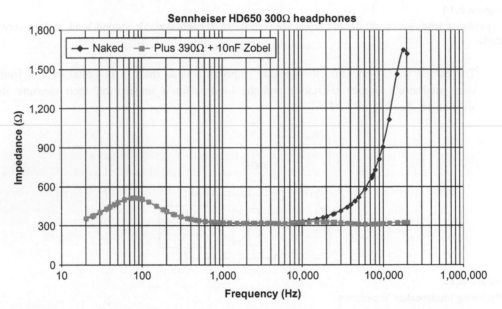

Figure 6.16

Impedance against frequency of Sennheiser HD650 headphones before and after compensation with a Zobel network.

The graph shows two plots: one naked, and one showing the effect of a parallel Zobel network consisting of a 390 Ω resistor and series 10 nF capacitor that converted the inductive impedance at high frequencies into a very nearly resistive (±4%) 320 Ω load below 200 kHz. Interestingly, the supplied (user-replaceable) cable has a surprisingly high 790 pF capacitance, perhaps explaining why there are reports of improved sound quality when it is replaced with a different cable – that 790 pF could easily have provoked marginal high frequency instability in a high-feedback headphone amplifier.

Pure resistive loads are desirable because:

- Non-resistive loads demand current that is out of phase with the applied voltage. Deviation from a perfectly straight loadline in the amplifier usually increases distortion.
- Reactive loads are likely to provoke instability in amplifiers requiring copious global negative feedback, such as output transformerless headphone amplifiers.
- Unless the amplifier has **zero** output impedance, load impedance that varies with frequency creates a potential divider having attenuation varying with frequency. Thus, the amplifier/load combination incurs an amplitude response error that varies with frequency (because the impedance of the load varies with frequency). Even if the errors are quite low (<1 dB), broadband errors are very noticeable, so they should be avoided.

Measuring the output resistance of a pre-amplifier

The output resistance of a pre-amplifier can be measured very easily by setting a convenient open-circuit output voltage at 1 kHz, then loading it with a variable resistance, and adjusting that resistance until the output voltage halves. At this point, the load resistance is equal to the source resistance, so the load resistance is removed and measured with a DVM, or if a resistance box was used as the load, the output resistance can be read directly. See Figure 6.17.

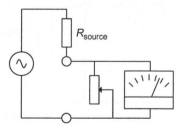

Figure 6.17
A variable resistor across the output of a pre-amplifier allows output resistance to be determined quickly and easily.

When used with a ×10 probe, an oscilloscope has an input resistance of 10 MΩ, which is certainly high enough to be able to measure the open-circuit voltage of any practical pre-amplifier (we're expecting an output resistance of <10 kΩ). Thus, we could apply a signal at the input of the pre-amplifier to produce a 6 V_{pk-pk} sine wave on an oscilloscope, with oscilloscope vertical sensitivity sct to 1 V/division.

We now connect the variable resistance across the output of the pre-amplifier and change oscilloscope sensitivity to 0.5 V/division. If we now adjust the variable resistor to restore the original deflection of six vertical divisions (now 3 V_{pk-pk}), then the output voltage of the pre-amplifier has been halved. The variable resistor can be removed and its value as measured by a DVM is equal to the output resistance of the pre-amplifier.

Measuring the output impedance of a power amplifier

Power amplifiers are designed to have very low output impedances (ideally a fraction of an ohm for conventional loudspeakers). If the previous method were used to measure the output impedance of a transistor power amplifier, the amplifier would be destroyed or would shut down. We need a different method. The way to measure a power amplifier's output impedance is to use a variation of the method used for determining loudspeaker impedance. See Figure 6.18.

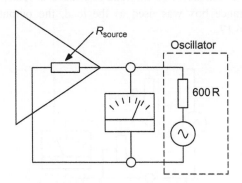

Figure 6.18
To measure a power amplifier's output resistance, we connect it across an oscillator, and see how well it can attenuate the oscillator.

We short-circuit the **input** of the amplifier to ensure that it cannot produce a signal. We then connect a meter directly across the output terminals, and also connect our oscillator across the output terminals. Because the amplifier is designed to have almost zero output impedance, it should be able to almost completely attenuate the output of the oscillator. We use the same equation for determining amplifier output resistance as we used for the loudspeaker:

$$Z_{\text{amplifier output impedance}} = \frac{R_{\text{oscillator}}}{\left(\frac{V_{\text{amplifier}}}{V_{\text{oscillator (open circuit)}}} - 1\right)}$$

Disappointing results tend to result from this measurement. See Figure 6.19.

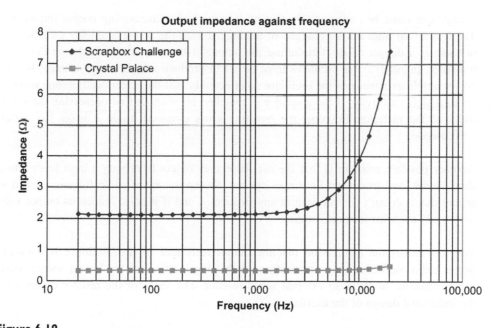

Figure 6.19
Output impedance against frequency for two power amplifiers. Lower: Crystal Palace push–pull amplifier with global feedback. Upper: Scrapbox Challenge single-ended amplifier with zero global feedback. Both amplifiers were configured to drive an 8 Ω load.

The output impedance is invariably higher and changes more with frequency than would be expected from a simple consideration of output valve r_a transformed by the square of the output transformer turns ratio. The reasons for this are:

- Valves in practical circuits rarely achieve the stunningly low values of r_a quoted by manufacturers on their data sheets (they tend to quote at $V_g = 0$ V and maximum anode dissipation).
- Output transformer secondary resistance and reflected primary resistance are in series with r_a.
- Output transformer leakage inductance is in series with r_a.
- Feedback amplifiers must reduce their feedback at high frequencies to maintain stability. Feedback reduces output impedance, so if feedback falls with frequency, its ability to reduce output impedance is reduced, and output impedance must rise with frequency.

Precautions must be observed when using this method of measuring output impedance. The test set-up must not be plugged or unplugged with the power amplifier switched on, or there is a danger of short-circuiting the output of the amplifier (which can destroy a transistor amplifier). The oscillator must be set to produce a high output voltage, but the power amplifier attenuates the oscillator so effectively that the meter/oscilloscope must be set to a high sensitivity. If the amplifier is switched off without first decreasing the sensitivity of the meter/oscilloscope, the resulting gross overload might damage the meter/oscilloscope.

Another possible problem is that the oscillator may object to driving a high level into a short circuit, and its current limit may operate. Sweep the entire audio frequency band to ensure that it doesn't current limit at any frequency, and if it does, reduce its output voltage or add a known series resistance.

The test makes the assumption that the output resistance of the oscillator is constant. Some audio oscillators may not conform to this ideal, so select the lowest source resistance available, and add an external series resistor. The value of the resistor depends on the individual design of the oscillator:

- Because function generators are designed to have a bandwidth of >10 MHz, they are designed to drive transmission lines and an output resistance of 50 Ω (constant at audio frequencies) may be assumed.

- Dedicated audio test sets that were designed to drive 600 Ω music lines had a constant output resistance with frequency as part of their design requirement. They do not need an external resistor.
- Modern bench audio oscillators tend to base their design on operational amplifiers, so they have an internal output resistance of zero, which is built out by a series resistor to give the specified output resistance. They are unlikely to need an external resistor.
- Old bench audio oscillators are likely to use a transformer to couple to their output terminals, so their output resistance is not necessarily constant. If in doubt, select the lowest output resistance, and add a metal film series resistor of ten times that value to swamp any variations.

Non-linear distortions

Non-linear distortion implies that the output is not simply a scaled replica of the input, but that additional frequencies have been added as a result of non-linearity within the DUT.

Maximum output power and distortion against level

One of the most important specifications of a power amplifier is the amount of power that it can deliver into a specified load impedance. There is plenty of scope for argument in defining what constitutes "maximum power". Possibilities for defining maximum power include:

1. The power developed at, or just before, the point when clipping of the output waveform can be observed on an oscilloscope.
2. The power developed when total harmonic distortion (THD) reaches an arbitrary value − often 0.1%, possibly 1%, perhaps even 3% or 10%.
3. The power developed just before the point when THD begins to rise sharply.

Definition 1 is quite useful for amplifiers that employ plenty of global feedback, as distortion tends to be very low until clipping occurs, whereupon it rises catastrophically. Even so, the definition is a little fluffy.

Definition 2 is popular when measuring zero-feedback single-ended amplifiers, although adjusting the THD criterion from 3% to 10% is a popular ploy for increasing the perceived value of an amplifier by substantially increasing its measured power output. At the opposite extreme, Leak specified amplifiers by the power they could deliver at 1 kHz with 0.1% THD.

Definition 3 requires the plotting of a graph of THD against level. This is easy with a modern automated audio test set, but harder with an older, manual test set.

Whichever definition is chosen, the amplifier is loaded with a dummy load, and the resulting THD is measured by the audio test set. In addition, it is useful if the monitoring output of the test set feeds the distortion waveform to an oscilloscope. See Figure 6.20.

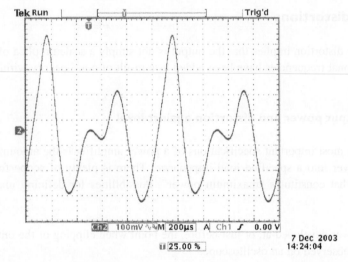

Figure 6.20
Distortion waveform 0.2 dB below grid current.

Fortunately, when testing valve amplifiers, there is a very distinct point that is very sharply defined, and therefore readily identifiable, so it can be used to define maximum output power. We are not looking for a particular distortion figure (after all, that can be adjusted by negative feedback), rather we are looking for an abrupt change in the distortion waveform that signifies grid current in the output stage. See Figure 6.21.

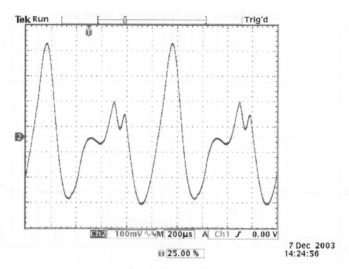

Figure 6.21
Distortion waveform at onset of grid current.

Comparing the two distortion waveforms, the second waveform has an additional valley at one of its peaks. This valley appears and disappears very sharply with level, so it is ideal for defining maximum power. In this particular test of the author's Crystal Palace amplifier, it defined maximum output power at 1 kHz to be 47 W with 0.56% THD. The amplifier delivered 52 W with 1.5% THD just below observable clipping on an oscilloscope.

Measuring THD against output power is fiddly with a manual test set, but manageable with care. Calculate the level in dBu for 1 W into the test load resistance. Adjust level until the power amplifier develops 1 W. Then, increase level in 1 dB steps, recording the amount of distortion at each point until gross distortion results. Finally, convert the levels in dBu into power, and plot a graph of distortion against output power. See Figure 6.22.

Once again, the reason for using a logarithmic scale for output power is that the ear perceives loudness logarithmically. Looking at the graph, we see a smooth curve with rising distortion and a kink at ≈50 W due to grid current. Considerable effort and number-crunching was required to produce this graph, yet it only tells us what we already knew,

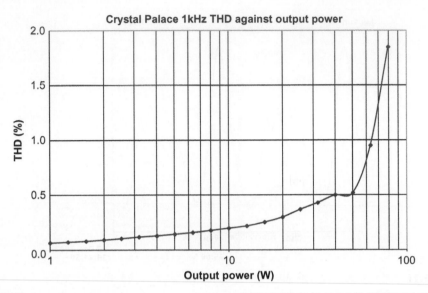

Figure 6.22
Distortion against output power at 1 kHz.

and does so with poorer accuracy. Manually plotting THD against output power is not useful for determining maximum output power, although ideal for an automated test set. Beware that an ill-chosen automated test can easily destroy a power amplifier – make sure there's a way of stopping the test if the amplifier shows signs of distress.

One justification for plotting THD against output power is that if the distortion rises at low levels, this can be an indicator of crossover distortion. Unfortunately, as all amplifiers produce noise, practical measurements of distortion are not so much THD as THD + N (total harmonic distortion plus noise). As output signal level falls, the noise becomes more significant, so all amplifiers suffer deteriorating THD + N at low levels, which may mask the measurement of crossover distortion.

A sudden change in the distortion **waveform** with level is a powerful indicator of a problem. At high levels, the sudden appearance of a single dip indicates grid current (valid for power amplifiers and small-signal amplifiers). Conversely, at low levels, the gradual appearance of triangular spikes as level falls indicates crossover distortion. If you designed the circuit you are testing, you already know which features to look for. If you're testing a single-ended amplifier, there is no need to look for crossover distortion because it is an exclusive failing of push—pull Class AB amplifiers.

THD against frequency

A popular test with feedback amplifiers is to test THD against frequency. The reason for this is that if the amplifier relies on global negative feedback to reduce distortion, the gain before feedback is applied must be maximised, but to maintain stability this gain must fall with frequency. But feedback can only correct by the ratio between the open-loop gain and the required closed-loop gain. This means that if the open-loop gain is 80 dB, and the required closed-loop gain is 20 dB, 60 dB of feedback is available for correction. But if at a higher frequency, only 40 dB of open-loop gain is available, and the required closed-loop gain is still 20 dB, only 20 dB of correction is available, so distortion must rise. Summarising, all feedback amplifiers must have distortion that rises with frequency. See Figure 6.23.

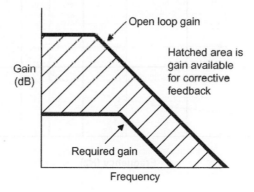

Figure 6.23
Open-loop gain and available corrective feedback.

Transistor power amplifiers often rely on global feedback to lower distortion, so plotting THD against output power can be very revealing. Valve power amplifiers almost invariably incorporate an output transformer, so they are typically limited to ≈ 30 dB of feedback before stability becomes distinctly questionable, tending to make the measurement less useful.

Measurement and interpretation of THD against frequency has to be done very carefully if it is to bear any relation to audible effects. If we measure THD at 1 kHz, then even the 20th harmonic (20 kHz) is theoretically within the audio band, but if we measure THD at 10 kHz, only the second harmonic is within the audio band. If the measurement is intended to correlate with audible effects, a 20 kHz low-pass filter must be used, but this would have the effect of making an amplifier producing predominantly third harmonic distortion appear to have falling THD with frequency after 6 kHz.

If we were measuring an amplifier with a view to possible improvement, we would not use the 20 kHz low-pass filter, as this colours the results. As an example, a 10 W valve amplifier employing substantial global feedback was tested at 10 W. See Figure 6.24.

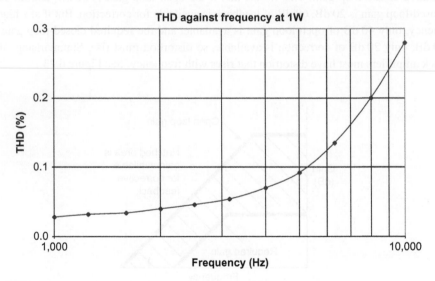

Figure 6.24

This 10 W high-feedback amplifier exhibits textbook rising THD with frequency.

As should be expected from a high-feedback amplifier, the THD is very low at 1 kHz, but rises with frequency in accordance with the falling open-loop gain required to maintain high frequency stability. In this example, the measured THD at 10 kHz was **exactly** ten times that at 1 kHz. Alternatively, we could state that the THD was 20 dB worse at a frequency one decade higher, or that THD rises at 20 dB/decade, which is **exactly** the slope produced by the CR compensation network required to maintain stability. Summarising, the graph of THD against frequency of a feedback amplifier always mirrors the graph of open-loop gain against frequency.

If we are already familiar with the amplifier design, then plotting THD against frequency tells us little that we didn't already know, but if we are forced to treat the amplifier as an unknown black box, then it becomes useful. One useful outcome of a plot of THD against frequency is that it reveals that high frequency compensation is a trade between stability and distortion. What the graph doesn't tell us very clearly is whether the amplifier might suffer from slewing distortion.

Slewing distortion

A capacitor *can* change its voltage instantaneously, but requires an infinite current to do so. At a more practical level, the faster we want to change the voltage across a capacitor, the more current is required. All amplifier stages have input capacitance, so when the preceding stage attempts to change the voltage at the input of the next stage, it must change the voltage across this shunt capacitance. In small-signal terms, the shunt capacitance forms a low-pass filter in conjunction with the output resistance of the preceding stage. See Figure 6.25.

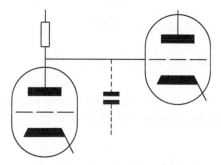

Figure 6.25
Shunt capacitance driven by inadequate current causes slewing distortion.

When we attempt to impose a large voltage change quickly, the amount of current that the capacitor can draw may be limited by the quiescent current in the preceding stage. Under this condition, the capacitor charges slower than the input waveform changes voltage, and this distortion is known as **slewing distortion**. Slewing distortion became particularly noticeable when early operational amplifiers were used for audio, but valve amplifiers are by no means immune to the problem. Fortunately, the problem is easily detected, if not quite so easily cured.

To test for slewing distortion, drive the amplifier to full power at 1 kHz, then increase frequency to 10 kHz and beyond. At some point, the sine wave will begin to look more like a triangular wave, and this is evidence of slewing distortion. See Figure 6.26.

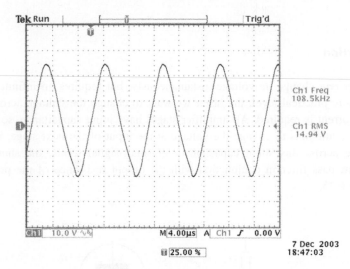

Figure 6.26
Slewing distortion turns a sine wave into a triangular wave.

Providing that the effect is not visible at 10 kHz, there probably isn't a problem, but if it occurs at a lower frequency, then the offending combination of stages should be identified and corrected.

The previous definition of whether slewing distortion is a problem is somewhat woolly, and this is because it all depends on what sort of music you listen to, and at what level.

Music played on acoustical instruments tends not to generate high amplitudes at 10 kHz, so it rarely provokes slewing distortion, whereas electronically generated music is more likely to provoke slewing distortion.

If the patient is a feedback amplifier, the global feedback should be removed, and the input level adjusted to restore full power at 1 kHz. Then, change the frequency to 10 kHz, and probe inside the amplifier to find the offending combination of stages. Slewing distortion can be tackled either by increasing available current or reducing capacitance:

- Substantially increasing the quiescent current in the preceding stage (probably by a factor of 5) forces a redesign, and a valve having similar amplification factor but higher mutual conductance will probably be required.
- Reducing the shunt capacitance. Usually the offending combination is the driver stage and output stage, because this is where the largest voltage swings occur, and output stages invariably have high input capacitance. Pentodes have far lower input capacitance than triodes, so converting an output stage from pentode to triode could cause slewing distortion! One way of reducing the input capacitance is to insert a cathode follower, but cathode followers are not immune to slewing distortion, so it will probably need to pass five times the current of the preceding stage. Another possibility for push–pull Class A amplifiers is partly to neutralise the output stage by connecting a small capacitor from each anode to the other valve's control grid. This is positive feedback, and needs to be used with great care.

Power bandwidth and transformer saturation

In addition to checking for slewing distortion, it can be useful to check the power bandwidth. This is essentially amplitude against frequency response at full power. The high frequency $f_{-3\,dB}$ point at full power is generally limited by slewing distortion and so long as it is beyond 20 kHz, it probably isn't a problem, but the low frequency power response is more significant.

The low frequency power response is almost totally governed by transformer core saturation. Rather than looking for the frequency at which output power halves (−3 dB), which would imply gross distortion, it is kinder to the output valves to measure the frequency at which distortion begins to rise. See Figure 6.27.

467

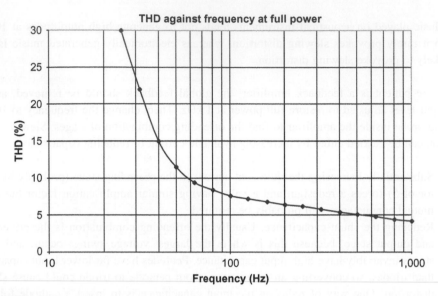

Figure 6.27
Transformer core saturation causes increased distortion at low frequencies. (This ill-suited transformer was deliberately chosen to ease measurement).

The graph shows the distortion of a zero feedback single-ended amplifier with an ill-suited output transformer having insufficient primary inductance. As can be seen, once frequency falls below 60 Hz, distortion rises dramatically.

Typical classic push–pull amplifiers could generally support full power down to 50 Hz before transformer saturation caused distortion that was visible on an oscilloscope. On test, the author's Crystal Palace amplifier produced 1% THD at full power at 22 Hz, and reducing frequency until transformer saturation became visible on the oscilloscope required 11 Hz, but produced pulsing orange flashes from within the output valves, so the test was hurriedly abandoned.

Investigating the distortion spectrum

It is very useful to perform a Fast Fourier Transform (FFT) on a distortion waveform to discern individual harmonic amplitudes. To find individual amplitudes, we need to make a sweeping assumption that (fortunately) is usually valid. The assumption is that **one** of the distortion harmonics is at a higher level than the others; >6 dB ($\times 2$) higher will do, but >10 dB ($\times 3$) is better, and it doesn't matter which harmonic it is. Provided that this assumption is true, then the THD measured by the audio test set is equal to the amplitude of that single harmonic, and because we have been able to assign an absolute level to this harmonic, if we can make relative measurements to other harmonics, we can assign them absolute levels.

As an example, the author tested one section of a 7N7 as the lower valve in a μ-follower amplifier and measured 0.175% THD, which is equivalent to -55 dB. See Table 6.2.

Table 6.2

Combining THD and FFT of its distortion residual to determine absolute harmonic levels

	Second	Third	Fourth	Fifth	Sixth
Levels measured by FFT or wave analyser (relative to -55 dB THD)	0	-33	-42	-50	-58
Levels relative to fundamental	-55	-88	-97	-105	-113

The absolute level at the monitoring output of the audio test set is unimportant. What is important is the relative amplitude between the dominant harmonic and other harmonics. Because the third harmonic is -33 dB compared to the second, the sweeping assumption that the level of the highest harmonic is equal to the level of the THD is justified, so (having previously converted THD into dB) we can immediately deem the level of the second harmonic to be equal to the THD. The higher harmonics were all measured relative to the level of the second harmonic, so their levels relative to the fundamental can now be found. In this example, the technique has produced some remarkably low levels for the fourth and higher harmonics, so it is best to check the distortion residual of the test equipment very carefully before inserting a DUT and believing impressive figures. The author's test equipment is generally reliable to ≈ -95 dB.

Investigating the distortion spectrum of an amplifier is a very powerful technique for choosing between different valves and bias points. In theory, it ought to be good for detecting power amplifier problems, but in practice observing the distortion waveform on an oscilloscope is faster and more sensitive.

There are many packages that allow computers to perform the FFT. Beware that when you use these, you rely on the linearity of your sound card. The built-in audio stages of a computer are not usually very good, and you need a recording-quality sound card, preferably 24 bit, 192 kHz sampling frequency, or better.

Measuring power supply ripple and hum

Since almost all valve electronics is mains powered, there is always the possibility of mains hum and its harmonics entering the audio signal. Full-power distortion spectra of power amplifiers often reveal power supply sum and difference frequencies around each distortion harmonic. Thus, a 1 kHz distortion measurement might have a 2 kHz harmonic plus lower-level 1.9 and 2.1 kHz frequencies because 1.9 kHz = 2 kHz − 2 × 50 Hz (the factor of 2 occurs because full-wave rectification in the power supply produces ripple at double mains frequency).

More commonly, we look for hum in the absence of signal, and we typically begin by measuring across the amplifier's output terminals with the input either short-circuited or terminated in the expected source resistance. The reason that a short circuit might be used is that we want to discriminate between hum generated within the amplifier and hum picked up by external cabling.

A digital oscilloscope is the best instrument for measuring hum, although its sensitivity may need to be increased by preceding it with a high-gain amplifier − an audio test set or AC millivoltmeter having a monitoring output is ideal. The reason that a digital oscilloscope is best is that good audio design/construction should have its hum buried in the noise, making it difficult to measure. However, if we trigger the oscilloscope from "line" and invoke "averaging", we can average out the random noise to leave the repetitive hum, then use an automated measurement to quantify that hum.

Having found hum (usually at a higher level than we would like), we want to reduce it, so we need to determine its cause.

If the hum is the same as mains frequency (50 or 60 Hz), it is probably caused electromagnetically, and due to transformer induction or a hum loop. If waving a sheet of steel between a power amplifier's mains transformer and audio transformer changes the waveform's shape, then induction is likely and the only real cure is to move the two transformers further apart. A less likely possibility is poor electrostatic screening at the amplifier's input, and if waving your hand near the input changes the waveform, improve the screening either by eliminating gaps in the screen or reducing the impedance bonding it to chassis.

If the hum is double the mains frequency (100 or 120 Hz), it is almost certainly power supply ripple due to inadequate filtering. Although an oscilloscope and associated probe can measure millivolts of HT ripple simply by engaging AC coupling at its input, we might need to measure microvolts of HT ripple without destroying our audio test set. There are various ways in which the sensitive input of the audio test set may be safely connected to the HT, and they all require the HT to be switched off **before** connection. See Figure 6.28.

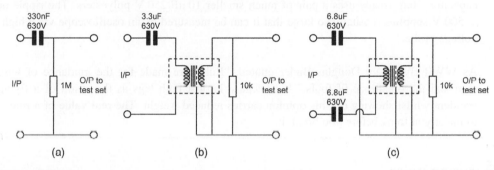

Figure 6.28
Connecting a delicate audio test set to the HT to measure hum.

- Capacitor coupling: This is a single-ended measurement and prone to inaccuracy due to earth loops on the input leads. The capacitor needs to have a DC rating greater than the HT and sufficient capacitance in conjunction with the tests set's input resistance to keep $f_{-3 \text{ dB}} < 20$ Hz. Given that the test set's input resistance might only be $100 \text{ k}\Omega$, we probably need a 330 nF 630 V capacitor. Some test sets have an input capacitor to protect them from DC, but the capacitor is likely to be rated at only 100 V. The 1 M resistor ensures that all the input DC appears across the capacitor that is rated for it.
- Single-pole capacitor coupling plus 1:1 line-level audio transformer: This breaks earth loops, and the transformer's common-mode rejection ratio (CMRR) assists in a quiet measurement. However, if inadvertently connected the wrong way round, although the HT is blocked from appearing across the primary (vaporising it), the HT could appear across the interwinding insulation, and break that down instead. The transformer

probably needs to be loaded by 10 kΩ to damp high frequency ringing, so the DC blocking capacitor now needs to be 3.3 μF 630 V.

- Double-pole capacitor coupling plus 1:1 line-level audio transformer: This avoids the cross-connection hazard but requires the primary to have a centre tap that can be earthed to prevent the HT appearing across the transformer's interwinding insulation. It also restores the earth loop. The two DC blocking capacitors are in series, so they each need double the previous capacitance − we now require a pair of 6.8 μF 630 V capacitors. These are big capacitors, engendering ingenious engineering justifications for why a particular application did not need such large capacitors and why the smaller (but available) capacitors were adequate.

When pressed, the author uses a pair of inconveniently large 6 μF 2 kV polypropylene capacitors, but usually uses a pair of much smaller 10 μF 250 V polyesters. The ripple on >500 V supplies is usually so large that it can be measured by an oscilloscope via a high-voltage probe.

As WWII flying ace Douglas Bader quoted, "Rules are made for the guidance of wise men and the obedience of fools," but because he lost both legs as the result of a flying accident whilst showing off, his opinion carries reduced weight. The real value of a rule is to make you think before you break it.

Measuring noise

We measure noise at the amplifier's output in the absence of an applied signal but under all other conditions of use. Thus, we either terminate the input of the amplifier with its expected source resistance or a short circuit, and to avoid gain errors due to non-zero output resistance, we load the amplifier's output with its expected load resistance. The classical assumptions are that noise is random, has a Gaussian distribution, and is white (constant power with frequency), so it doesn't matter whether we measure a noise bandwidth of 20 kHz at 500 kHz or 500 MHz. $1/f$ noise means that the white assumption may not be exactly true for audio, but it is a good place to start.

Remembering that:

- The mean is the DC component.
- Any deviation from the mean is the AC component.

- RMS stands for root of the mean of the squares (of the deviation from the mean) — which we otherwise know as standard deviation.

Thus, an RMS rectifier calculates the standard deviation from the mean and this is the correct statistical calculation for expressing the deviation of a Gaussian distribution. Summarising, RMS is the only rectifier that is mathematically justifiable for measuring random noise.

Practical RMS measurement

Imagine that we choose to calculate RMS by sampling the signal at a sufficiently high frequency that Nyquist is satisfied and we take many samples at short intervals. To calculate RMS, we must take the mean of the **squares** of individual samples, so any error in measuring the amplitude of narrow high-amplitude spikes is greatly magnified. We need an instrument that can cope with a high crest factor. We briefly mentioned crest factor when investigating DVMs in Chapter 4, but it is noise measurement that really exercises this parameter.

There are various ways of determining RMS:

- Thermocouple: This was the traditional (and expensive) method. A small resistor and thermocouple are thermally bonded in an evacuated glass envelope having minimal thermal loss to the environment. The signal to be measured is applied to the resistor and the heating effect is proportional to the square of the signal ($P = V^2/R$). We measure the mean temperature using the thermocouple, and take the square root of this DC term to give a true RMS measurement. A certain amount of calibration is necessary using simple waveforms (sine and square) at different amplitudes, but once these are made, correct noise measurement is a given. Measurement bandwidth and maximum crest factor are determined purely by the bandwidth and clipping voltage of the amplifier feeding the resistor, so these meters tend to give near-ideal performance. Commercial examples were generally intended for wide-bandwidth measurement at video or radio frequencies rather than audio and this tends to be reflected by their input connectors (BNC, MUSA, etc.). Thermal mass means that the measurement may take several seconds to stabilise, so such a slow response on an unknown instrument is a useful indicator of the technology.
- Analogue multiplier: The signal to be measured is applied to both inputs of an analogue multiplier, resulting in its square, and we take the mean and the square root as before. Maximum crest factor is determined by the self-noise, linearity, and clipping voltage of

the multiplier, but multiplier linearity is never ideal and the squared input voltage hits clipping rather quickly, so this method is not ideal. Commercial examples include hand-held true RMS DVMs and analogue audio test sets.

- Digital sampling: We sample at a high frequency and provided that we do not overload the analogue to digital converter (ADC), we faithfully capture the amplitudes of all peaks. We perform the entire RMS calculation in the digital domain to arbitrary accuracy, so the only limitation is the resolution and sample frequency of the ADC. Commercial examples include oscilloscope automated measurements, specialised audio software in conjunction with a recording quality sound card, and 6½ digit bench DVMs.

The gold standard used to be the thermocouple meter, but ADCs are now so fast, accurate, and cheap that digital sampling is now by far the best technique. Buying all the qualities we need as a single audio instrument tends to be expensive, but connecting a good digital oscilloscope to the monitoring output of an analogue audio test set substitutes the oscilloscope's presumed accurate RMS computation for an imperfect analogue rectifier and meter to give excellent accuracy. Alternatively, a recording quality computer sound card plus dedicated software can do the job.

Bandwidth limiting filters

Remember that the noise produced by a resistor is:

$$v_n = \sqrt{4kTBR}$$

where:

v_n = RMS noise voltage
k = Boltzmann's constant (1.3806×10^{-23} J/K)
T = absolute temperature (°C + 273.16)
B = bandwidth in Hz
R = resistance in ohms.

Although we may have the correct rectifier to measure white noise, the preceding equation reminds us that we must always define the noise measurement bandwidth. The required measurement bandwidth is the bandwidth of the human ear − not the amplifier. Thus, all audio noise measurements **must** be preceded by a 20 Hz−20 kHz band-pass filter, although 22 Hz–22 kHz is also common (this slightly increased bandwidth causes it to read an almost unnoticeable 0.4 dB high when measuring the same white noise source).

Many audio test sets include a 400 Hz high-pass filter to remove hum when measuring noise. The reduction in noise bandwidth from 19,980 to 19,580 Hz causes an entirely negligible 0.08 dB error unless there is $1/f$ noise — which there will be in an RIAA stage (due to the equalisation, even if the input device doesn't produce it). Think carefully before invoking the 400 Hz high-pass filter.

Noise weighting filters

Although we can make an engineering measurement, we might want to correlate it to its subjective effect, and the usual technique is to insert a noise weighting filter before the rectifier. See Figure 6.29.

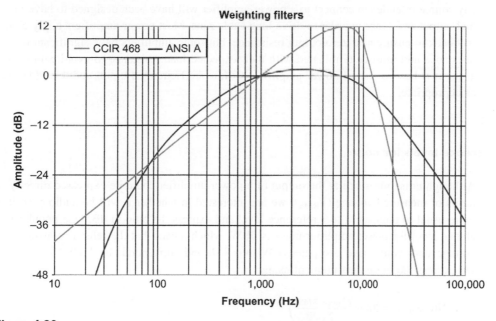

Figure 6.29
Amplitude against frequency response of two noise weighting filters.

Sometimes alleged to correlate with average human hearing at 50 dB(A), the ANSI "A" weighting filter is popular with manufacturers because not only does it reduce noise bandwidth but it also attenuates mains hum, guaranteeing good marketing figures. As another example, CCIR468-2 was designed to be sensitive to telegraph induction onto analogue telephone circuits, but analogue telephone circuits other than local ends between subscriber and exchange are obsolete, and so is telegraph, so it is now only used (rarely) for weighting distortion harmonics.

In short, noise weighting filters were designed for obsolete technologies or their basic premise is questionable, allowing very little engineering justification for weighted noise measurements; the author only makes unweighted measurements.

Power amplifier noise

It is legitimate to measure a power amplifier's noise with its input short-circuited because any source intended to connect to a power amplifier will have been designed to have minimal output resistance in order to drive long cable runs. Most power amplifiers apply global feedback to ensure near-zero output resistance, but some amplifiers have significant output resistance and must be terminated with their expected load resistance to avoid gain errors. Thus, it is best to test all power amplifiers with a short-circuited input, and 4 or 8 Ω load as appropriate.

Reference levels for noise

Although noise measured at the output of a power amplifier could be expressed directly in terms of measured voltage (V_{RMS}), we are interested in whether it will be audible, so it is more usual to compare it to a reference level and express the ratio in dB. One popular reference level for power amplifier noise is dBW, which means that the reference level is the voltage that would have dissipated 1 W into 8 Ω, and corresponds to 2.828 V_{RMS}. Thus, we express the noise measurement using:

$$\text{Noise}_{dBW} = 20\log\left(\frac{v_{\text{noise(RMS)}}}{2.828}\right)$$

We expect the noise voltage to be small, so a typical result from the above calculation might be −70 dBW. The virtue of the dBW reference level is that if two amplifiers are connected to the same loudspeaker, the one having the more negative number will be the quieter.

Manufacturers and designers often adapt a fundamental design for different powers, so a 50 W amplifier might share its driving circuitry with a 200 W variant. If both amplifiers have been designed to achieve full power from 2 V_{RMS}, the 200 W amplifier must have 6 dB more voltage gain, and if the input stage is the same, 6 dB more noise. The 200 W amplifier is invariably the more expensive, so much engineering effort is expended in reducing its noise to make it quieter than the cheaper 50 W model, but is this necessary?

We normally match an amplifier's power to its partnering loudspeakers, so it would be unusual to drive loudspeakers having a maximum power rating of 250 W from a 10 W amplifier. Conversely, it would be foolish in a domestic setting to drive loudspeakers having a sensitivity of 110 dB/W from a 250 W amplifier. Thus, we might choose maximum power as the reference level because the resulting number would be the achievable electrical dynamic range, and with the presumption of appropriate loudspeakers, an indication of audible dynamic range.

Despite the previous minor flaw, the dBW reference level is an accepted standard for power amplifiers and electrical dynamic range can easily be calculated by subtracting the dBW noise measurement from the maximum power (also referred to dBW).

Digital audio embraces the dynamic range concept and invariably refers noise to maximum level dBFS (dB full scale).

Older analogue standards accept that their maximum level is somewhat ill-defined and choose a line-up level (such as 0 dBu), subsequently referring noise measurements to that level. Thus, a twin track ¼″ analogue tape machine might have noise at −58 dBu, and a maximum operating level (MOL, usually specified for 3% THD + N) of +10 dBu, giving a dynamic range of 68 dB.

Vinyl defines a line-up velocity of 5 cm/s at 1 kHz and a very ill-defined peak level (post-RIAA) perhaps 12 dB higher. But cartridges have varying sensitivities, ranging from 100 μV to 10 mV at 5 cm/s, making it necessary to specify the RIAA stage's sensitivity for a standard output voltage. A better technique is to specify the RIAA stage's input-referred noise as an absolute voltage because this allows genuine comparison — the smaller the voltage, the quieter the stage.

Measuring RIAA stage noise

We expect (hope) that the noise of an RIAA stage is very low, so it has to be measured carefully if the correct result is to be obtained, and we must consider how we terminate the input of the RIAA stage. A short circuit would give an optimistic figure because it would remove:

- Thermal noise generated by the cartridge's internal resistance and arm wiring resistance.
- Voltage noise developed across the cartridge's source impedance by current noise emanating from the RIAA stage's input device.

Fortunately, low output moving coil cartridges can be treated as pure resistances <20 kHz. See Figure 6.30.

Thus, a moving coil RIAA stage's input may be terminated with a resistance equal to the intended cartridge's specified internal resistance (r_s) plus (if significant) arm wiring resistance. See Figure 6.31.

Happily, an RIAA stage employing a valve or JFET input stage produces negligible current noise, making it permissible to measure such a stage's noise terminated by a resistor corresponding to the cartridge's resistance. Sadly, applying the same technique to a bipolar input moving magnet RIAA stage is likely to result in a signal-to-noise measurement that is optimistic by between 1 and 5 dB because the RIAA stage's current noise develops an additional noise voltage across the cartridge's source impedance that varies significantly with frequency. See Figure 6.32.

It would be convenient to substitute a defined series inductance and resistance for noise measurements. Sadly, despite manufacturers' published figures, the impedance of a moving magnet cartridge cannot be simulated accurately by a simple inductance in series with resistance, and a six-component model becomes necessary to describe their 20 Hz−200 kHz impedance undulations even to ±5% accuracy. See Table 6.3.

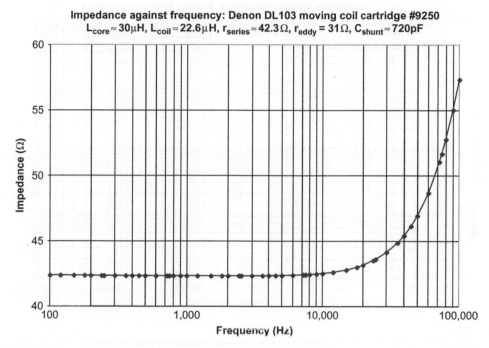

Figure 6.30

Impedance against frequency of Denon DL103 moving coil cartridge.

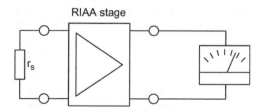

Figure 6.31

Measuring RIAA stage noise.

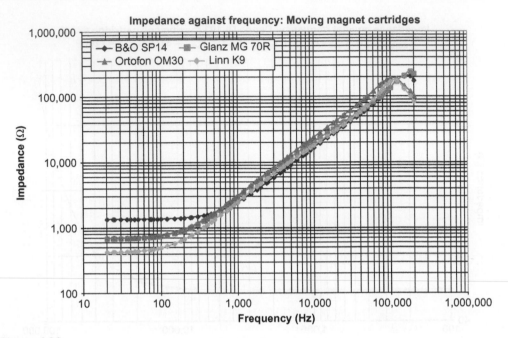

Figure 6.32

Moving magnet cartridges have an impedance that changes significantly with frequency.

Table 6.3

Six-component models of four moving magnet cartridges

	B&O SP14	Glanz MG-70R	Linn K9	Ortofon OM30
C_{shunt} (pF)	6.5	2.4	11	8.5
R_{shunt} (kΩ)	240	240	190	200
R_{eddy} (kΩ)	17	11	9.7	11.8
L_{core} (mH)	166	230	205	290
L_{coil} (mH)	134	210	175	180
L_{total}(mH)	**300**	**440**	**380**	**470**
R_{DC}(Ω)	**1360**	**684**	**380**	**703**

The practical solution in the vinyl heyday was to measure moving magnet RIAA stage noise with the input terminated by a Shure M75ED cartridge within a heavily screened box. Few cartridges had quite such high inductance ($\approx$600 mH) as the M75ED, but the technique was valid because the only other cartridge to come even close in popularity to the Shure M75ED was the electrically similar Shure V15/III.

It is essential when measuring noise of an RIAA stage to check that the waveform genuinely is noise and not hum. The noise measurement is best made using an audio test set with a digital oscilloscope connected to its monitoring output, triggered from "line", and with "RMS" selected on the automated measurements. If averaging (over as many samples as possible, certainly >32) is invoked on the oscilloscope, the noise will disappear, possibly to leave a hum residual. Provided that the residual hum waveform has an RMS amplitude less than one-third of the noise waveform's RMS amplitude, the noise measurement is plausible.

It is the difficulty of securing a hum-free noise measurement using the fundamentally flawed unbalanced phono interface standard that leads to the engineering desire to substitute a pure resistance for the real cartridge in arm on turntable when measuring RIAA stage noise. Turntable rumble spectra are effectively system noise measurements that necessarily include a cartridge in an arm on a turntable, and published spectra invariably include peaks at mains frequency and its harmonics, even when the motor cogging frequency is known to be unrelated to mains, implying that these peaks are not vibration at all, but electrical interference.

Having made a 20 Hz–20 kHz bandwidth-limited noise measurement at the output of the RIAA stage, we need to express the measurement in a way that can be understood. We might have configured the RIAA stage's gain so that when used with its intended cartridge, +12 dB (ref. 5 cm/s at 1 kHz) produces 2 V_{RMS} in order to minimise volume changes when switching between digital and analogue sources. Our noise measuring instrument is likely to use a reference level of 0 dBu (perhaps incorrectly specified as 0 dBm), and we might measure noise at −68 dBu. If we wanted to express this noise in terms of dynamic range, we would need to express the maximum level of 2 V_{RMS} in dBu:

$$ dBu = 20 \log \left(\frac{V}{V_{reference}} \right) = 20 \log \left(\frac{2}{0.775} \right) = +8.2 \ dBu $$

Subtracting the −68 dB noise, we would have a dynamic range of 76 dB.

Alternatively, we might note that the RIAA stage had a measured 1 kHz gain of 70 dB (0.25 mV input, 775 mV output), and noise referred to the input could be expressed as:

$$v_{noise(input)} = v_{noise(output)} - \text{Gain (dB)} = -68 \ \text{dBu} - 70 \ \text{dB} = -138 \ \text{dBu}$$

If we wanted to compare this input-referred noise with that derived from a semiconductor manufacturer's data sheet, we would probably convert it into a voltage, resulting in a figure of 98 nV$_{RMS}$.

Using the same RIAA stage as in the previous worked example, we might instead have used an oscilloscope preceded by additional gain and 20 Hz−20 kHz bandwidth filtering to make the noise measurement directly in V$_{RMS}$. Perhaps we added a ×1000 (60 dB) amplifier and were able to configure the oscilloscope's probe menu to compensate, resulting in an oscilloscope automated measurement of 309 μV$_{RMS}$ noise at the output of the RIAA stage. We know the RIAA stage has a 1 kHz gain of 70 dB (×3160), so the noise at its input must be:

$$v_{noise(input)} = \frac{v_{noise(output)}}{\text{Gain}} = \frac{309 \ \mu V_{RMS}}{3160} = 98 \ \text{nV}_{RMS}$$

We might prefer to express the oscilloscope measurement in terms of dynamic range from the maximum output of 2 V$_{RMS}$, in which case:

$$\text{Dynamic range} = 20\log\left(\frac{306 \ \mu V_{RMS}}{2 \ V_{RMS}}\right) = 76 \ \text{dB}$$

Cartridges and magnetisation

We saw in Chapter 4 that using a DVM to measure winding resistance inevitably magnetised the core of a push−pull output transformer, necessitating explicit demagnetisation if primary inductance was to be measured correctly. But what about cartridges, and will a winding resistance measurement impair the cartridge?

The author used a DVM to measure the DC resistance of a Glanz MG 70R moving magnet cartridge (656 Ω), thereby magnetising it. At very low frequencies (<1 Hz), inductive reactance is negligible, so current will be limited by the DC resistance. Thus, even applying the maximum signal of 7 V_{RMS} from his 50 Ω function generator could only result in a current of 10 mA, and dissipate 71 mW. This didn't seem likely to damage the wire, so he demagnetised the cartridge by applying 7 V_{RMS} directly across one coil, sweeping frequency from 0.1 Hz to 10 kHz. He then measured coil inductance at 360 Hz (528.85 mH), measured DC resistance again (to magnetise the core), and immediately repeated the inductance measurement (528.66 mH). The effect of explicit magnetisation was to reduce inductance by $\approx 0.04\%$.

The reason that the inductance change was so small is that moving magnet cartridges are deliberately wound on soft magnetic cores having very low remanence. We need not worry about permanently magnetising a moving magnet cartridge's soft magnetic core by measuring its DC resistance.

If a moving coil cartridge had its coils wound on an iron armature, measuring DC resistance would magnetise it. Further, it would be difficult for the coils to impose sufficient alternating flux to demagnetise the armature. However, given that the armature is immersed in a magnetic gap imposed by pole pieces that are almost certainly operated at saturation, a little direct current has no effect. We need not worry about magnetising a moving coil cartridge's armature by measuring its DC resistance.

References

1. IEC standard 268 (Sound System Equipment), Part 10: Peak programme level meters.
2. J. Hagerman, On reference RIAA networks. *Audio Electronics*. Available at: http://www.hagtech.com/iriaa.html.
3. S.P. Lipshitz, On RIAA equalization networks, *Journal of the Audio Engineering Society*, 1979, **27** (6), 458–481.
4. A.N. Thiele and R.H. Small, A series of papers in the *Journal of the Audio Engineering Society*, starting from May 1971 (Vol. 19, No. 5) to January 1972 (Vol. 20, No. 1).

SECTION 3

EXAMPLES

CHAPTER 7

PRACTICAL PROJECTS

In this final chapter we will put all of the previous chapters to use by following the construction of three very different projects:

- Bulwer-Lytton power amplifier designed to drive the author's Arpeggio loudspeaker (see the "Articles" section of the diyAudio website).
- RIAA stage remote power supply.
- Hybrid balanced RIAA stage designed to allow a Denon DL103 moving coil cartridge to drive the Bulwer-Lytton power amplifier.

Detailed discussion of audio design featured in *Valve Amplifiers* (fourth edition), so our primary concern here is practical implementation and EMC (electromagnetic compatibility). The intention is not to provide a set of construction instructions that must be slavishly followed, but to highlight practical problems that occur using different chassis techniques and show how they may be overcome.

Bulwer-Lytton power amplifier

The high signal level makes it easier to render a power amplifier hum-free than an RIAA stage, and we can always listen to calming music from a digital source whilst making the RIAA stage, so it makes sense to tackle the power amplifier first. Despite the Bulwer-Lytton amplifier being rated at a mere 4 W per channel, it still has the problems of a larger amplifier, just to a less intimidating scale. See Figure 7.1.

Building Valve Amplifiers. DOI: http://dx.doi.org/10.1016/B978-0-08-096638-0.00007-2

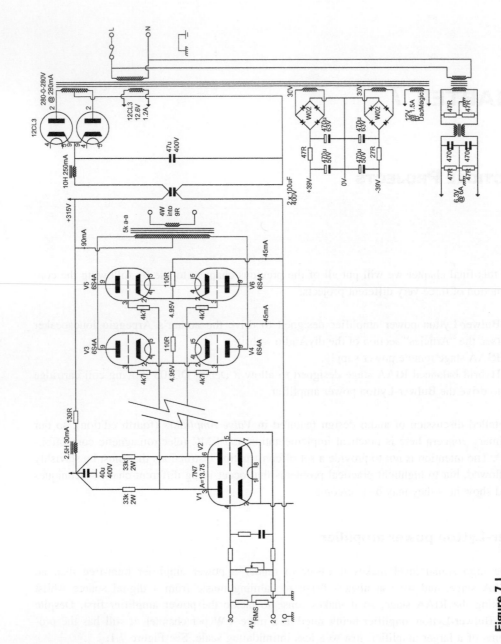

Figure 7.1

Simplified Bulwer-Lytton circuit diagram.

From the simplified diagram we see that the amplifier is a stereo push—pull design operating from a shared power supply and having a volume control plus baffle step equaliser dedicated to the Arpeggio loudspeaker, but no input switching. The inclusion of a volume control is highly significant because it immediately places the amplifier at table height in plain view, so it needs to be domestically acceptable.

Planning

Power amplifiers invariably require a strong chassis, so the rigidity conferred by the author's traditional picture frame technique is ideal. Having chosen our chassis type, we know that we have extreme flexibility on size, but we must first consider the factors that will ultimately determine chassis size and proportions.

Planning invariably starts with the largest heaviest lumps, and that means transformers and chokes. Referring to the diagram, we see that two HT chokes have been specified. In addition, we know that we should not power heaters from a transformer feeding rectifiers, so we need two mains transformers. We therefore have the following lumps of iron:

Sowter 0491s mains transformer for DC supplies (quite large)
Heater transformer providing 6.3 V_{AC} at 6 A for signal valves (probably reasonably small)
Two Hammond 1615 output transformers
5 H 250 mA choke (medium size)
2.5 H 30 mA choke (tiny).

Some of these lumps are large and all either emit or are sensitive to stray magnetic fields so leakage flux considerations probably override all others.

The mains transformer for the DC supplies is inevitably quite large, so it was specified to have steel shrouds (reduces leakage flux at coil edges) and also to have a Faraday shield.

The heater transformer for the signal valves is a compromise. A split bobbin EI transformer would have minimised interwinding capacitance C_{ps} but substantially increased leakage flux (especially in such a small transformer). Whilst it is possible to partially compensate for increased C_{ps} by adding common-mode filtering, shielding a low frequency magnetic field is almost impossible. Because the heater transformer is on an audio chassis (so separation from sensitive circuitry is limited) a toroidal transformer was chosen to minimise leakage flux, and common-mode filtering added.

We choose output transformers solely by their electrical characteristics (usually maximising the L_p/L_l ratio) and must live with any leakage. Because the Hammond 1615 transformers have steel shrouds, leakage from coil edges will be attenuated.

Fortunately, the supply is not choke input, so the first smoothing choke has minimal alternating voltage across it and its leakage flux will be correspondingly low. The second choke has even less alternating voltage across it and its leakage is entirely negligible.

Continuing in a power supply vein, the symbol for a capacitor gives no indication as to a capacitor's physical size, but HT capacitors are invariably large, so we should check all along the HT line. At the transformer end, we find a pair of 6CL3 valve rectifiers, so we make a mental note to check their size (1.188″ or 30.2 mm diameter), we then find a 47 μF 600 V reservoir capacitor, followed by a parallel pair of 100 μF 400 V four-wire capacitors, and finally a 40 μF 400 V capacitor. The diagram's capacitor symbols tell us that none of these capacitors has been specified to be electrolytic, so they will all be metallised polypropylene (polypropylene is the dielectric of choice for high voltage industrial capacitors).

Surprising as it might seem, custom plastic capacitor manufacture is a cottage industry, so the largest capacitors were specially made for the author by Suppression Devices of Clitheroe. The ideal extended foil capacitor has a large diameter but is short because this shape minimises capacitor inductance, and short four-wire tails at each end reduce lead inductance. Thus, the ideal implementation of the 100 μF 400 V capacitors is short and fat, but the largest practical winding diameter was 60 mm, so these capacitors are ≈ 62 mm diameter including their sleeving and ≈ 85 mm long. Because the tails exit at each end, safety considerations and minimising tail inductance require the capacitors to go inside the chassis, and **that** means we have just specified a minimum internal chassis height of 62 mm. But a capacitor this large certainly cannot be supported by its tails and must be clamped securely in place. Capacitor clamps invariably consume extra height, so an internal height of 75 mm (3″) is required.

You will see from later photographs that the author did **not** mount the capacitors as recommended, and there is a very simple reason for this. At the time of this amplifier's planning, the author had yet to realise the advantages of four-wire capacitor connection and had already made and cut all the holes in a 2″ chassis for the capacitors he had in stock, so he flatly refused to repeat all that careful metalwork. Nevertheless, measurement showed that the advantage of the four-wire capacitor connection was so great that it was worth having a pair of four-wire capacitors specially made that were drop-in replacements for the original types. They might look pretty, but they were expensive and slightly inferior, partly because winding them from 100 mm film marginally increased inductance, but mainly because forcing all the tails to exit from one end greatly increased lead inductance. Learn from the author's mistake by using the cheaper and superior version wound from 75 mm film, with two tails at each end, neatly encapsulated in heatshrink sleeving. The ideal position for each 100 µF 400 V four-wire capacitor is directly under its corresponding output transformer, thus allowing the output valves to be reasonably close to their transformer.

We have a number of hot valves, so we can use the cooling factor concept introduced in Chapter 1 to determine cooling provisions. See Table 7.1.

Table 7.1

Estimated cooling factors for the Bulwer-Lytton valves

	Estimated cooling factor
6S4A	151
6CL3	322
7N7	424

As should be expected, only the 6S4A output valves merit any real consideration, and even they are on the borderline of the weakest cooling provision of a twice envelope diameter exclusion zone, so our chassis does not need perforated sheet, and we can place the valves almost anywhere.

We have a capacitor input power supply, so the loop formed by the mains transformer, rectifier, and reservoir capacitor carries sharp pulses at 4−6 times the DC load current of 250 mA. We minimise unwanted voltage drops along the wires carrying these currents by minimising their length. Further, to minimise electromagnetic emissions we must minimise the loop area included by the wires. Taken together, these requirements mean that the rectifier valves and reservoir capacitor must be close to their mains transformer and **all three** connections to the HT winding must be tight together, preferably twisted.

Transformers are invariably heavy, so they should always be towards the edges of any chassis (where it is most rigid). Placing output transformers close to the loudspeaker terminals minimises the length of secondary wiring and associated resistance.

The previously mentioned loop area rule applies equally to magnetic susceptibility, but more importantly at valve impedances, capacitance to interference sources must be minimised, so input valves, volume control, input sockets, and associated audio wiring need to be close together and as far away as possible from the mains transformers and associated wiring. As a general rule of thumb, the signal connection to an output valve grid should add less than a tenth of output stage input capacitance (not forgetting Miller capacitance), so this path usually needs to be reasonably short.

Purely considering ergonomics, the volume control must be on the front panel, as should be the mains on/off switch. But putting a mains switch on the same panel as a volume control implies that the two controls are close and capacitive coupling becomes likely. There are three possible ways of avoiding the potential hum problem, and they all aim to reduce capacitance between the two circuits:

- Make the front panel sufficiently wide that the mains switch and volume control can be adequately separated.
- Put the mains switch towards the back of the amplifier but add linkage to bring its control to the front. The linkage is usually mechanical, but could be electrical − a low-voltage switch controlling a power relay. Positioning the mains switch at the back is safest because it confines mains wiring to a small area, and this technique is commonly used on commercial equipment.
- Fit an electrostatic screen between the two circuits.

The author didn't have any switches that would accept a mechanical linkage and relay control necessitates the constant power consumption of a standby supply, so he chose a wide front panel and, if necessary, a screen.

Microphony is inversely proportional to control grid signal amplitude, so we should always check the input valve. The indirectly heated 7N7 has a rigid cathode, so microphony is purely due to grid movement, but the low mutual conductance implies a reasonable grid-to-cathode spacing, making any grid movement a small proportion of grid-to-cathode spacing. With a grid sensitivity of $\approx 1\ V_{RMS}$, microphony should not be an issue, making anti-vibration valve sockets unnecessary.

Finally, we want the amplifier to look nice, and that means maximising the visibility of all that glowing glass. Because the amplifier needs to be at a convenient height for easy adjustment of controls, it must not be too deep to fit on a small table.

Taking all the previous considerations together, we find that our picture frame chassis needs $3'' \times 1''$ extruded aluminium channel plus a plain 1.6 mm top plate. Loudspeaker terminals will be on the back panel, output transformers will be nearby, and the 6S4A output valves will be in front of the transformers. The 7N7 will be in front of the 6S4A and the volume control will be nearby on the front panel. In order to separate the front panel mains switch from the volume control, the front panel must be wide, and amplifier depth minimised to allow it to fit neatly on a table. We can draw a preliminary mechanical layout. See Figure 7.2.

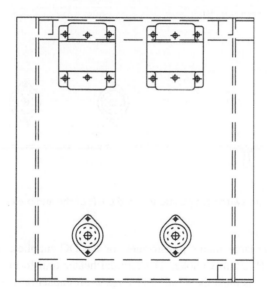

Figure 7.2
Power amplifier layout starts with the output transformers.

Although the placement of audio circuitry seems hopeful, we have not yet included the power supply. We could add the power supply at the back or to one side. If we add it at the back, it will make the chassis deeper (undesirable) and lengthen wires between output transformers and loudspeaker terminals (also undesirable). If we add the power supply to one side, depth will remain the same (desirable) but the wider chassis will increase separation between mains switch and audio circuitry (very desirable). We will widen the chassis. Motorcycle throttles are on the right, so the author favours volume throttles on the right, suggesting that the power supply be added to the left of the chassis. See Figure 7.3.

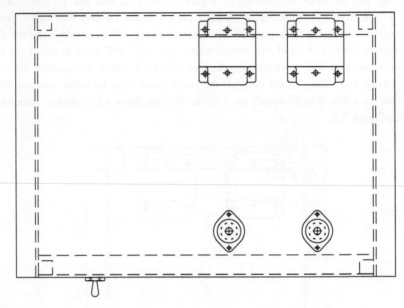

Figure 7.3
The power supply and mains switch are positioned to the left of the audio electronics.

We have to fit the large mains transformer for the DC supplies, the two 6CL3 rectifier valves, and the 5 H 250 mA choke. The toroidal heater transformer must be fitted within the chassis (safety), as must the 47 μF 600 V reservoir capacitor. (The author fitted his reservoir capacitor externally because he already had one in an expensive can, but it would be better and cheaper to fit a four-wire polypropylene capacitor internally directly under the large mains transformer.) To minimise wiring loop area, the rectifiers and reservoir capacitor should be between the mains transformer and the 5 H 250 mA choke. The chassis has sufficient depth to place the mains transformer and choke one behind the other, so the decision is between the mains transformer being at the front or back of chassis. The mains switch is at the front, and the choke connects to the 100 μF 400 V four-wire capacitors directly under each output transformer, so it makes sense to place the mains transformer at the front.

We know the rectifier valves must be between the mains transformer and choke, but they're quite pretty, so we put them in line at the edge of the chassis where they can be seen, which will help cooling. See Figure 7.4.

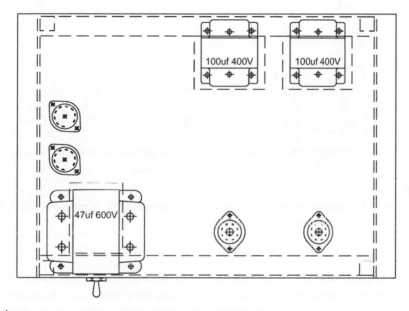

Figure 7.4
The HT transformer, rectification and filtering are generally close together.

Now that we know the positions of the major magnetic components, we can determine their optimum coil orientation. The most important orientation is that of the large mains transformer — its coil axis must not point towards the input valves, so we align it front to back. There is still a leakage field due to coil edge effects, but this tends to cancel 90° off the coil axis, so the input valves are best positioned on this cancellation line. The stepped attenuator volume control would be most conveniently wired if it was between the two input valves. We don't want to pass sensitive input wiring past output valves, so the input sockets could be on the side of the amplifier, towards the front, and because the amplifier has balanced inputs, they will be three-pin XLRs. See Figure 7.5.

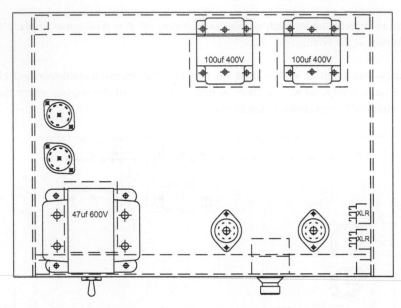

Figure 7.5
HT transformer coil orientation is crucial, so not only is it pointed at 90° to the input valves, but they are positioned on the null line from the centre of its core.

It's not a good idea to direct even the small leakage flux of an output transformer towards output valves, so most manufacturers bring the primary and secondary wires out on opposite sides of the core, enabling more direct wiring when the coil axis is in the preferred orientation. Although aligning coil axes parallel to the rear edge of the chassis is best, and allows the secondary tails to drop through the chassis directly above the loudspeaker terminals, it has the disadvantage of aligning the two transformer coils, potentially allowing coupling that would degrade stereo crosstalk. Later considerations resulted in a 35 mm gap between the two output transformers that should render this problem negligible.

The author likes big loudspeaker terminals that can be firmly tightened to make a gastight connection (preventing the formation of distortion-producing copper-oxide rectifiers in series with the signal). The Belling Lee 32A L309 range is 1″ in diameter, so you can get a good grip on them — provided there's enough surrounding room for your fingers.

It makes aesthetic sense to place terminals symmetrically either side of an output trans-former centre line, but there's no point in allowing a large separation between the left channel's two terminals if one is squeezed tight against a right channel terminal — the four terminals need equal spacing.

The Hammond 1615's steel shrouds shield coil edge effects, so there isn't a magnetic problem in moving the output valves close to their output transformer, but there isn't much advantage either, and thermal caution would suggest that a twice envelope diameter thermal exclusion zone be retained for each 6S4A — the author used 50 mm. Note that the dull black textured finish of the Hammond 1615 output transformer shrouds minimises reflection of radiant heat back to the output valves (unlike chrome-plated shrouds). Just like the loudspeaker terminals, we want equal spacing between valves, and the thermal exclusion zones force two rows of four 6S4A output valves.

Each 7N7 input valve should be reasonably close to their respective output valves but not close enough to be heated by them, and this criterion sets the depth of the final amplifier because the distance from the rear panel to the first row of 6S4A has been set by the size of the output transformers and the 6S4A thermal exclusion zones, whereas the distance from the front panel to the centre of the 7N7 was set by the centre line of the mains trans-former. See Figure 7.6.

We know that the chassis will be fabricated from extruded channel and a 1.6 mm top plate, and because the front and rear panels have controls or terminals on their surface, their channels must face inwards. The amplifier's weight will not be trivial, so facing the side channels outwards avoids difficult joints whilst providing convenient handles. The author prefers to mount transformers directly above the channel because it prevents their weight from bending the top plate and accepts the resulting increased metalwork. Thus, the chassis depth we derived earlier can be assumed not to have been altered by considera-tions of channel orientation.

Having fixed chassis depth, we can consider width. The spacing of the 6S4A has been set, and although we would be happy for the inside of the channel to come very close to a 6S4A socket, doing so pushes the XLR input sockets a little too close to the nearest 7N7 for comfortable wiring — a spacing of ≈ 60 mm from the centre of the nearest 7N7 to the inside face of the channel seemed better. Incidentally, right angle XLR cable plugs are available, enabling the external input cables to be discretely routed to the back of the amplifier within the channel if required. All of our chassis considerations so far have been

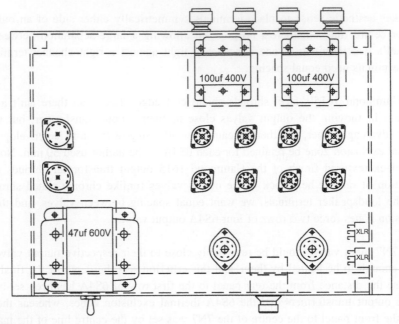

Figure 7.6
Output transformers always have their coils oriented away from the output valves, thermal considerations dictate output valve positioning, whereas wiring convenience generally determines input valve positioning.

referenced to the inside faces of the channel, so we must remember to add the width of the channel either side of the top plate.

The toroidal heater transformer's tape covering is not robust so it must be enclosed by the (conductive) chassis. The accepted way of securing toroidal transformers is to pass a large screw through their centre and use this to clamp the transformer between chassis and the large rigidised steel washer supplied with the transformer, but this deforms a thin aluminium chassis because the area around the screw is unsupported. A simple solution might fit a >30 mm diameter steel washer above the top plate to spread the load, but this is scarcely pretty. A more elegant solution fits a cylindrical distance piece within the toroid having a slightly smaller diameter than the toroid's internal diameter – preventing side to side movement. The distance piece is fastened to the top plate using three small screws near its circumference, minimising the top plate's unsupported distance.

Provided that the length of the distance piece is made such that the toroid's rigidised washer touches it when the foam or rubber washers under the toroid are lightly compressed, an axial screw can be tightened securely (so that it does not loosen with vibration) yet avoid deforming the top plate. See Figure 7.7.

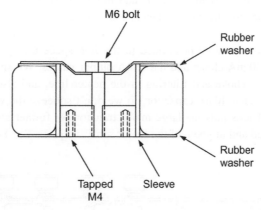

Figure 7.7
A distance piece made of a tubing sleeve and press-fitted solid insert prevents the toroid loosening over the years.

If all that seems a lot of work, it's because it was. But it made for a neat job and that toroid won't loosen over the years.

The drop-through 5 H 250 mA choke is best placed at the back of the chassis, exactly on the output transformer coil axis but with its coil axis at right angles. Although the choke can't couple hum into the output transformers, its coil is aligned with the large mains transformer, but there's a good distance separating them, so this shouldn't be a problem.

We need a fused IEC mains inlet and it should be as far away as possible from the loudspeaker terminals on the back panel. The author decided it would be nice to fit a Curtis 700 series hours counter to log valve use (and be reset at valve replacement), but this doesn't need to be highly visible, and it could be powered from the raw HT, so it made sense to put it near the 5 H 250 mA choke. Burying the hours counter between the choke and an output transformer would make it difficult to read, so the hours counter moved

499

towards the edge of the chassis, together with the IEC mains inlet, allowing twisted pair mains wiring to be pushed firmly into the corner of the channel and top plate.

Because of their relative coil orientations, the 5 H 250 mA choke could slide along the output transformer centre line quite close to the nearest output transformer, but it is better to position it close to the hours counter, and move the two away from the output transformer to leave space for the toroidal heater transformer.

The heater toroid can now be positioned in the free space under the chassis somewhere between the 5 H 250 mA choke and the 6S4A output valves. We should allow some room for a common-mode choke and filtering in the heater line, and also for the 2.5 H 30 mA choke and its associated 40 μF capacitor, so we don't squeeze the width too hard — especially as a wider chassis puts the large mains transformer further away from the 7N7. We have now positioned and aligned all the major sensitive parts. See Figure 7.8.

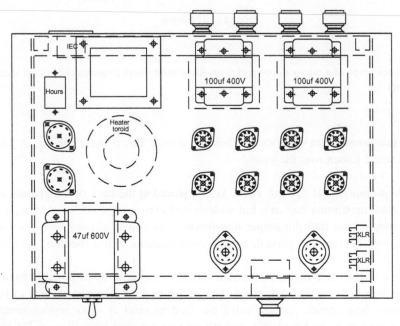

Figure 7.8
Final layout of major components.

The simplified diagram of the Bulwer-Lytton amplifier showed a housekeeping block, and this now needs to be considered. See Figure 7.9.

The housekeeping circuitry consists mostly of silicon and small components, so hard-wiring or tag strips are inappropriate — only strip board or a PCB make sense. If you want to lay that circuit out neatly and efficiently on strip board, then good luck to you — the author made a pair of PCBs. Separate PCBs were chosen not because of misguided stereo crosstalk considerations, but because mounting individual PCBs vertically between the 6S4A valve sockets enabled efficient use of chassis space. See Figure 7.10.

The housekeeping block has an MJE340 in each constant current source having its tab electrically tied to its collector, which is connected to the audio signal, and the 6–8 pF shunt capacitance that would be suffered by thermally bonding the MJE340 to the chassis would expend our entire 10% allowance of output stage 60 pF input capacitance. Fortunately, the four MJE340 transistors only dissipate 1.6 W in total, so we could avoid the unwanted capacitance by losing that heat via free-standing heatsinks and accept the minor rise in internal air temperature.

The greater a simple equaliser's correction, the wider the frequency band encompassed by the sloping section of its response. The significance of the previous statement is that a misplaced slope causes amplitude errors for all frequencies encompassed by that slope. Although slope positioning is determined by the combination of resistors and capacitors, capacitors have a much larger thermal coefficient than resistors, so high-loss equalisers such as RIAA cannot be the exact inverse of recorded response unless care is taken to mini-mise capacitor temperature changes. Conversely, the 2.4 dB baffle step equaliser within the Bulwer-Lytton amplifier is relatively insensitive to temperature, and this is why we could tolerate the increased internal air temperature caused by free-standing MJE340 heatsinks.

The FQP1N50 source followers each dissipate 1.1 W, but their tabs are connected to their drains, which are connected to the $+39$ V supply, so we're happy to incur extra capaci-tance by heatsinking them to the chassis via an insulating kit — which is just as well because 4×1.1 W $= 4.4$ W would raise internal air temperature appreciably. The PCBs were laid out so as to put the FQP1N50 at the edge of the board where they could bond to a heatsink bracket. See Figure 7.11.

The heatsink brackets were cut from $1'' \times 1'' \times \frac{1}{8}''$ extruded aluminium angle, with most of the material cut away on one side to clear PCB components. The bulk of the material was

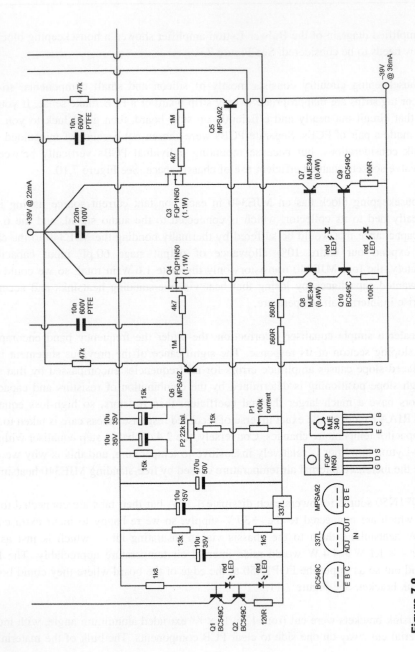

Figure 7.9

Like many contemporary valve amplifiers, the Bulwer-Lytton contains silicon for housekeeping.

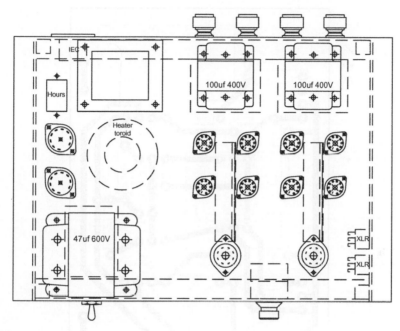

Figure 7.10
The housekeeping PCBs will fit between output valves.

sawn away and the remainder filed, although the author cut a later version of the bracket on the mill. Note that the two bias adjustment resistors point through the bracket, allowing external adjustment, but beware that the author's (surplus) resistors had an unusual pin-out, so a PCB layout change is probably necessary. Short M3 screws into tapped holes secure the two FQP1N50Ps to the heatsink bracket. See Figure 7.12.

Once made, the housekeeping boards were powered from a bench supply and tested:

- Supply voltage was gently increased from 0 to ± 39 V and load current monitored ($+39$ V at 22 mA, -39 V at 36 mA expected). Had current exceeded expectations by >5 mA, applied voltage would have been quickly reduced and the fault investigated, hopefully without any silicon being destroyed.
- -12 V regulator checked for correct output voltage (± 1 V).
- Balance control adjusted to send matched bias voltages to output valve grids.

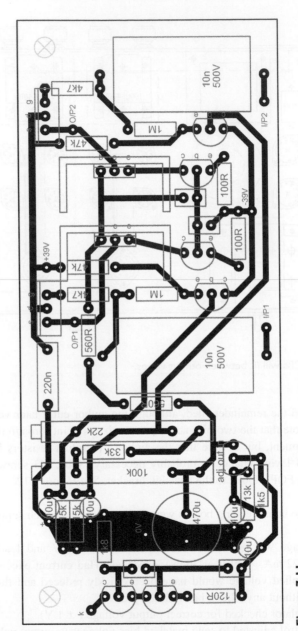

Figure 7.11
Housekeeping PCB layout (view from foil side).

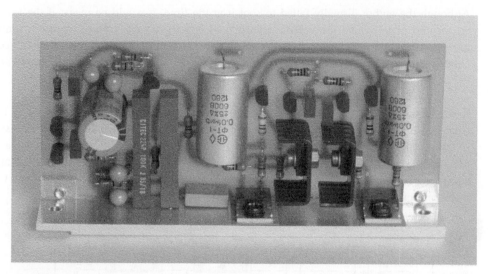

Figure 7.12
The housekeeping PCB fits to an aluminium bracket that also serves as a heatsink for the FQP1N50.

- Current control adjusted to set output valve grid bias voltage to -5 V. This should allow substantially correct operation.
- Source followers tested with a 10 kHz 10 V_{pk-pk} square wave monitored by a >100 MHz oscilloscope. A square wave is more onerous than a sine because the 100% feedback in a source follower tends to mask faults, and the sharp edges of the square wave are more likely to provoke any incipient instability (power FETs sometimes oscillate at tens of MHz).
- Note that without an external load, the input CCS cannot operate correctly and its LEDs did not light.

Having been successfully tested, the housekeeping boards were put to one side.

Returning to chassis metalwork, we know where we need to drill/cut the top plate for the major components, but we must secure the top plate to the picture frame formed by the aluminium channel. A number of the transformers have screws that pass through the top plate into the channel, so it's a matter of positioning an appropriate number of screws evenly spaced along the remaining edges.

Although the author dismally failed to spot it at the time, it would be nicest if the screw on the top plate behind the volume control was exactly on the volume control's centre line, so that's one screw positioned. We need a screw at each chassis corner, and we need another midway between the screw behind the volume control and the corner screw, and the addition of **that** screw sets the screw spacing along the front.

It would look odd if screw spacing on the sides was radically different to that along the front and back, so the sides need similar spacing to the front spacing. The author's chassis is fractionally deeper than yours because of his exposed 100 μF 400 V four-wire capacitors, and he needed six screws (including the ones at the ends). The back didn't need any additional screws because the output transformers and 5 H 250 mA choke secured the top plate perfectly well. See Figure 7.13.

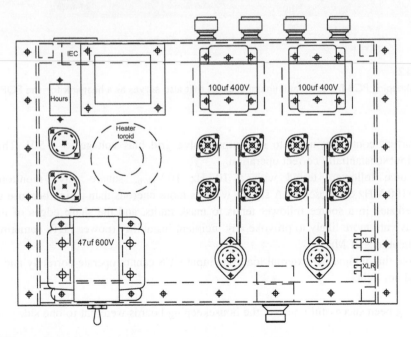

Figure 7.13
The top plate looks best with equally spaced screws around its edge.

The planning phase is over — we can now mark out and cut metal.

Making the Bulwer-Lytton chassis

The author first cut the aluminium channel and fastened the picture frame at the corners using 16 mm square bar, then placed it at one corner of a sheet of 1.6 mm aluminium, then marked the two cutting lines with a scriber, donned earmuffs, and cut the metal with a power jigsaw. Once cut, the fit was confirmed, and the rough cut edges smoothed first with 160 grade emery cloth backed by a piece of flat wood. In woodworking, longer planes average deviations over a longer distance, so longer planes are preferred when flatness is important. Similarly, emery cloth backed by a 4″ block of wood has superior averaging compared to a 1″ wide file, resulting in straighter edges. Coarse emery cloth tends to throw up quite a burr, so this was carefully removed using 320 grade silicon carbide paper backed by the wood block. Finally, the cut edges were smoothed with a medium file, followed by a gentle deburr using a flat needle file (had the needle file been used on the first deburr, it would probably have dug in). The top plate was again checked for fit on the picture frame.

The chassis is quite large, and the author has a standard printer, so his layout had to be printed in several overlapping sections, aligned, and stuck together to form a single printed sheet having the same size as the top plate. At this point, it's a good idea to check a few dimensions that cross from one sheet of paper to another to make sure that nothing has stretched or moved. Errors are easily fixed at this stage.

Satisfied that the paper template was correct, it was lightly glued to the aluminium top plate and centres of holes were marked through using a sharp scriber. The 5 H 250 mA choke and hours counter both needed square holes and these were marked by **very** lightly spotting through their corners, again using the sharp scriber. (It's essential to spot the corners of square holes as lightly as possible because three-quarters of each dimple remains after the hole is cut.) It's easy to omit a hole position, so check carefully before peeling away (and probably damaging) the paper template.

Marked drill holes were centre-punched and part-drilled to leave a shallow cone using a 3 mm centre drill lubricated at each hole before drilling by a drop of methylated spirits from a child's paintbrush. Because the top plate could not be clamped when drilling, a small burr resulted on the reverse side of each hole as the centre drill's pilot broke through, and this needed to be deburred after each hole to allow the top plate to lie firmly on the wooden drilling block — tedious, but essential for accurately placed holes. Once all the holes were drilled, the top plate was compared against the paper template and the previously unnoticed scriber dimples centre-punched and drilled.

It's best to drill the smallest holes first (that way, if you accidentally drill a hole to incorrect size it can always be opened out to the correct size), so a lubricated 2.8 mm drill was used for the 6BA clearance holes needed for the 6S4A valve sockets, and these were deburred. Centre distances were checked across each valve socket, and as they seemed correctly aligned, they were redrilled using a new 2.9 mm drill carefully lubricated and running appreciably faster than the 2.8 mm drill. The point of this seeming complication is two-fold:

- Had a hole wandered, a needle file could have corrected it and the final hole would still have been the correct size in the correct place.
- A fast-running 2.9 mm drill removing very little material produces hardly any burr, and deburring almost always removes a little material at the edge of the hole. This loss of material becomes proportionately significant when a small nut (such as 6BA or M3) must bear upon it, so we take extra care with small holes to minimise both the burr and subsequent deburring to minimise this loss.

The 6S4A valve sockets need ¾″ holes, and the 7N7 need 1¹⁄₁₆″ holes, so the pilot sizes for these punches were checked and their holes drilled appropriately. The remaining top plate holes were drilled, deburred, and checked for accuracy. The chassis punches were freshly greased to enable clean cutting and the valve socket holes cut, taking care to ensure that the punch distorted the top plate's inside rather than outside face. Punched holes were deburred using a wiggly deburring tool.

Having checked that all was well, the top plate was aligned and clamped to the picture frame using four small G-clamps. Cardboard spacers 1 mm thick protected the comparatively soft aluminium from marking by the G-clamps. Once aligned, a transfer punch marked the picture frame holes that needed tapping for the top plate-securing screws.

An open picture frame as large as this tends to tilt when held under the drill because it can't always be supported properly by the table, so you might want to disassemble it for drilling. Before you disassemble, make sure you uniquely identify each piece's position so that it goes back together correctly. Once under the drill, a ⅛″ centre drill started the holes, prior to drilling them 4.3 mm for tapping to M5. Methylated spirits lubricated each drilling stage.

The M5 screws and their washers easily mark the top plate, so the washers were fitted to the screws with their sharp edges away from the top plate. Once drilled and tapped, the picture frame was test-fitted to the top plate. All was well, and the two square holes for the 5 H 250 mA choke and hours counter needed to be cut, so the top plate was removed. The hole for the choke could have been cut using a power jigsaw, but a better job can be achieved by hand using a tension file. The top plate needs to be held solidly whilst cutting, so it was clamped vertically between the jaws of a Workmate, making it easy to see where the cut needed to go. The top plate was rotated for each of the four cuts so that cutting was always vertical, making it easier to obtain a precise cut. Once cut, the edges were filed, deburred, and checked by carefully dropping the choke into place. The hole for the hours counter was much harder because it is smaller and rear-mounting leaves the cut edge visible, but the technique was the same. Cutting this hole took a long time because of the care required to obtain a neat snug fit.

The top plate was fitted to the picture frame again and checked for areas where holes in the top plate required the picture frame to be cut away. The 5 H 250 mA choke required a large rectangular cut, so this was marked through onto the rear section using a scriber, and this section of channel unscrewed. This required rectangular cutout was slightly awkward because it was in only one of the edges of the channel, so the other edge was only 2″ away (remember that the author **didn't** use the recommended 3″ chassis depth), making it difficult to saw and file. Ordinarily, the author would have solved the problem in a trice using his milling machine, but at that point the mill was still crated following a house move. There was nothing for it but to cut the rectangular cutout by hand. The bulk of the material to be removed was sawn away (vertical cuts at each end, then a horizontal cut with the saw blade angled to clear the channel's far side), followed by filing. Although the sliding stainless steel cover taken from a 3½″ floppy diskette served as a guard on the channel's far side, care and patience were still required. See Figure 7.14.

The output transformer secondary leads pass through the top plate, needing grommets to prevent chafing, so the corresponding holes in the channel below needed ⅜″ diameter holes to allow the grommets to sit inside them. These holes were carefully drilled ½″ with the work firmly clamped to the drilling machine's table. Once drilled, the holes were enlarged and converted into slots by sawing and filing so that the rear section would not be held captive by output transformer secondary leads. See Figure 7.15.

Holes in the rear section's face side were then drilled for the loudspeaker terminals. The hole for the IEC inlet was marked out, accurately cut using a tension file in hacksaw frame, tidied with appropriate files, and deburred using a needle file. The inlet was inserted, securing holes marked through, then drilled and tapped M3.5. A large circular

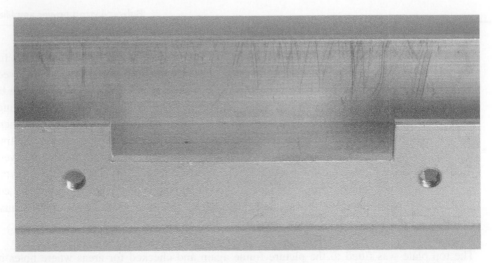

Figure 7.14
This slot for the 5 H 250 mA choke was cut and filed by hand.

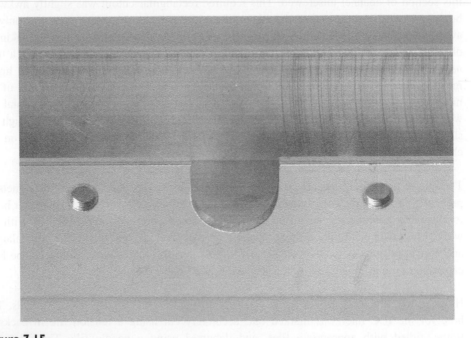

Figure 7.15
Rather than leaving closed holes, the holes for output transformer leads were converted into slots to enable the rear channel to be withdrawn if necessary.

hole was needed for the four-pin XLR socket that supplies $12\,V_{AC}$ to the author's DacMagic (allowing him to dispense with its loathsome wall wart), so this was marked using a pair of dividers, and a ring of 3 mm holes drilled just inside it to weaken the ⅛″ aluminium sufficiently to enable a chassis punch to cut that thickness of aluminium without damage. See Figure 7.16.

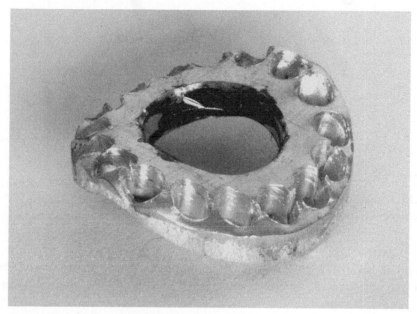

Figure 7.16
Chain drilling a series of 3 mm holes just inside where the punch would cut weakened the metal, enabling the punch to cut through ⅛″ aluminium.

The volume control and equaliser module

The volume control consists of a switched variable resistor plus additional components at input and output, and minimising capacitively coupled hum requires that these small components are as close to their variable resistor as possible. Each variable resistor is a thirty-position switch wafer having surface-mount resistors soldered directly to it, so the ideal solution was to mount the other components on a PCB between switch wafers. Thus, the

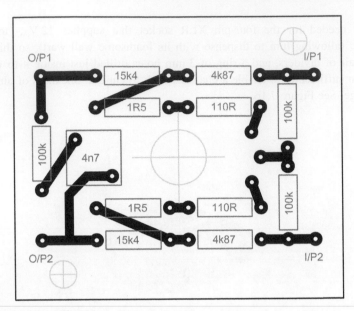

Figure 7.17
Volume control PCB layout (view from foil side).

baffle step equaliser and volume control tail resistance live on a pair of subsidiary PCBs that were dimensioned and drilled so that they could be stacked between wafers on the Type 72 switch that formed the volume control. See Figure 7.17.

The 4n7 capacitor was carefully positioned so as not to foul the protruding 8BA fasteners on the switch rotor as it rotates. See Figure 7.18.

As he has traditionally done, the author soldered the volume control's 1206 surface-mount resistors directly between the Type 72 switch pins. See Figure 7.19.

Unsurprisingly, given their age, the switch contacts needed cleaning (brass polish), and a test joint showed that the solder pins were reluctant to solder. Surface-mount components are quite fragile, so to enable damage-free soldering to the resistors, each of the switch pins was scraped clean with a scalpel, tinned, cleaned with desolder wick, then defluxed.

Figure 7.18
This PCB carries the baffle step equaliser, grid-leak resistors, and volume control tail resistance. Note that the volume control tail resistance (221 Ω ± 0.1% in series with 2 Ω ± 1%) was broken into a pair of balanced tails (110 Ω ± 0.1% in series with 1R5 ± 1%).

Figure 7.19
1206 surface-mount resistors just fit between the pins of a Type 72 switch.

With the switch still assembled, its control was turned to minimum, **then** the wafers removed, taking care not to disturb the position of the moving contact. Since the resistors are on the back of the wafer, as volume is increased, total resistance increases as the resistors are added **counter**clockwise. It's easy to make a mistake when soldering many small resistors, so the author started soldering from the end having the lowest value resistors (the position of the moving contact) and checked total resistance as each resistor was added – so much easier than correcting an error later.

Heater wiring

Once the subsidiary PCBs had been made and the resistors soldered to the switch wafers, the entire stereo volume control could be assembled and test fitted. At this point, it became clear that the 7N7 heater wiring would need careful routing to keep it away from the switch wafers, so it was pushed firmly into the channel's internal corner at the very front of the chassis. Even so, the author was uneasy about capacitive coupling between heater wiring and studs on the switch, and wasn't reassured by a back-of-envelope estimate of capacitance and resultant coupling, so he cut and folded a screen out of 0.058 mm aluminium foil to cover the heater wiring, then trapped that foil between channel and top plate to make electrical contact to the chassis. See Figure 7.20.

Each of the ten signal valves has its own twisted pair going back to the heater transformer, with the 7N7 heaters being distinguished by red/black rather than blue/black. The twisted pairs were laced together in a loom and glued to the chassis where necessary with hot melt glue. The 7N7 heaters were fed via the common-mode filter, whereas the 6S4A received raw power from the 230 V: 6 V 50 VA toroid.

Because each valve has its own pair going back to the heater transformer, and they are of differing lengths, they suffer differing voltage drops, so each heater voltage was measured and an average taken. On test, from regulated mains via a Variac, the heater transformer required $235\,V_{RMS}$ to set $6.3\,V_{RMS} \pm 2\%$ on all ten signal valves. Heater transformer mains current was $180\,mA_{RMS}$, so 5 V could be dropped by a $27\,\Omega$ resistor dissipating 0.9 W, enabling 240 V operation (measured UK mains remains 240 V despite the European paper harmonisation of 230 V). Since this resistance would be in series with the mains, insulation from resistive element to chassis capable of withstanding a $500\,V_{RMS}$ LN to E portable appliance insulation test is required, so the 560 V WH25 type was chosen even though its power rating greatly exceeds what is needed.

Figure 7.20
The volume control and baffle step equaliser assembly have been test fitted and heater wiring put in place.

To minimise the number of visible screws, and optimise cooling, a pair of holes were tapped M3 to secure the resistor to the nearby aluminium channel rather than the top plate. See Figure 7.21.

Installing the housekeeping PCBs

Once the heater wiring was in place, the housekeeping PCBs on their brackets could be fitted. The PCBs carry the CCS for the differential pair input stage, and this needs to be close to the valve pins, implying that the PCB and its bracket should butt up against the socket's PTFE base, requiring the bracket to cover the socket's steel flange. Thus, part of the rear face of the PCB bracket was relieved by fractionally less than the thickness of the Loktal flange so that a screw passing through the chassis and flange into the bracket could clamp the Loktal flange without distorting the top plate. See Figure 7.22.

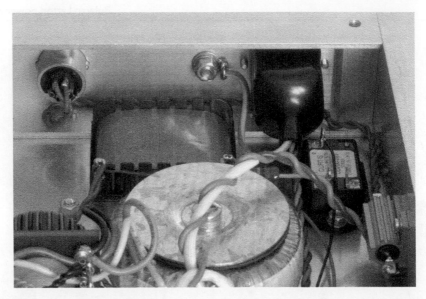

Figure 7.21
Counterclockwise from bottom right: Heater transformer 27 Ω series resistor, hours counter, booted IEC inlet, mains earth bond using M6 fasteners, 5 H 250 mA choke, four-pin XLR 12 V$_{AC}$ outlet to DacMagic.

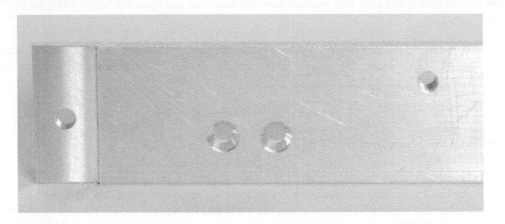

Figure 7.22
The rear face of the housekeeping PCB bracket has been relieved to allow it to clamp the valve socket's flange. Note also that the holes leading to the bias adjustment resistors have been countersunk using a centre drill to guide a fine screwdriver into the correct position.

An alarming problem surfaced when the author decided that as he had bought twenty 33 kΩ ± 5% 2 W anode load resistors for the 7N7, it wouldn't hurt to select matched pairs. Removed from their bandolier and measured, all but one turned out to be 12 kΩ, despite all being clearly marked orange, orange, orange, gold. Such poor quality control might be handy for deterring reverse engineering, but it's not so useful when you want a circuit to work as per the diagram. Moral: It's worth checking component values before soldering.

Loudspeaker terminals and output transformer wiring

The 32A Belling Lee screw terminals for loudspeaker connection require some of their thread to be removed to enable direct soldering. See Figure 7.23.

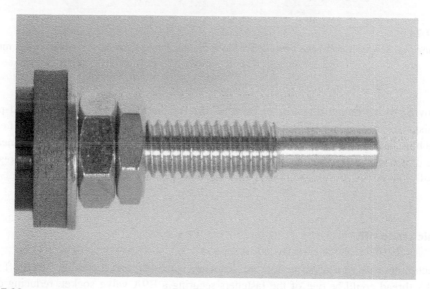

Figure 7.23
The far end of this 32A Belling Lee screw terminal has been machined to expose the brass and make for easier soldering. Note the flimsy metal washer to the left of the nuts.

Fitting the loudspeaker terminals requires a hole of ½″ diameter for the central boss and another of 3.1 mm diameter for the anti-rotation spigot, with ¹³⁄₃₂″ (10.3 mm) between their centres. The author drilled the ½″ holes with the rear panel firmly clamped to the drill table. Sadly, the spigot is quite weak, and if the washer bearing onto the chassis applies insufficient friction, all the torque when securing the terminal or tightening onto cable is transferred to the spigot, shearing it. Having sheared one spigot and identified the problem, the author replaced the terminal's easily distorted ⅜″ plated brass washer (seen in Figure 7.23) with a 1″ diameter thick steel washer. See Figure 7.24.

Figure 7.24
The Belling Lee 32A terminals have been fitted. Note the much more substantial replacement metal washer.

Moving to the output transformer primary, the Hammond 1615 has unequal half-primary resistances, but adding 15 Ω in series with the brown half-primary (81 Ω) made it equal to the blue half-primary (96 Ω), enabling output stage current balance to be set by measuring the direct voltage (no signal) between the anodes and adjusting for 0 V. A 15 Ω resistor is negligible in terms of the AC load of 5 k$\Omega_{a\text{-}a}$ and need not be bypassed.

Ceramic stand-offs

Each output valve needed an adjacent insulated stand-off, and a ceramic stand-off using a 6BA thread could be one of the fasteners securing a B9A valve socket, reducing drilling. The best positioning was immediately adjacent to the heater pins, which would have made heater wiring very difficult, so they were fitted **after** heater wiring. Given the likely corrosion, scarcity, and consequent price of NOS stand-offs, all of these stand-offs were salvaged; the solder was wicked off and bulk flux cracked away with a scalpel before scrubbing with defluxer followed by isopropyl alcohol. Two additional stand-offs were needed for the junction between each 100 μF 400 V Kelvin capacitor and output transformer centre tap. Take extra care with ceramic stand-offs requiring a nut to secure them in place as this implies the use of a pair of spanners, easily applying enough torque to shear the brass thread — which is only fractionally stronger than the ceramic shaft.

Harmonic equaliser resistor

The Bulwer-Lytton amplifier ordinarily uses the Hammond 1615's 8 Ω tap to match all four 6S4As to the 9 Ω load of the Arpeggio loudspeaker, but connecting the 9 Ω load to the 4 Ω tap allows the amplifier to operate at half power using only one pair of output valves (this is why each pair has its own harmonic equaliser resistor). Prior to building this pretty version, the author selected thirty 6S4A that were close to the mean of $I_a = 22.5$ mA at $V_a = 320$ V, $V_{gk} = -11$ V, then tested each against a reference valve by temporarily substituting his AC constant current sink (Figure 4.80) into the 6S4A cathode circuit (set to 45 mA). Optimum harmonic equaliser resistor value against the reference valve was determined by minimising third harmonic amplitude compared to second at a signal level 1 dB below 6S4A grid current. Out of these thirty valves, eleven were rejected for THD $\geq$ 0.5% or disproportionately high amplitudes of higher harmonics. The mean value of optimum harmonic equaliser resistor for the remaining nineteen was 111 Ω ($1\sigma = 21$ Ω), so the original 91 Ω harmonic equaliser was increased to the E24 value of 110 Ω. Annoyingly, this change caused the housekeeping board on one channel to be unable to set correct output stage current without a resistor change, which is why Figure 7.9 has a 15k series resistor feeding the 22k balance control rather than the original 33k.

Rather than select valves to match a fixed resistance, you might prefer to fit a 100 Ω variable resistor in series with a 47 Ω fixed resistor as a variable harmonic equaliser resistance, but beware that adjusting this resistance during operation also changes DC conditions.

The fact that the valves had to be selected twice, first for DC conditions, then for AC conditions, resulting in 50% wastage, underlines why the otherwise elegant harmonic equaliser distortion cancellation technique is not commonplace. So far as commercial production is concerned, selecting devices is always expensive (in time, even if the devices are cheap and wastage is low), and negative feedback is a cheaper and more reliable way of achieving low distortion.

It's obvious with hindsight, but valves requiring the same value of harmonic equaliser resistor when compared against a reference valve must be matched, so when the author took two valves that had individually needed a 135 Ω harmonic equaliser and tried them as a push—pull pair with an optimised harmonic equaliser resistor, he achieved near-perfect second harmonic cancellation and 2 W with only 0.1% THD.

Mains wiring

Unlike the toroidal heater transformer, the large mains transformer has taps for different mains voltages, so these flying leads were taken to a turret tag board, each entering its turret coaxially from the back of the board. Turning to the front of the board, the heater transformer (via its 27 Ω resistor) was soldered using a wrapped mechanical joint to the 0 and 240 V taps towards the bottom of their turrets. Incoming switched mains was tack-soldered to the top of the turrets so that 220, 230, 240, or 250 V mains could be accommodated by judicious repositioning of these incoming wires — the large mains transformer would then act as an autotransformer to the heater transformer, always applying 240 V and thus ensuring correct heater voltages.

A short thick wire connected the IEC mains inlet's earth pin to chassis via an M6 solder tag with a serrated washer between tag and chassis to ensure a gas-tight low-resistance bond. Once the wiring from the IEC mains connector was in place, the amplifier's earth bond resistance and high voltage leakage current could be pass/fail tested using a PAT tester, and it passed. A more scientific earth loop test using Kelvin probes measured 45 mΩ resistance from the earth pin of a standard 1.2 m IEC lead to the amplifier chassis. Mains safety having been established, subcircuits on transformer secondaries could be connected and tested in situ.

Low voltage power supplies

The large mains transformer secondaries were on flying leads, which was convenient for the nearby rectifier valves, but the housekeeping and DacMagic windings led to destinations that were too far away, so another turret tag board was needed, again with the transformer connections entering turrets coaxially from the back of the board.

The power supply for the housekeeping boards was the first PCB to be fitted, connected to the mains transformer, and tested. See Figure 7.25.

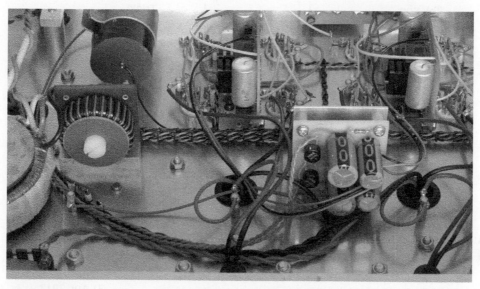

Figure 7.25
The power supply for the housekeeping boards can be seen to the right, the heater common-mode choke to the left.

The housekeeping board's two outputs were individually monitored, and a Variac gently increased mains voltage from zero, immediately causing both meters to rise from zero, eventually reaching 43 V unloaded. Had either supply failed to immediately rise, power would have been cut and the fault investigated.

When the supply was connected to the housekeeping boards and tested, each supply produced its nominal value of 39 V with 240 V mains applied, and the two LEDs near the power devices lit, but the two LEDs in the cascode CCS did not because the CCS was not yet connected to its load. Briefly connecting an ammeter between each CCS output and 0 V lit the CCS LEDs and ≈ 8 mA was registered (the exact current is uncritical, and 8 ± 1 mA is fine). Under quiescent conditions, the -39 V supply is loaded more heavily than the $+39$ V, and this is reflected by their ripple amplitudes. See Figure 7.26.

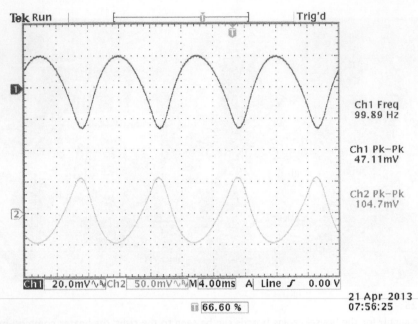

Figure 7.26

± 39 V ripple waveforms: Ch1 (upper) +39 V; Ch2 (lower) −39 V.

The rounded ripple waveforms suggest minimal high frequency content, and an FFT confirmed that harmonics >500 Hz were below 1%.

In the time since the fourth edition of *Valve Amplifiers* was written and work began on this second edition of *Building Valve Amplifiers*, the author has developed his measurement techniques to the point where he can obtain a plausible model for mains transformer secondary leakage inductance. Further work established that rather than fitting a capacitor across each diode, a single snubber network across each transformer winding was more effective at preventing ringing. Naturally, once an improvement was possible, it had to be made, so all the rectified supplies in this chapter use this snubbing technique [1]. Having established that the ± 39 V supplies worked correctly, 1 kΩ + 1 nF snubbers were connected across their transformer secondaries on the turret tag boards to suppress rectifier ringing.

With power applied to the housekeeping boards, grid bias voltages were investigated for mains ripple, which was of the order of $2\,\text{mV}_{\text{pk-pk}}$ or less. Given that we expect a maximum of $20\,\text{V}_{\text{pk-pk}}$ signal at each grid, $2\,\text{mV}_{\text{pk-pk}}$ ripple is $-80\,\text{dB}$. Push–pull output stages do not achieve the $\approx 50\,\text{dB}$ common-mode rejection of a differential pair, but $>20\,\text{dB}$ rejection can be expected, which would improve the $-80\,\text{dB}$ hum due to this single cause to an acceptable $-100\,\text{dB}$.

Although placing the housekeeping boards between output valves was mechanically convenient, it necessitated rebating a channel in their brackets to enable sensibly short wires to pass between valve pairs.

Power to the DacMagic

The Sowter 0491s large mains transformer includes a 12 V 1.5 A winding for powering the author's DacMagic outboard digital to analogue converter. The DacMagic uses voltage doublers followed by 7815/7915 linear regulators to provide its ± 15 V analogue supplies, requiring one side of its $12\,\text{V}_{\text{AC}}$ incoming supply to be tied to 0 V common, necessitating half-wave rectification for the digital supplies, which is why it imposes a DC component (CycMean) of $\approx 150\,\text{mA}$ as part of its peak current of 1.6 A. See Figure 7.27.

Noting the DacMagic's current waveform, the author connected between transformer secondary and four-pin chassis XLR socket using solid-core wire having a 0.8 mm diameter conductor, twisted the pair tightly, and buried it as far into chassis corners as possible. A $1\,\text{k}\Omega + 1\,\text{nF}$ snubber was connected across the transformer secondary at the turret tags.

It seemed a good idea to protect external cabling from the DacMagic's current waveform, so the author used pins 2 and 3 of the four-pin XLR for the $12\,\text{V}_{\text{AC}}$, tied pin 1 to chassis, then made a cable having an earthed double braid screen (pin 1), around the tightly twisted pair (pins 2, 3) that took the $12\,\text{V}_{\text{AC}}$ to the DacMagic. The tinned copper braid screen was sleeved with nylon braid to prevent crackles if it touched metalwork.

The cable screen was insulated (heatshrink sleeving) at the 2.1 mm coaxial power connector end so that only the twisted pair made electrical connections. With all this sleeving and shielding the final cable was far too large to pass through the coaxial connector's outer shell, so this was discarded and a final sleeve of glue-lined heatshrink

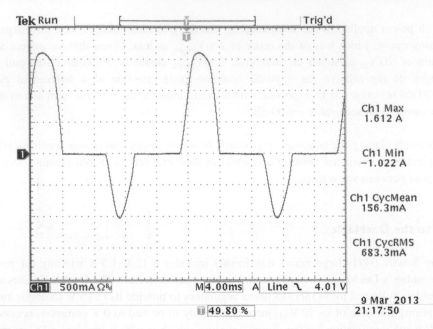

Figure 7.27
DacMagic current waveform. Note the asymmetry that results in the 156 mA DC component (Ch1 CycMean 156.3 mA).

was shrunk over the connections, cable and (most importantly) the threaded section of the connector.

The DacMagic power cable was tested for end-to-end continuity and insulation between all three conductors. 12 V_{AC} was tested first at the chassis XLR, then the DacMagic power cable was plugged in and power confirmed at the cable's far end.

HT wiring and audio testing

Having installed and tested all the low voltage circuitry, attention turned to the HT. DC for the hours counter was taken from the 47 μF 600 V reservoir capacitor via a 100 kΩ 2 W series resistor. In this way, the hours counter not only logs the time that HT is present, but also ensures that high voltage capacitors are safely discharged when power is removed. The individual wires leading to the two 100 μF 400 V Kelvin capacitors from the 5 H choke are effectively a hum loop, but by twisting them together, their loop area becomes zero, so no problem occurs. The same technique was used for their 0 V returns to the 47 μF reservoir. There is no particular significance in the star solder tag other than that a lot of 0 V connections were needed at one point. See Figure 7.28.

Figure 7.28
Clockwise from top right: Heater transformer 27 Ω series resistor, I2CL3 rectifiers and I kΩ + I nF HT snubbers to star washer, I kΩ + I nF snubbers on turret board, 40 μF 400 V capacitor, heater common-mode filter and distribution, heater transformer.

The outputs from the Kelvin capacitors passed to their respective output stages and, when connected to the two housekeeping boards, this eventually formed a hum loop of non-zero area, but it is further away from the mains transformers and signal levels at the push–pull output stage are quite high (and it has some common-mode rejection), so this proved not to be a problem. A possibly superior alternative might have brought all 0 V connections back to a single earth bus-bar passing between the two housekeeping PCBs.

Hum loops are always a problem in stereo amplifiers having a common power supply, and there are three ways of dealing with them:

- Star earthing
- Single bus-bar for ground follows signal
- Make the loops irrelevant.

The problem is essentially mechanical. A stereo amplifier has twice the chassis area of a mono amplifier, invariably lengthening connections to the common point (whether a star or bus-bar), and parallelled output valves further increase chassis area. Neither star earthing nor ground follows signal could be implemented neatly in the Bulwer-Lytton, so it is fortunate that the amplifier had been designed from the start to be tolerant of interference on the 0 V rail — the balanced input passing to a differential pair and thence push—pull output stage confers considerable immunity. So much so, that when the author first powered the amplifier and was disappointed to find a clear $1.7\,\text{mV}_{\text{pk-pk}}$ hum waveform at the loudspeaker terminals, he then realised that this was actually a very good result for an amplifier where he had unaccountably forgotten to bond the 0 V rail to the chassis. Bonding the 0 V rail to chassis reduced hum to $112\,\mu\text{V}_{\text{RMS}}$ on one channel and $85\,\mu\text{V}_{\text{RMS}}$ on the other, equating to -95 and -97 dB respectively referred to 4 W. See Figure 7.29.

Note that the previous hum waveform was averaged over 512 waveforms to remove random noise, and that the residual is very distorted — if it isn't very distorted then it's probably due to a single cause, and susceptible to improvement.

One point always worth checking when using a full-wave valve rectifier is the ripple at rectifier cathodes. If one rectifier should have low emission, it causes the ripple waveform to have alternating high and low amplitudes. Fortunately, individual rectifiers such as the 12CL3 allow this problem to be quickly resolved by replacing the offending single rectifier, rather than replacing an entire (and much more expensive) GZ34. Conversely, one could argue that the common cathode of the GZ34 **ensures** equal ripple amplitudes.

Output power and grid current

The volume control's 1 dB intervals enabled convenient measurement of THD as a function of output power into a $9\,\Omega$ load. See Figure 7.30.

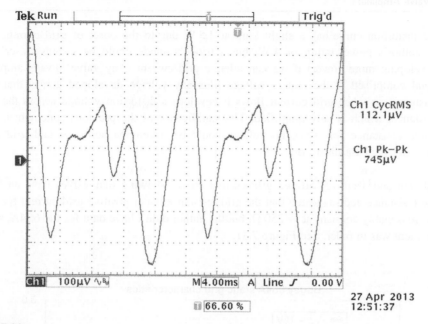

Figure 7.29
Hum waveform from the poorer channel. Note that the waveform is very definitely not a clean 50 or 100 Hz sine wave.

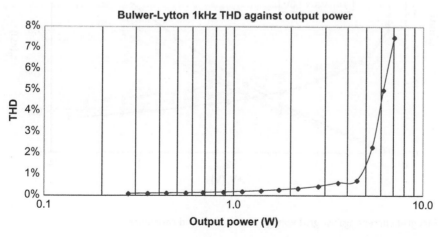

Figure 7.30
Bulwer-Lytton distortion against output power.

The distortion curve has a slight kink at 4.5 W due to the onset of grid current, hence the author's power rating of 4 W, but a triode's anode could be swung closer to 0 V (developing more power) if we can tolerate grid current. Any valve always amplifies a signal comprised of the source voltage **plus** any voltage developed across that source resistance, so when grid current flows it develops a distortion voltage across the source resistance, which is duly amplified. The significance of the previous statement is that if source resistance was zero, no voltage could be developed across it and grid current could not increase distortion.

Millman and Taub [2] offered grid current curves showing that valves have an internal grid resistance and suggested that the grid resistance under positive grid current for a 5965 was reasonably constant at $\approx 250\ \Omega$, but the author didn't have data for the 6S4A, so measurement was in order. See Figure 7.31.

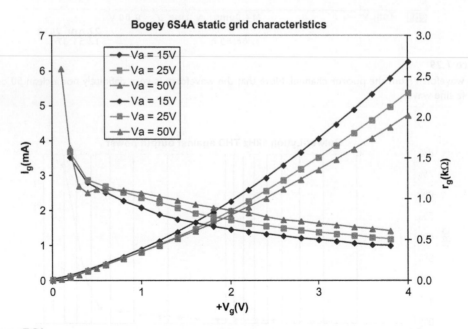

Figure 7.31
Static 6S4A grid current against grid voltage and internal grid resistance.

As the grid current curves rise from 0 V, their gradient (r_g) changes, so this was calculated and also plotted. The significance of r_g is that it adds to source resistance and is high at the onset of grid current (which is why distortion due to grid current starts so sharply) but is only very approximately constant at 800 Ω. If r_g **were** constant, its effect on distortion would be minimal, but it changes, and the kink in the $V_a = 50$ V curve at $V_g = 0.4$ V is due to the anode being briefly able to attract electrons away from the positive grid. Thus, we should also expect kinks at higher anode voltages (measurement foreshortened because the resulting anode current was way over $I_{a\ max}$). More significantly, we could note that this grid/anode perturbation can occur only when the electric field strength at the grid tips in favour of the anode rather than the grid, and that because the anode swing is inverted, this occurs for only a very brief proportion of one cycle. Grid current is complex.

Source resistance driving each 6S4A was comprised of:

- Source follower output resistance = $1/g_m \approx 30$ Ω
- Grid stopper resistance = 4700 Ω
- 6S4A internal grid resistance $r_g \approx 800$ Ω.

Although r_g prevents source resistance from being reduced to zero, reducing the value of grid-stopper resistors should be worthwhile if stability can be maintained. The author first tried reducing grid-stopper resistances by a factor of 10, and was sufficiently encouraged by the results that he removed them entirely, whereupon he fully expected ringing or outright oscillation because these stoppers are designed to keep both their 6S4A and its preceding source follower stable. Nevertheless, careful probing at each of the 6S4A anodes and grids with a 300 MHz oscilloscope failed to show any radio frequency instability even when the amplifier was driven by a 10 kHz square wave, so he measured 1 kHz distortion again. See Figure 7.32.

The curves speak for themselves, with maximum output power almost trebling at 3% THD to 16 W, but this improvement should come as no great surprise; after all, George Anderson (tubelab at diyAudio.com) has been championing the principle for years.

If we're going to drive grid current, we ought to know how much we're driving, so the author applied a current probe to one 6S4A grid and measured grid current at each of the power levels requiring grid current (1 dB steps of input signal level above grid current). See Figure 7.33.

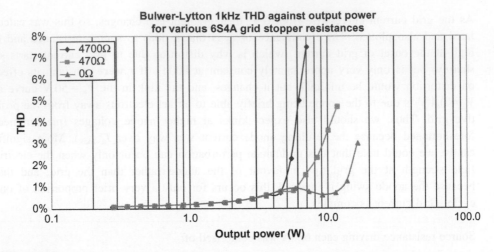

Figure 7.32
Bulwer-Lytton distortion against output power for various grid-stopper resistances.

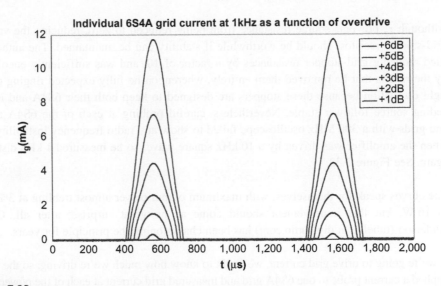

Figure 7.33
6S4A grid current under 1 kHz sine wave drive in 1 dB steps above the onset of grid current.

At the onset of grid current, $\approx 12\,V_{pk}$ is required to drive each grid, yet at 16 W the 12 mA_{pk} demanded by each grid would develop an equal voltage across a 1 kΩ source resistance – demonstrating why low source resistance is crucial when distortion is to be minimised. We observed earlier that grid current has a complex dependency on both grid and anode voltage, which accounts for the flame-like pulses changing shape with amplitude, and noting that these pulses occupy roughly one-fifth of each cycle, we should expect grid current to produce a varying distortion structure starting at H5.

Individual harmonic amplitudes were measured and initially plotted for all harmonics up to 20 kHz (H20) at power intervals of 1 dB, but this produced a graphical mess, so H2 and H3 due to valve linearity were neglected, and only the dominant harmonics attributed to grid current are shown. See Figure 7.34.

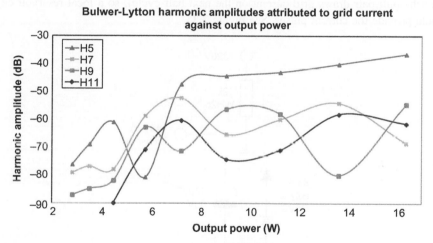

Figure 7.34

Bulwer-Lytton dominant grid current harmonics against output power.

We see that as grid current approaches, H5 rises rapidly, then dips abruptly before resuming its earlier rise. Meanwhile, the previously negligible H7 and H11 rise sharply at the onset of grid current, dip, then remain significant. H9 is indescribable. A distortion spectrum consisting of a spray of harmonics bouncing up and down with signal amplitude like the stops on a fairground steam organ cannot be expected to sound nice, although arguably preferable to the solid crunch caused by clipping at a lower power. Achieving 16 W necessitated driving the 6S4A grids to $+11$ V, and with a $+39$ V supply this implies FQP1N50 $V_{ds} = 28$ V, which is a fraction less than the >30 V minimum distortion criterion laid down in the fourth edition of *Valve Amplifiers*, so it is conceivable that some of the distortion at 16 W is due to the FQP1N50.

It's worth noting that if grid current **is** to be driven, this requires either power FET source followers or cathode followers that can source tens of milliamps without themselves hitting grid current. Whichever is chosen, the followers need a positive rail capable of providing those grid current pulses without sagging, ideally a dedicated supply (as in the Bulwer-Lytton). Alternatively, if the followers must be supplied from the main HT, we can note that although the peak grid current is large, its duration is short and charge low, and in the Bulwer-Lytton, each 11.7 mA_{pk} grid current pulse averaged to only 4.5 mA_{DC} over the 500 μs of each half-cycle (we must consider half-cycles because the other half of a push–pull pair draws grid current on the next half-cycle), so a local reservoir capacitor could provide the current pulses. See Figure 7.35.

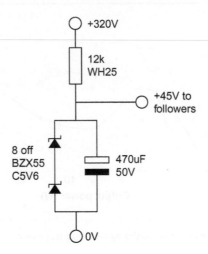

Figure 7.35
A variation on this simple supply would allow the followers to be powered from the main HT.

Considering one channel only, 2×4.5 mA_{DC} (9 mA) is required at peak grid current, and this is in addition to the 10 mA quiescent current demanded by the cathode CCS. A parallel Zener drawing an average current of 4 mA maintains a constant voltage in the absence of grid current. If an oscilloscope automated measurement is used to obtain the average grid current at one valve, that result must be doubled to account for the other half of the push—pull pair. Alternatively, the followers could be powered from a bench supply and their supply current measured whilst maximum grid current is being supplied — directly providing the necessary grid current averaging and quiescent current addition. Note that the 12k resistor dissipates 6.3 W — this circuit might be cheap and retrofittable, but it isn't efficient, and we still need the −39 V supply. Even if cathode followers are used (necessitating a higher voltage positive supply), the series resistor is still needed to attenuate back-feeding of current pulses.

Summarising the Bulwer-Lytton, grid current distortion is proportional to source resistance, and there is no justification for excess source resistance when a valve is driven into grid current, so the 6S4A grid-stoppers were removed. The design purpose of the source followers was to prevent blocking distortion and permit instant recovery from overload, so removing the 6S4A grid-stoppers has not magically converted the Bulwer-Lytton into a 16 W amplifier, just one that tolerates momentary overload even better than before. If you have the ability to measure distortion spectra, optimisation of harmonic equaliser resistance can significantly improve distortion character below 4 W.

Heat

Having been powered for nine hours continuously, the surface temperature of the hottest parts was measured. See Table 7.2.

Table 7.2
Bulwer-Lytton surface temperatures

	Temperature (°C)
Large mains transformer	41
HT choke	32
6S4A	90
7N7	52
12CL3	78

All of these are perfectly acceptable temperatures, suggesting that a long service life may be expected.

Mass

At 14.8 kg (31 lb) the Bulwer-Lytton is half the mass of the author's "Scrapbox Challenge" and slightly more than a quarter that of his "Crystal Palace", making it rather more manageable. See Figure 7.36.

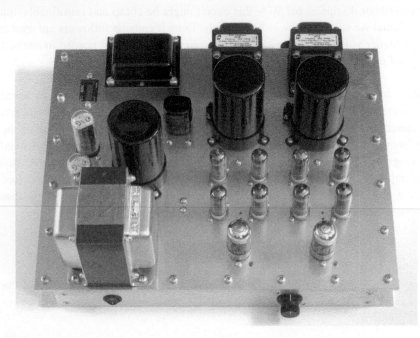

Figure 7.36
Bulwer-Lytton overall view.

RIAA power supply (a.k.a. the "Thing")

Moving coil RIAA stages are so sensitive that their power supply must be remote to prevent leakage flux from mains transformers inducing hum currents into signal wiring. Once the power supply has been moved away, low leakage flux transformers such as toroids are no longer enforced and types having reduced interwinding capacitance (but possibly higher leakage flux) become tenable — which is fortunate because RIAA stages are susceptible to common-mode interference via their heater supplies.

The +195 V HT supply was set in stone by the match between the subminiature tag panels (RS: 433−680) that could accommodate 35 off 5V6 Zeners and the independent requirement for +195 V arising from the 33 kΩ anode loads. Conversely, the low voltage supplies for the valve heaters and the subsidiary negative supply are much more flexible. For his pretty prototype (as opposed to the horrible proof-of-concept lash-up used to prove the design), the author chose to use the 8416 variant of the E88CC, which requires 12.6 V heaters — easily provided by rectification, smoothing, and 317 IC regulation from a 12 V transformer winding.

Construction of this power supply required many trips between mechanical workshop and laboratory, all past the withering gaze of She Who Must Be Obeyed [3], who dubbed it the Thing. In a post-factual world where very ordinary electronics can be packaged in CNC-machined bling and sold for the price of a small car, the Thing is quite the most unassuming piece of electronics the author has made.

The box, screening, and testing the salvaged iron

Although the supply will be remote, there's no virtue in allowing it to radiate or pick up any more interference than necessary, so a 275 mm × 175 mm × 60 mm die-cast aluminium box was chosen for the gap-less electrostatic screening that it offered. Similarly, the HT transformer and HT choke were of oil-filled Admiralty pattern because the necessary containment of the oil (thin steel can) confers electrostatic screening at all frequencies and electromagnetic shielding above 200 kHz. Since the choke was known to be 50 years old, and the HT transformer presumed to be of a similar vintage, both were tested to check that their cores and insulation hadn't deteriorated over the decades. Thus, the mains transformer was connected to the mains via a Variac, its magnetising current measured at regular voltage intervals, and inductance calculated and plotted as a function of applied voltage. See Figure 7.37.

The primary inductance at the expected mains voltage (240 V_{RMS}) is only 16% down on its maximum, and this would be creditable for a new transformer, let alone an old one. Having established that the core was fine, the author investigated the magnetising current waveform with 240 V_{RMS} applied. See Figure 7.38.

As can be seen, the 35 mA_{RMS} magnetising current waveform is **not** a pure sine wave, having symmetrical distortion implying that the FFT will contain odd harmonics. See Figure 7.39.

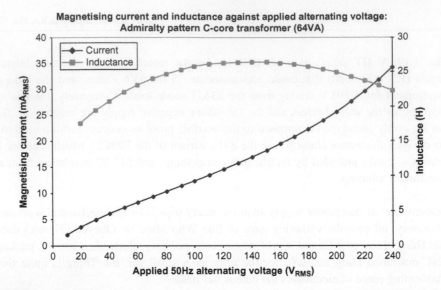

Figure 7.37
Provided inductance doesn't fall significantly at the intended operating voltage, a transformer's core is healthy; this is perfectly satisfactory.

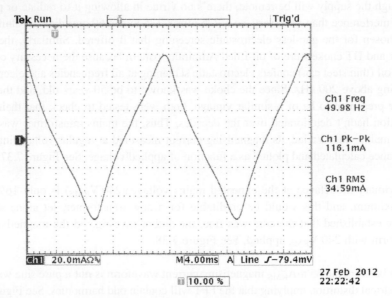

Figure 7.38
Even a good transformer draws a non-sinusoidal magnetising current.

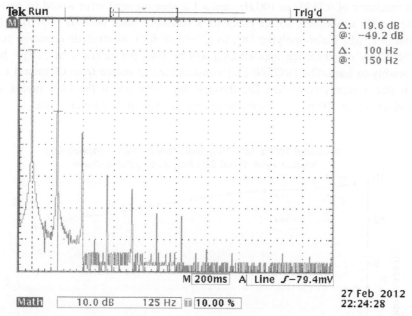

Figure 7.39

An FFT of magnetising current shows that the 10% distortion is dominated by third harmonic (150 Hz) and this is characteristic of a transformer core approaching saturation. (Mains transformers are invariably designed to operate just below the knee of saturation).

The magnetising current spectrum shows the fundamental at 50 Hz followed by a decaying series of odd harmonics dominated by the third harmonic (150 Hz) at -19.6 dB (10.5% distortion). Sadly, this demonstrates that even a **good** small transformer (and most are worse) draws a distorted magnetising current, and this is why we must separate power cables from signal cables behind our completed electronics.

Turning to the choke's core, it must withstand the magnetic flux associated with 68 mA$_{DC}$, so its inductance needs to be measured as a function of DC, but that DC must be prevented from reaching and damaging the LCR bridge. A series capacitor will block DC from the bridge, but (to minimise measurement errors) its reactance at the test frequency must be a small proportion of the inductive reactance to be measured. We measure smoothing chokes at twice mains frequency (their fundamental operating frequency); 10 H has a reactance of 6.28 kΩ at 100 Hz, and a 1% measurement error would be perfectly tolerable in this application, so the capacitor's reactance must be <63 Ω at 100 Hz, which implies >25 μF. We're applying DC, so a >33 μF 63 V electrolytic capacitor (connected the right way round for the applied DC) will be fine; the author used 220 μF because it was readily to hand. The DN2540 FET constant current source from Chapter 4 fed from a bench power supply controlled DC through the choke whilst the LCR bridge measured AC inductance at 100 Hz as the current source was varied. See Figure 7.40.

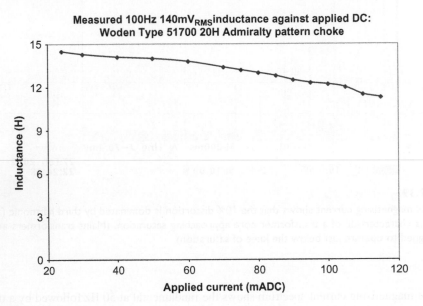

Figure 7.40
Measuring an unknown choke's inductance whilst varying the current through it enables its current rating to be found.

As can be seen, the choke never achieves the 20 H claimed on its label, and there are two possible reasons for this:

- The choke could be a swinging choke deliberately designed to have a high inductance at low current and reduced inductance at higher current, although such chokes usually state inductance at maximum and minimum currents.

- Noting the variation of inductance with AC excitation of the (similar pattern) mains transformer, the AC excitation of only 140 mV$_{RMS}$ was insufficient to achieve the design inductance. This seems highly likely.

Fortunately, we are not so much concerned with absolute inductance as curve shape, and although inductance falls gently with applied DC, it does not suddenly collapse, suggesting that the choke is good for at least 100 mA$_{DC}$, which is fine.

Finally, the mains transformer and choke had their insulation from coils to chassis tested by an insulation tester applying 500 V$_{DC}$. The choke immediately read >50 GΩ, which is excellent, but capacitance to the nearby E/S screen meant that the mains transformer's primary winding took 20 seconds to reach its asymptotic value of 30 GΩ, which is fine (even 500 MΩ would imply an entirely negligible 1 μA leakage current).

The low voltage transformer was a typical modern part of open-frame construction, but a shroud salvaged from an old transformer enabled its conversion to a drop-through type and electrostatically screened it from the outside world. Because the output of the HT choke was predicted by PSUD2 to have quite low ripple, its coil orientation was determined using a search coil and oscilloscope (see Chapter 4) and its coil oriented at right angles to the HT transformer's coil. Although three coils can be arranged mutually at right angles, it requires one coil to be pointing vertically, and this wasn't possible. Thus, the low voltage transformer's coil was aligned with the HT transformer's coil in order to avoid inducing currents into the HT choke. See Figure 7.41.

Valve rectification having been specified, positioning the rectifier in the gap between HT transformer and choke increased the distance between their centres and allowed compact wiring. Moving the EZ81 to the edge of the chassis allowed best cooling, although an EZ81 is barely stressed when passing only 68 mA$_{DC}$. The Kelvin HT filter capacitor had to be mounted within the die-cast box because it was a large wire-ended axial component. A very large multipole plug was chosen for the umbilical cable not because it was entirely necessary, but because it was robust, had large pins (suggesting low series resistance), and the author had a matching chassis socket.

Die-cast box metalworking issues

Be warned that the aluminium alloy needed for die-casting is horrible — it is soft and it crumbles. The lack of side support causes drills to wander, so it is essential that the

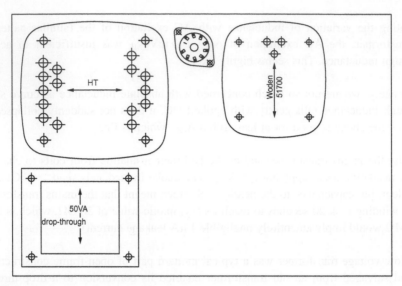

Figure 7.41
Aligning coils on the chassis appropriately minimises hum.

drilled surface is truly horizontal and unable to move. The horizontal requirement might seem obvious, but placing a die-cast box on a drill table and trying to drill into the side compounds the 1° draw to put the upper surface 2° off horizontal. The author solved the problem by clamping the box bottom to an adjustable angle plate set to put the drilled surface horizontal, but an alternative is to swing the drill table sideways and support the drilled surface from directly underneath using an overhanging scrap of wood clamped to the table.

Another problem is that the horrible casting alloy is far too soft to take a thread. But don't die-cast boxes secure their lids by screws tapped into cast pillars? Yes, they do, and those threads survive two or three tightenings with their supplied fasteners before stripping. The author checked the thread gauge (4BA for old Eddystone, 8-32 UNC for current Hammond), ran a 4BA tap into the pillars, thoroughly cleared the swarf, and fitted clean new 4BA fasteners.

The die-cast box had internal guides that allowed a 2 mm thick aluminium screen to be fitted between the (presumed clean) DC output connector and the (presumed dirty) AC compartment. To be effective, the screen needed to be a close fit, otherwise interference would simply leak through gaps from the dirty to clean compartment, so it was cut and filed carefully. Making the screen the correct width was not a problem, although allowing for the 1° draw on the "verticals" of a die-cast box slightly complicated matters, but ensuring that there was minimum gap when the lid was fitted involved a good deal of cut-and-try work.

Heater supply for the 8416 variant of the E88CC

As shown in the original diagram, the supply requires two mains transformers, one for the HT and one for the low voltage supplies. Choosing to use the 8416 variant of the E88CC means that 12.6 V heaters are needed, so the low voltage supplies could be supplied from a 2×12 V transformer and the halved (600 mA) heater current requirement is easily catered for by the 317T IC regulator. The steep rising edges on the sawtooth waveform produced by reservoir capacitor smoothing imply rectification harmonics, so it is best to regulate the heater supply within the remote power supply and carry it over a screened twisted pair to the RIAA stage.

Even 600 mA of heater current causes the 317T to dissipate >1 W, so it clearly needs to be thermally bonded to the die-cast box, but what of the preceding bridge rectifier?

Bridge rectifiers, power dissipation, and cooling

Calculating the heat dissipated in a bridge rectifier is awkward because we ideally need to split a mains cycle into many time divisions and calculate instantaneous power at each point by multiplying instantaneous rectifier current by instantaneous rectifier voltage drop, then calculate the mean power over one cycle. Nevertheless, a rough estimate suggested that for the required 600 mA load current, 1.25 W was likely to be dissipated in the chosen W02 rectifier. Rough estimate or not, such power in a small package explains why so many W02s char their PCBs, so the author planned to add a T05 transistor heatsink and gave the whole device a little more cooling room. See Figure 7.42.

The heater supply is so simple that it fits on a PCB small enough to be supported by the reservoir capacitor's M5 pillars. The thermal bond to chassis from the 317T was achieved by a 35 mm square of ¼″ aluminium plate bonded to the chassis by a long screw also

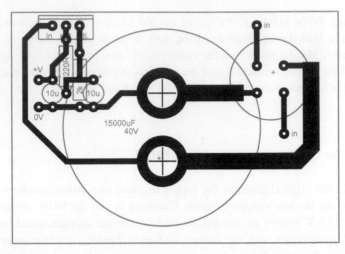

Figure 7.42
PCB layout of the 12.6 V heater regulator (view from foil side).

securing the capacitor clamp going into a 15 mm deep hole drilled and tapped 4BA into one edge. At the first attempt, the square would not quite reach the capacitor clamp screw without fouling the PCB, but a little work with a round needle file elongated the 3.6 mm circular chassis hole into a slot that matched the slot in the capacitor clamp and allowed the square to be snugged up neatly to the PCB. Once the PCB was fastened to the capacitor using its M5 terminal screws, and the aluminium square firmly fastened to the chassis, it was easy to use a right-angled scriber to mark through the 317T's mounting hole onto the aluminium square, remove it, drill and tap M3, then refit with the usual insulating kit. See Figure 7.43.

Test fitting of the PCB and aluminium square was done without serrated washers between the PCB tracks and reservoir capacitor because repeated fitting and removal would cut the (thin) copper tracks. Once the washers are fitted (and they are needed to obtain a long-term low resistance bond), the PCB lifts by ≈ 1 mm, preventing the aluminium square from thermally bonding to the chassis. Fortunately, the problem is easily solved by loosening the capacitor in its clamp and sliding it until the aluminium square just kisses the chassis. Note that this method of mounting the PCB requires a little of the capacitor body to

Figure 7.43
The very simple 12.6 V heater regulator mounts directly on its reservoir capacitor terminals.

enter the box in order for the foil face of the PCB to clear the (conductive) chassis, necessitating a capacitor clamp without an internal flange.

Once made, the 12.6 V heater supply was connected to its 12 V transformer and tested by loading it with the author's dummy loudspeaker load set to its maximum value of 19.5 Ω, which drew 629 mA, and is near enough to the expected load current of 600 mA for preliminary testing. Loaded regulator output voltage was stable at 12.595 V (0.04% low), and regulator temperature measured using a Type K thermocouple and DVM was an entirely acceptable 10°C above ambient. However, the rectifier temperature rise above ambient was an unacceptable 30°C. The author can't imagine why he was surprised to find that the T05 heatsink was ineffective.

Because room had been allowed for the rejected heatsink, there was room to fit a second ¼″ aluminium plate thermally bonded to the first and having a 9 mm hole for the rectifier. The plate needed to be clamped to the rectifier to transfer heat effectively, so a slot was sawn across the hole, and a 6BA screw at its open end tightened the hole onto the rectifier. When making clamps of this sort, the hole before the slot needs to be clearance, but after the slot, tapped. The construction order for the 9 mm hole and its subsequent clamping arrangement was:

- Drill 9 mm hole. (Clamp the work to the drill table, drill an 8.5 mm pilot hole, then use a well-lubricated sharp 9 mm drill and, once drilled, withdraw it gently to leave the smoothest possible hole.)
- Drill tapping size for clamping screw to full depth.
- Drill most of the clearance depth.
- Tap the hole (the preceding clearance helps align the tap).
- Saw the slot.
- Drill the remaining clearance. If a thin piece of tin plate is inserted in the slot and held in fingers, when the drill touches it, vibration is felt and drilling can stop at exactly the right depth.
- Deburr the slot using a square needle file for open ends and riffler file for closed ends.

In addition, a hole was needed along the plate to allow a screw to fasten it to the first. Drilling such a long hole and expecting it to remain true, exiting in the right place, requires careful alignment in a vice clamped to the drill table, followed by gentle drilling, using a sharp drill, with frequent withdrawals to clear swarf. Provided those precautions are taken, it's easier than you would think. Once made, the new plate was slid over the rectifier, and a line scribed on the edge of the first plate where the two met. The first plate was removed and the scribed line used to determine where the tapped hole needed to be drilled. Drilling and tapping this hole was comparatively easy. Finally, all the parts were fastened together, then to the chassis, and the rectifier clamped.

On test, the new rectifier temperature rise above ambient was an acceptable 13.5°C. Making the W02 heatsink was a lot of fiddly work, and you may prefer to take the easier option of screwing a 4 A rectifier directly to the chassis and accept the very slightly increased mains interference due to the increased off capacitance of the larger diodes.

Alternatively, you might choose to make up a bridge rectifier out of individual 1 A diodes spaced 10 mm from the board to allow cooling.

Testing the heater supply

Having established that the regulator didn't seem likely to explode in the near future, the oscilloscope was warmed up and the regulator's voltage drop probed. Connecting an oscilloscope probe's earth clip to the output of a supply would ordinarily be courting disaster because the regulator's 0 V is usually tied to earth, so the earth clip applies a short circuit. But at this point, the heater supply was still floating, with no connections to chassis/mains earth, so the earth clip **could** be safely connected to the output of the heater supply and the probe tip to the reservoir capacitor, enabling measurement of the voltage across the regulator whilst under load. See Figure 7.44.

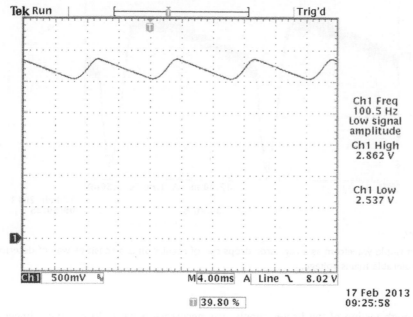

Figure 7.44
Voltage drop **across** the 12.6 V heater regulator. Note that it falls to a minimum of 2.537 V (Ch1 Low).

As can be seen, the minimum voltage across the regulator is just over 2.5 V, which although a little lower than the author likes to see, is still comfortably above the 1.8 V drop-out voltage suggested by the 317 data sheet at 600 mA. The definitive test is to monitor ripple voltage across the output of the regulator with the oscilloscope AC coupled and sensitivity set at or near maximum, trigger from line, then reduce mains voltage using a Variac – the moment 100 Hz ripple rises from the noise, the regulator is dropping out. The author tried this test and drop-out occurred when mains was reduced to 231 V_{RMS}, which was fine because the author's mains voltage is typically between 240 and 245 V_{RMS}. See Figure 7.45.

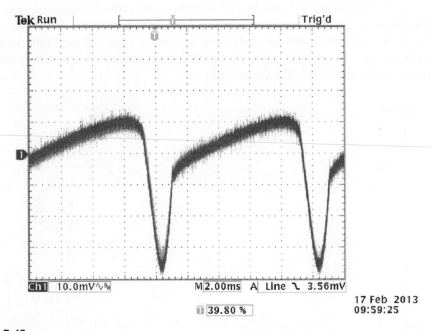

Figure 7.45
The sudden ripple waveform as a regulator drops out of regulation is a sensitive way of determining minimum tolerable mains voltage.

Thorough testing of the heater supply was necessary because low voltage linear supplies are always a balance between minimising heat dissipation but avoiding drop-out at the minimum expected mains voltage. The more efficient the design, the smaller the small operating window, and discrepancies between theory (PSUD2, LTspice, etc.) and practice become more significant, making rigorous testing essential.

Voltage doubler vs. bridge rectification for the subsidiary negative supply

The subsidiary negative supply feeds the constant current sink that biases an input stage driven by a cartridge specified to have a 1 kHz sensitivity of 0.3 mV — so it needs to be quiet. The author was never entirely happy with the original 337 regulator in the proof-of-concept construction because IC negative regulators tend to be poorer than positive, and the 8416 decision had already demanded a different transformer, encouraging the subsidiary negative supply and its regulator to be revisited.

Counterintuitively, for the same load voltage and current, PSUD2 suggested that RMS capacitor ripple current would be lower in a voltage doubler than a bridge rectifier. RMS ripple current is significant because it is the DC that would have the same heating effect, and reduced self-heating implies longer capacitor life. Some comparative measurements were clearly needed.

The voltage doubler was first tested from a single 12 V winding, using 1N4002 diodes and a pair of matched reservoir capacitors (492 and 493 μF, both at 100 Hz), feeding a 47 Ω series resistor leading to a nominal 470 μF filtering capacitor and 220 Ω load resistance (chosen to be representative of the expected load). Reservoir capacitor ripple current was measured using a Tektronix P6302 Hall effect current probe and, because capacitances had been matched, was identical for both capacitors. See Figure 7.46.

Note that capacitor ripple frequency is ≈ 50 Hz (not 100 Hz) — this is an inevitable consequence of voltage doubling. Capacitor ripple current was 336.9 mA$_{RMS}$, whilst load voltage as measured by a DVM was 27 V (impressively, measurement and PSUD2 prediction agreed within $\pm 1\%$).

The test circuit was then reconfigured, connecting two 12 V windings in series to feed a W02 bridge rectifier, one 492 μF reservoir capacitor, and the previous filter and load circuit. See Figure 7.47.

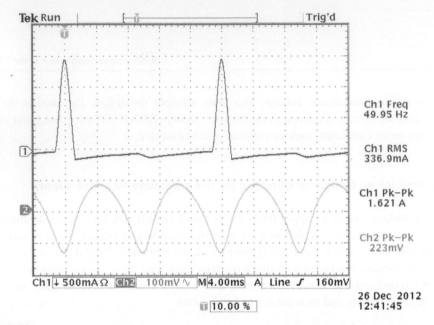

Figure 7.46

Voltage multiplier. Ch1: Individual capacitor ripple current. Ch2: Ripple voltage at the output of the voltage multiplier.

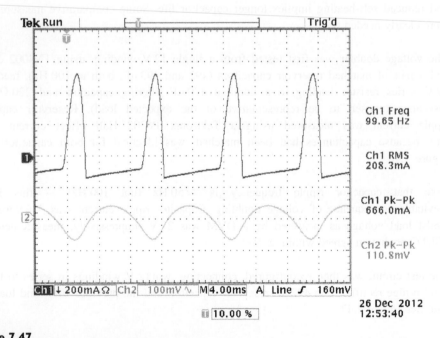

Figure 7.47

Bridge rectifier. Ch1: Capacitor ripple current. Ch2: Ripple voltage at the output. Compare with Figure 7.46 and note the reduced ripple current (Ch1 Pk–Pk) and ripple voltage (Ch2 Pk–Pk) due to capacitor charge rate of 100 Hz rather than 50 Hz.

Ripple frequency has reverted to the expected 100 Hz (double mains frequency), so the reservoir capacitor is being recharged twice as frequently, halving ripple voltage, and causing capacitor ripple current to fall to 208.3 mA$_{RMS}$ or 666 mA$_{pk-pk}$, whilst load voltage rises to 28.3 V. Unfortunately, PSUD2's predicted capacitor ripple current of 341 mA$_{RMS}$ or 1.58 A$_{pk-pk}$ was not supported by these measurements. The practical rule of thumb is that capacitor ripple current peaks at 4−6 times DC load current, and with 28.3 V across the 220 Ω load, load current was 129 mA, so 5 × 129 mA = 643 mA, agreeing with measurement and suggesting that this particular PSUD2 prediction is incorrect.

The value of modelling is that it eliminates poor solutions faster than construction and testing, thereby saving hours of wasted effort, but it is vital to remember that any model is always a simplification of reality. Once poor solutions have been eliminated, any chosen solution should always be tested to verify its quality. PSUD2 has some issues regarding rectifier currents but that does not detract from its undoubted utility − it simply reinforces the need to check models against measurement.

The preceding measurements showed that a voltage doubler imposes a significantly higher ripple current upon its reservoir capacitors than a bridge rectifier, and this seems intuitively correct because each capacitor is being recharged half as frequently yet must supply the same DC load current. Thus, all factors being equal, we would always prefer a bridge rectifier over a voltage doubler (and benefit from halved ripple voltage). But all factors were **not** equal, and a voltage doubler would allow a standard 2 × 12 V transformer to power heaters **and** a superior subsidiary negative regulator.

Refining the subsidiary negative supply regulator

Given that a 12 V voltage doubler provides plenty of voltage for regulation, it suggests a variation on the CCS/Zener theme used in the Statistical Regulator. Although there is no problem in simply using a pair of 5.6 V Zeners to set a −11.2 V supply (the exact voltage isn't critical), the 88 mA required by the Toaster (increased by those additional LED chains) plus 20 mA of Zener current means that we must source 108 mA, and that causes problems for the two-FET CCS that reduce its power supply rejection ratio (PSRR), which we wish to maximise. See Figure 7.48.

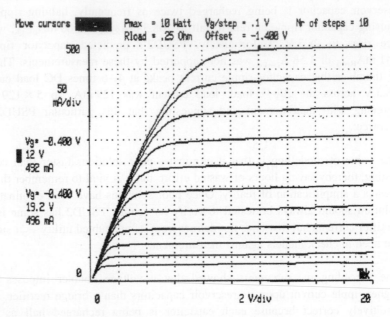

Figure 7.48

Output characteristics of the DN2540N5 FET. Note that minimum required V_{DS} increases as I_{DS} increases.

As can be seen, if V_{DS} is low and constant, as I_D increases there comes a point when the curves are no longer horizontal and r_{ds} falls significantly. This region is known as the resistive region or, very occasionally, the triode region because of its slight similarity to triode curves. Entering the resistive region causes the two-FET floating CCS's output resistance to fall, and it hits a maximum usable current limit (typically around 90 mA for DN2540N5) rather sharply because two factors act simultaneously upon the lower FET:

- Increased current requires V_{GS} for the outer device to fall, thereby directly reducing the inner device's V_{DS}, moving its operating point to the left, towards the resistive region.
- That same increased current drives the inner device's operating point up, again towards the resistive region.

Paralleling the outer device with a second reduces the effect of the first factor, whereas paralleling the inner device reduces the effect of the second. Paralleling both is the same as putting two CCSs in parallel (halving their output resistance, and thereby reducing PSRR by 6 dB), but it is even worse than that because halving the current through each CCS significantly reduces g_m for each FET, reducing μ and thence CCS output resistance. Thus, although initially counterintuitive, LTspice modelling suggests that if devices are to be paralleled, it should be the inner device only. See Figure 7.49a.

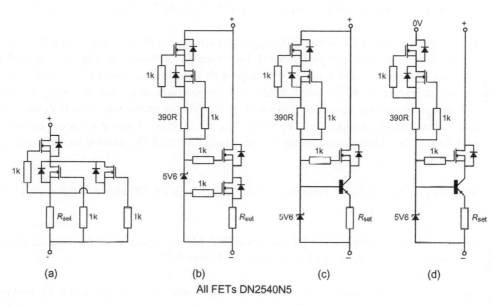

All FETs DN2540N5

Figure 7.49
The evolution of an increased current CCS.

An alternative strategy might lift the outer DN2540N5's gate voltage using a 5.6 V Zener, thereby increasing the inner device's V_{DS}. The Zener needs to be powered, perhaps by giving it its own two-FET CCS returned to the load, so we now need four expensive DN2540N5s, yet modelled PSRR was 16 dB worse than by simply paralleling the inner device. See Figure 7.49b.

PSRR is dominated by the need for the inner device to be clear of the resistive region and to have maximum g_m, suggesting that a bipolar junction transistor (BJT) might be a better choice than an FET. The BJT needs to have its base lifted to force a programming current into its emitter resistor, so this can be done as before using a Zener and two-FET CCS returned to the load, and modelling suggested 2 dB better PSRR than the four-FET solution. See Figure 7.49c.

Passing the same current, we expect far more g_m from a BJT than an FET, so such a small improvement suggests that something is still wrong. Relinquishing the two-terminal requirement and returning the reference chain current to 0 V rather than the load finally brought low frequency PSRR to within 1 dB of the three-FET solution. See Figure 7.49d.

Given that the three-FET solution is simple and effective, why struggle to match its performance with a more complex circuit? Unfortunately, although simple, the three-FET solution is expensive and not terribly stable with temperature. The CCS current in the circuit of Figure 7.49d is more stable with temperature than the preceding circuits because programming current is dominated by the (almost unchanging) 5.6 V Zener voltage rather than the temperature-dependent V_{GS} of an FET. A more complex circuit is justifiable if it's cheaper and better, so we should minimise the number of those expensive FETs.

We can lose two FETs by replacing the reference chain's CCS with a Williams ring-of-two. See Figure 7.50.

We now need to compare the LTspice modelled AC performance of the original three-FET solution to the new BJT/FET solution. See Figure 7.51.

Although the graph shows two possible BJT/FET solutions, we will begin by looking <1 MHz, where the two BJT/FET solutions are virtually identical. Firstly, we see that a BJT/FET solution is 8.1 dB better at low frequencies, but significantly worse than the three-FET between 15.43 and 569 kHz. However, it pays to look very closely at what the simulation is telling us. The BJT/FET solutions have attenuation that rises at 6 dB/octave up to $\approx$1 MHz, implying that output impedance <1 MHz is dominated by a single reactive component (DN2540 output capacitance). Conversely, the three-FET solution's slope always rises at 12 dB/octave, implying two reactive components and therefore questionable stability even though it seems fine in the simulation. Finally, the BD135/FET solution maintains its 6 dB/octave slope to 10 MHz, making it the best choice.

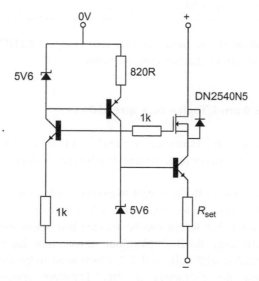

Figure 7.50
Replacing the Zener's FET CCS with a Williams ring-of-two improves performance and is cheaper.

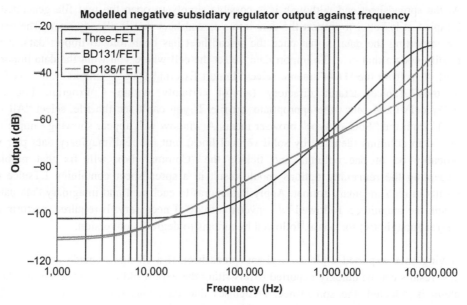

Figure 7.51
Comparison of modelled performance of the three CCSs.

There's no doubt about it, the more complex (but cheaper) BJT/FET solutions are genuinely better (and a purely bipolar solution was worse).

Exporting raw results from LTspice to a spreadsheet

Exporting raw results from LTspice into a spreadsheet is useful because once the data is in a spreadsheet, it is much easier to determine details such as slopes and $f_{-3\,dB}$ frequencies.

In LTspice, first plot the results you want to export, click anywhere on the results graph, then from the "File" menu, select "Export". A box will appear with the plotted results highlighted, and the number format can be selected just below the "Browse" button. The default is "Polar: (dB, deg)" for Bode plots, but for each plot this produces a comma separated data pair appended with "dB" and "°", which need to be removed before a spreadsheet can recognise the characters as data. Different spreadsheets have different commands for truncating data characters, but a universal solution is to select "Cartesian: re, im" and export.

At the spreadsheet end, although it is possible simply to open the text file generated by LTspice, it is better to import the data into a spreadsheet that is already open. The reason for importing the data is that once the spreadsheet has been told to import data, it will readily update that data. To import data, select the cell where you want the data import to start, then from the "Data" menu, select "Import External Data", "Import Data", navigate to the LTspice default directory (almost certainly within C:\Program Files\LTC \LTspiceIV) and select the appropriate .txt file. If you can't see the file, select "All files (*.*)" from the "Files of type" browser menu. A window will appear showing sample data placed in columns (denoted by solid vertical lines), but the real/imaginary data pair will probably be in one column, and ticking the "Comma" box will fix that problem. Accepting the remaining options should result in a spreadsheet containing LTspice raw results in real/imaginary format. Apply Pythagoras to each real and imaginary data pair to obtain the magnitude, followed by conversion to dB if necessary. If required, perform arctan(real/imaginary) for phase, followed by radians to degrees conversion.

Having once exported a plot to a given spreadsheet, if a change is made in LTspice, the new results can be quickly exported (overwriting the export old file), but when the "Data" menu is selected, the spreadsheet recognises that data was imported and offers "Edit External Data", which allows the spreadsheet to be updated with a minimum of fuss.

Why a power BJT?

Diligent readers will already have wondered about the choice of a power BJT in the CCS cascode. Annoyingly, 105 mA exceeds a BC549C's 100 mA $I_{C(max)}$, and although it is below a 2N3904's 200 mA $I_{C(max)}$, we are perilously close to the 2N3904's $P_{(max)}$ of 300 mW. Moreover, although h_{FE} for a BC549C would be >420 as specified on the data sheet at low currents, it falls as I_C approaches $I_{C(max)}$, evidenced by reducing vertical spacing between curves. See Figure 7.52.

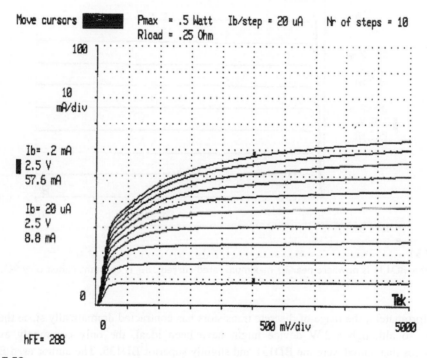

Figure 7.52
Like all bipolar junction transistors, this BC549C suffers decreasing h_{FE} as I_C approaches its rated maximum.

Not only does h_{FE} fall as I_C approaches $I_{C(max)}$, but the Early effect means that $1/h_{oe}$ also falls (the curves became less horizontal), and neither of these effects is desirable in a constant current sink. We must keep well clear of $I_{C(max)}$, and choose a higher-rated transistor. Although the h_{FE} of a typical BD131 is only 100, it maintains that value over the current range we require, thus offering higher h_{FE} at the operating point than a smaller and notionally superior device. See Figure 7.53.

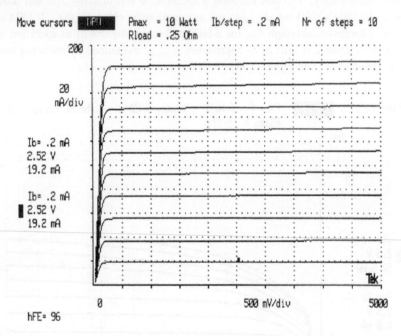

Figure 7.53
Because this BD131 is nowhere near its maximum rated current, h_{FE} is constant (albeit only 96).

Unfortunately, the range of discrete transistors has contracted dramatically since their heyday, so although a 2 W device might have been ideal, the only commonly available devices that suited were the BD131 and slightly superior BD135. The author used BD135.

Sanity checks

Having established that the BD135/FET solution's output impedance looks like resistance in parallel with capacitance (just like the original Statistical Regulator), we know that suitable choice of capacitance across the load Zeners will make attenuation constant with frequency — LTspice suggested that 10 μF was required for a very nearly flat response. The significance of iteratively adjusting capacitance to obtain a flat response is that, combined with the modelled 112 dB attenuation (2.79×10^{-6}), we know that C_1/C_2 must be in the same ratio, so the CCS output capacitance must be:

$$C_{\text{out}} = \frac{V_{\text{out}}}{V_{\text{in}}} \cdot C_{\text{Zener}} = 2.79 \times 10^{-6} \times 10 \times 10^{-6} \approx 24 \ \text{pF}$$

The author has previously measured 12 pF (including strays) for other DN2540N5 CCSs, so a modelled capacitance of 24 pF seems possible if somewhat pessimistic. (We will revisit this capacitance later.)

If we know the output capacitance and find the frequency where the attenuation rises 3 dB from its maximum (C_{Zener} removed), we can estimate the CCS's output resistance, r_{out}. The easy way to do this is to change the units of C_{Zener} in LTspice from μF to pF (effectively setting it to zero), then note the attenuation at low frequencies (112.88 dB in this model), and look for the frequency where attenuation is 3 dB worse (109.88 dB), which is 1.725 kHz. We then find r_{out} using:

$$r_{\text{out}} = \frac{1}{2\pi f C_{\text{out}}} = \frac{1}{2 \times 3.14 \times 1725 \times 24 \times 10^{-12}} \approx 3.9 \ \text{M}\Omega$$

This doesn't sound much, but remember that we're passing 108 mA, so a genuine resistance would drop 420 kV and dissipate 45 kW — it's actually a very good CCS.

Knowing that, for a flat response, the time constant of the load resistance (160 Ω) in parallel with the Zener's slope resistance, and bypass capacitance must be the same as the CCS's time constant, we can also estimate Zener slope resistance. The CCS time constant (CR) is 3.9 MΩ × 24 pF = 92 μs, and since the bypass capacitance for a flat response is 10 μF, Zener slope resistance must be ≈ 10 Ω. The Zeners passed 20 mA in the simulation, but the author's BXZ55 C5V6 measurements gave 5.9 Ω at 20 mA, so two in series would be 12 Ω, suggesting that although the posited 10 Ω slope resistance is in the right ballpark, it is a little optimistic.

It is essential to perform these sanity checks because complex simulations easily produce results having high precision but zero accuracy. Seemingly to prove precisely this point, at one iteration, LTspice objected to tying the FET's gate-stopper resistor directly to the BD135 base, despite the fact that such a circuit was working on a breadboard nearby. But LTspice didn't throw up an error message, it just predicted that the modification improved attenuation by almost 10 dB — an unexpected improvement. The truth was whispered by better attenuation below 1 kHz, then shouted out by DC conditions claiming fA of current through the load, rather than mA.

Having decided that the simulation is probably not far from the truth, we can consider fine detail.

Noise and Zener bypassing

Unlike the HT Statistical Regulator, the Zener bypass capacitor in this regulator only has to bypass 11.2 V, making an electrolytic capacitor the obvious choice, especially since we don't need the capacitor to be a precise value, just large enough to maintain the low-frequency attenuation to high frequencies. However, we expect this form of regulator not only to have good ripple attenuation, but also to be low noise, yet we know that electrolytic capacitors must pass DC leakage current to maintain anodisation in the face of their corrosive electrolyte, and this leakage current has a noise component.

The author's measurements suggest that a typical 470 μF 50 V aluminium electrolytic capacitor passes a noise current of $\approx 50\,nA_{RMS}$, so a 10 μF 50 V capacitor could be expected to have a proportionately smaller plate area and be expected to pass $1\,nA_{RMS}$. The output resistance of the regulator is defined by the slope resistance of the two Zeners, and passing 20 mA is likely to be $\approx 12\,\Omega$, which is orders of magnitude higher than a feedback regulator would achieve, but is the price paid for simplicity and low noise. Coupling the two factors together, we could expect $1\,nA_{RMS}$ to develop $12\,nV_{RMS}$ noise voltage across 12 Ω, which is certainly small, but is it small enough?

Using a 22 Hz–22 kHz measurement bandwidth, the author found noise at −103 dBu ($5.5\,\mu V_{RMS}$) across a string of 35 BZX55C5V6 Zeners passing 25 mA. Remembering that the noise sources are uncorrelated, noise rises by the square root of the number of devices, so individual Zener noise must be:

$$v_{n(individual)} = \frac{v_{n(total)}}{\sqrt{n}} = \frac{5.5\ \mu V_{RMS}}{\sqrt{35}} = 0.9\ \mu V_{RMS} = 900\ nV_{RMS}$$

We have two Zeners in series, so we expect their noise to increase by a factor of $\sqrt{2}$ to $1300\,nV_{RMS}$, which is two orders of magnitude greater than the $12\,nV_{RMS}$ posited noise due to an electrolytic capacitor's leakage current. We need not fear the electrolytic capacitor's noise.

Having calculated this regulator's noise, it's useful to put it into context by comparing it to the displaced IC regulator. As was remarked earlier, negative IC regulators are invariably poorer than positive. Rather than producing noise at 0.001% of output voltage (317), the 337 produces 0.003%, so if set to -11.2 V, a 337 would produce 336 μV_{RMS} for a 10 Hz–10 kHz bandwidth. Correcting to a 22 Hz–22 kHz bandwidth increases the raw 337 noise to ≈ 0.5 mV$_{RMS}$. However, we would certainly bypass the 337's adjust pin with 10 μF to improve ripple rejection, and this reduces the bandwidth of its amplifier to 100 Hz, reducing output noise to ≈ 34 μV_{RMS}, which is 28 dB worse than the expected 1.3 μV_{RMS} of the discrete component subsidiary negative regulator.

Returning to the Zener bypass capacitor, a typical 10 μF 50 V electrolytic capacitor has ESR ≈ 0.5 Ω and perhaps 20 nH series inductance, so this was compared to a four-wire 10 μF film capacitor in the model. See Figure 7.54.

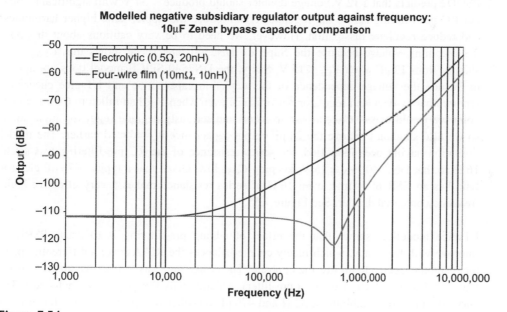

Figure 7.54
The significant ESR of a 10 μF electrolytic capacitor impairs bypassing between 10 kHz and 3 MHz compared to a 10 μF plastic capacitor, and the reduced inductance of a four-wire termination is a bonus.

Just as in the fourth edition of *Valve Amplifiers*, when comparing between film and electrolytic capacitors in HT supplies, the deciding factor is ESR (and to a lesser extent, series inductance), with the film capacitor exhibiting superior interference rejection >20 kHz. Thus, in addition to the expected 10 μF Kelvin capacitor for the HT, the author required space within the RIAA stage for a 10 μF Kelvin capacitor for its subsidiary negative supply.

Thermal issues and pre-filtering

All circuits have weaknesses, and the CCS's are only hinted at by LTspice. When measured, the BD135 passed 108 mA and had ≈ 2.7 V across it, dissipating 286 mW, but a good DN2540N5 might have a higher V_{GS} for $I_{DS} = 108$ mA, increasing BJT dissipation, and the BD135 temperature rise without heatsink in free air was 28°C, so it needs cooling. The DN2540N5 definitely needs cooling.

PSUD2 predicts that a 12 V voltage doubler should produce -34 V with significant ripple but RC pre-filtering before the regulator could reduce the ripple, tame higher harmonics, and reduce regulator dissipation. However, we need to be very cautious about dropping the incoming voltage because the Supertex DN2540 data sheet shows that C_{OSS} rises sharply from 12 pF when $V_{DS} < 15$ V. Sadly, the DN2540 SPICE model does not appear to include the voltage dependence of this depletion capacitance and LTspice cheerfully reported barely degraded regulator performance even when V_{DS} had fallen to 6 V, at which point we would expect C_{OSS} to rise to 39 pF and seriously degrade regulator attenuation, so we also shouldn't believe the 24 pF output capacitance it predicted earlier. The model limitation having been identified, the series resistance of the RC pre-filtering was set to 16 Ω so that DN2540 $V_{DS} \geq 15$ V, keeping C_{OSS} low. Given that a typical 470 μF electrolytic has an ESR of 50 mΩ, even 16 Ω of series resistance makes a very effective high-frequency potential divider. See Figure 7.55.

LTspice modelling shows that the effect of adding pre-filtering is to cause PSRR to improve with frequency, which nicely counterbalances the RIAA stage's differential pairs that can be expected to do just the opposite. The design aim was to improve upon the original 337 design (which also had a pre-filter), which is why the graph also included a 337 curve obtained by combining a 33 Ω and 470 μF pre-filter response with data taken from the -10 V ripple rejection graph on the LT337 data sheet. As Figure 7.55 shows, the new circuit is a minimum of 31 dB better, and attenuates ripple by >100 dB from DC to 1 MHz. That's pretty good. See Figure 7.56.

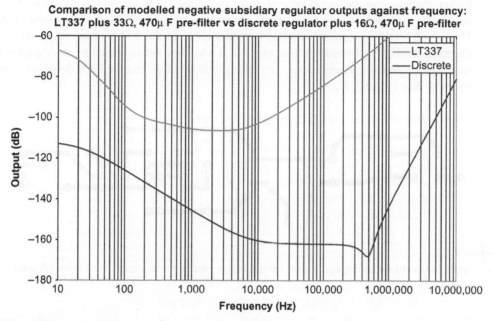

Figure 7.55

Because it was optimised for these particular requirements, the discrete component regulator is superior to the general-purpose 337.

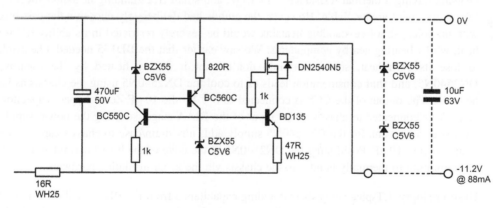

Figure 7.56

Circuit of the subsidiary negative regulator.

Rather than pollute the comparatively quiet umbilical cable and definitely quiet RIAA stage with the ripple entering the subsidiary negative supply's CCS, the CCS lives in the Thing, and feeds its current via RG174 coaxial cable to the RIAA stage where the two Zeners and their 10 μF capacitor live, and this is why they are separated by dashed lines in the previous drawing. Rectification, smoothing, pre-filtering, and the CCS occupied a single PCB. See Figure 7.57.

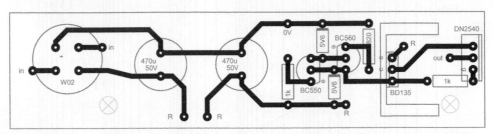

Figure 7.57
PCB layout of the subsidiary negative supply using a bridge rectifier — see later (view from foil side).

The significance of splitting the supply into two parts is that we must now consider CCS cooling within the unventilated case (die-cast aluminium box) of the remote supply.

The DN2540N5 dissipates ≈ 1.5 W, so to keep its temperature rise below 20°C, it needs a heatsink having a thermal resistance <13°C/W, and small free-standing heatsinks meeting this requirement are available. However, the author had deliberately chosen a die-cast box with no holes, so a free-standing heatsink would be severely restricted in its ability to lose heat, whilst heating nearby components. We saw earlier that the BD135 needed a heatsink to lose its own heat, so we certainly don't want it further heated by the (nearby) DN2540N5, and **that** consideration leads us to cool the DN2540N5 using the chassis as its heatsink. The output of the CCS is connected to 0 V by the 10 μF Zener bypass capacitor, and 0 V is connected to chassis at the input of the RIAA stage, so using the power supply chassis as a heatsink for the DN2540N5 simply adds any drain/case to chassis capacitance to an existing 10 μF. Positioning the DN2540N5 at the edge of the board permitted a block of aluminium to thermally bond it to the chassis via the usual insulating hardware.

Unsurprisingly, LTspice suggests that adding capacitance from the BD135 collector to 0 V degrades regulator attenuation at high frequencies, so it needs a free-standing heatsink, which is not a problem, and on test a nominal 19°C/W heatsink just above the breadboard reduced measured temperature rise from 28 to 10°C. We might have expected 5.4°C (19°C/W × 0.286 W = 5.4°C), but the breadboard plane underneath restricted air flow — heatsink specifications are always given in free air, yet we rarely operate them this way and invariably suffer degraded performance. Enclosing the transistor and heatsink within the die-cast box further increases device temperature, but that's the price we must pay for minimising capacitance to chassis. Note that because we do not tolerate the thermal resistance of an insulating kit between the BD135 and its heatsink, the heatsink is electrically connected to the collector so we must ensure that it is not close enough to chassis (or any other wiring) to add stray capacitance.

The 47 Ω current programming resistor dissipates >0.5 W, and the easiest way to lose its heat is to choose a WH** aluminium-clad type heatsunk to the chassis; the typical 13.5 pF capacitance to case of a WH** resistor combined with its 47 Ω resistance results in $f_{-3\,dB} = 250$ MHz, which is certainly high enough not to be a problem in this application. The 16 Ω pre-filter resistor dissipates >0.2 W, so the author used WH25 for both resistors because they were easier to fit and no more expensive than WH5, and thermally bonded them to the chassis. The remainder of the rectification and CCS components stayed on the PCB.

The subsidiary negative supply's PCB and two WH25 resistors were mounted on the 2 mm aluminium dividing screen mentioned previously, along with the block thermally bonding the DN2540N5. The advantage of this construction was that a number of components became a single module that was easily constructed and tested outside the die-cast box, then dropped in to require an absolute minimum of installation wiring. See Figure 7.58.

Testing the subsidiary negative supply

The subsidiary supply's rectification and CCS were tested using a 100 Ω WH15 resistor across its output in lieu of Zeners and RIAA stage. DVMs monitored voltages either side of the 16 Ω pre-filter resistor whilst the oscilloscope monitored AC ripple at the same points. A third DVM across the 100 Ω load resistor monitored output voltage — you really can't have too many DVMs. Mains voltage was gently increased from zero to nominal using a Variac. Rectifier temperatures were 9°C above ambient in free air (perfectly acceptable) and voltages were as expected, so attention turned to the oscilloscope. See Figure 7.59.

Figure 7.58
The −11.2 V subsidiary supply is mounted on the divider panel. Note the six feed-through capacitors to the right.

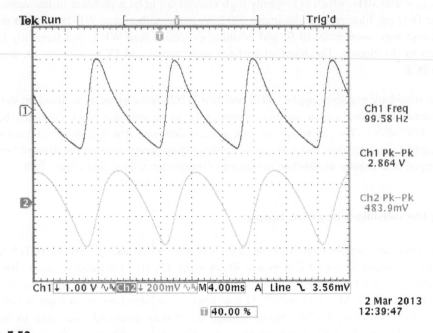

Figure 7.59
Ripple on the subsidiary negative supply measured before and after the pre-filter. Ch1: Before pre-filter. Ch2: After pre-filter.

The author is more used to looking at positive voltages, so both waveforms were inverted for clarity, and the ripple leaving the doubler (Ch1) is ≈ 2.864 $V_{pk\text{-}pk}$, whilst after the pre-filter (Ch2) it is ≈ 483.9 $mV_{pk\text{-}pk}$, so the pre-filter attenuates ripple by ≈ 15 dB. More significantly, the pre-filter waveform is rounded, implying reduced amplitude higher harmonics, but the difference only becomes clear when we use the oscilloscope's FFT to obtain spectra for both signals, measure each harmonic amplitude, and compare the data graphically. See Figure 7.60.

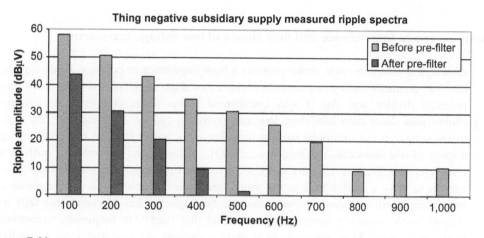

Figure 7.60
Measured spectra showing the efficacy of the pre-filter in the subsidiary supply.

Harmonic amplitudes are all referred to 1 μV_{RMS}, and it is now clear that the pre-filter becomes progressively more effective at attenuating ripple harmonics as frequency rises. This is exactly what we expected from modelling and demonstrates that a pre-filter before a regulator is very useful.

As with the heater supply, mains voltage was reduced until ripple suddenly appeared across the load, at an applied voltage of 146 V_{RMS}. Such a low voltage makes the CCS appear very good (albeit inefficient), but remembering that its DN2540N5 needs $V_{DG} > 15$ V to keep its output capacitance low (and therefore attenuate high frequency interference), a more valid mains drop-out test was to determine when $V_{DG} < 15$ V, and this required 176 V_{RMS}. Note that whenever we measure an FET's V_{DG} we should not attempt to measure directly at the gate, but must measure on the far side of the gate-stopper resistor to avoid DVM input capacitance provoking oscillation. Provided that there is no gate current (and we don't expect any) the two measurements give the same voltage.

A change of mains voltage from 176 to 240 V changed the voltage across the 100 Ω load from 10.202 to 10.222 V (equivalent to 0.03% line regulation), so the author concluded that the CCS was working as intended.

Common-mode interference and final choice of low voltage transformer

Although a common-mode choke presents a high impedance to common-mode signals, it can only attenuate them when combined with a low shunt impedance to earth that forms a potential divider, and this is why commercial mains filters comprise common-mode chokes plus shunt capacitors from line and neutral to chassis. Unfortunately, the typical mains voltage is not sinusoidal because capacitor-input power supplies cause flat tops and a spray of odd harmonics reaching beyond 1 kHz. Since neutral is bonded to earth by the electricity supplier, the undesirable effect of connecting a shunt capacitor from line to earth is to drive a current back to the neutral/earth bond, developing a voltage across any intervening earth impedance. Worse, because the capacitor has reactance that falls with frequency, the resulting current has a spectrum that rises with frequency, exacerbating mains harmonics. Thus, although such shunt capacitors are essential from a common-mode point of view, they may cause problems elsewhere by imposing buzzy interference upon the earth.

An additional consideration is that there is usually a residual current device (RCD) that detects earth leakage currents at the interface between the electricity supplier and the consumer. RCDs typically trip at 20 mA and multiple shunt capacitors cause leakage current that quickly eats into this 20 mA limit, increasing the likelihood of nuisance tripping.

On balance, the author is not in favour of shotgun mains filters and prefers a more targeted approach that directly attacks a specific and known problem whilst minimising side-effects. The only direct path for mains-borne common-mode interference to enter valve audio is via a valve's C_{hk}, and provided that the cathode has a low-impedance bypass, even this path is closed, but cascodes and differential pairs cannot bypass their cathodes, leaving them vulnerable.

DC heaters are mandatory in a moving coil RIAA stage, yet the single most effective barrier to mains-borne common-mode interference is minimised C_{ps} in the heater transformer, so the author initially fitted an off-the-shelf dual chamber transformer having a pair of 12 V secondaries. Unfortunately, it transpired that the two secondaries were wound bifilar, and connecting the other secondary to a voltage doubler for the subsidiary negative supply superimposed very nasty common-mode 100 Hz switching spikes onto the heater supply due to the ≈ 2.5 nF secondary interwinding capacitance. The capacitance plus following impedance to chassis formed a loop containing a potential divider. See Figure 7.61.

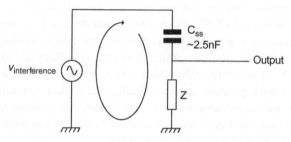

Figure 7.61
Transformer interwinding capacitance in conjunction with impedance to chassis forms a high-pass filter that allows rectifier switching spikes to pass through.

Now that the circuit is drawn, we see that making $Z = 0$ would eliminate interference, and since $Z = R + jX_L$, we must minimise both resistance and inductance. Thus, rather than attempting common-mode filtering, a carefully positioned chassis bond should minimise interference, and so it proved. Experimentally determined using a short clip lead, the best position for the heater supply's 0 V chassis bond was from its low ESR reservoir capacitor, so a short low-inductance link made of aluminium foil was fitted.

Unfortunately, although the carefully positioned chassis bond substantially attenuated the spikes, it couldn't eliminate them and the author reluctantly conceded that the low voltage transformer's second bifilar winding was unusable for the subsidiary negative supply. There were three possible solutions:

- Use the two spare 6.3 V 2 A windings on the HT transformer (but risk unwanted coupling between the HT and subsidiary negative supplies).

- Add another transformer solely for the subsidiary supply (probably requiring the metal-work to be started afresh).
- Replace the nasty bifilar secondary transformer with a proper one.

Expressing the problem and possible solutions properly made the choice obvious. Unfortunately, there was no reason to suppose that **any** contemporary off-the-shelf transformers would not be wound bifilar, so a custom transformer was required. Having accepted a custom transformer, the author looked to see if he could justify its considerable expense by gaining a further advantage. The basic premise of minimising interwinding capacitance still holds, suggesting earthed foil electrostatic screens not only between mains primary and secondaries, but also between the two secondaries, typically reducing interwinding capacitance to 5 pF or less. Freed from the 2×12 V restriction, but needing to justify commissioning a batch of transformers, the heater winding became 2×6.5 V (enabling 6.3 V_{AC} heaters on other projects) and the subsidiary negative supply's winding increased to 20 V, enabling reversion from voltage doubler to bridge rectifier, improving smoothing and reducing ripple current. Finally, the author requested a low core flux (0.9 T) to minimise magnetising current. The custom low voltage transformer made by Majestic Transformers of Poole was far superior to the off-the-shelf dross initially fitted and eliminated cross-coupling of switching spikes.

On the positive side, the original transformer's crippling interwinding capacitance forced determination of the best way of attenuating rectifier switching spikes coupled via inter-winding capacitance.

Testing HT rectification and smoothing

These tests were left until last because the author didn't fancy poking around inside a box having a 100 μF film capacitor charged to 260 V_{DC}. Thus, although the HT had been wired all the time, the EZ81 rectifier was removed until needed — this makes valve recti-fiers very user-friendly.

The HT rectification and smoothing were tested by loading with the DN2540N5 CCS seen in Chapter 4 (Figure 7.80). Although not the most stable load in terms of maintaining a set current, it approximated to the conditions of use (constant current load) and allowed test-ing with a minimum of excitement. Once the EZ81 had warmed up, 206 V_{pk-pk} ripple volt-age was seen across the 2 μF "reservoir" capacitor. See Figure 7.62.

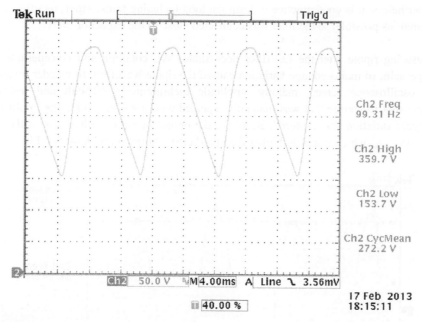

Figure 7.62
Voltage across the 2 μF "reservoir" capacitor. Note that the oscilloscope is DC coupled, with 0 V at the bottom of the screen.

The "CycMean" voltage is the DC component, and at 272 V is roughly as expected, so attention turned to the output of the LC filter, which a DVM measured to be slightly lower at 248 V_{DC} than the modelled voltage of 260 V_{DC} but perfectly acceptable.

A four-wire Kelvin connection is ideal for a reservoir capacitor because it maximises separation between incoming and outgoing currents, but this is far less significant for an intermediate supply because:

- The ripple current (and therefore voltage dropped across any common impedance) is much lower than a true capacitor input supply.
- The ripple voltage across the "reservoir" is so large that any unwanted voltage drops become proportionately smaller.

Nevertheless, it is good practice to keep the loop including transformer, rectifier, capacitor as small as possible, so the load was connected close to the capacitor body.

Measuring ripple after the LC filter necessitated AC coupling and considerable oscilloscope gain, so mains voltage variation caused the ripple waveform to wander up and down the oscilloscope screen, making amplitude measurements difficult and necessitating single-shot triggering that was manually repeated until a waveform having minimal cycle-to-cycle variation was captured. Ripple at the output of the LC filter was of the order of 285 mV$_{\text{pk-pk}}$, which is a little higher than predicted, but not a problem. See Figure 7.63.

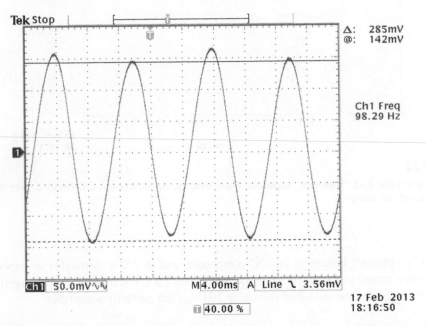

Figure 7.63
Ripple voltage after the LC filter. Ripple is of the order of 285 mV$_{\text{pk-pk}}$ (as measured by cursor Δ).

As with the subsidiary negative supply, input and output ripple spectra allow the filter's effect to be fully appreciated. See Figure 7.64.

The Thing's LC filter is sufficiently effective that it was necessary to reference the vertical scaling to 1 nV$_{\text{RMS}}$. The LC filter achieves a respectable 56 dB of attenuation at 100 Hz,

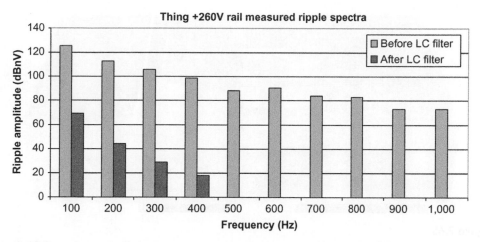

Figure 7.64
Measured spectra showing the efficacy of the LC filter in the 250 V HT supply.

but its 12 dB/octave slope means that the 400 Hz harmonic was barely out of the noise. Remembering that the following Statistical Regulator is likely to offer >90 dB PSRR at all frequencies up to 1 MHz, the +195 V to the RIAA stage will be dominated by noise, and there won't even be very much of that.

Six connections (+260, 0, −11.2, 0, +12.6, 0 V) need to pass through the screen, so these were taken via 1 nF 500 V feed-through capacitors. Feed-through capacitors look like a power diode with a wire at each end and a threaded stud to hold them in place, but the wire is continuous and a ceramic capacitor is connected between it and the body. These devices are popular in radio frequency work for decoupling power as it enters a screened compartment because they have very low inductance to their body and minimise entry hole size, thereby minimising the passage of interference. Because the feed-through capacitors made the screen an explicit electrical connection, a low-resistance, low-inductance bond was needed to the die-cast box and this was provided by a small right-angled bracket between screen and box near to the capacitors. As with the feed-through capacitors, serrated washers were used to ensure reliable gas-tight connections.

The author provided a switched IEC mains outlet to feed the turntable, but other than that, the Thing is fairly simple, hence all that fresh air inside. See Figure 7.65.

Figure 7.65
General view of the Thing from underneath with bottom cover removed. Clockwise from IEC inlet: HT transformer, low voltage transformer (note snubbers), heater regulator (note W02 heatsink clamp, foil chassis bond), subsidiary negative supply (bridge rectifier variant), 100 μF 400 V Kelvin capacitor, umbilical connector, HT choke, EZ81 rectifier, IEC outlet.

Note that the custom mains transformer means that the subsidiary negative supply's regulator board now has a bridge rectifier rather than the initial voltage doubler, and feeding the heater rectifier from 13 V rather than 12 V allows operation down to 220 V mains. See Figure 7.66.

It probably wasn't obvious from the internal photographs, but the author sought to minimise the number of screws passing through the case and this accounts for the uncluttered external appearance of the Thing. See Figure 7.67.

Hybrid balanced RIAA stage (a.k.a. "Toaster")

The hybrid balanced RIAA stage was designed for use with the Denon DL103 low output moving coil cartridge (0.3 mV at 5 cm/s, 1 kHz), which translates to only 42 μV at 50 Hz, so minimising hum is a huge practical concern.

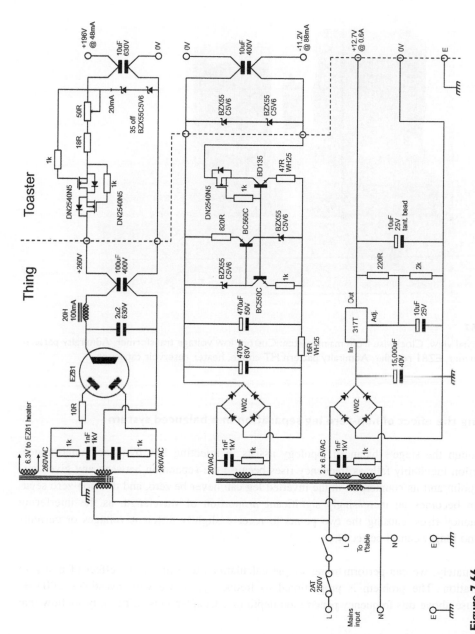

Figure 7.66
Thing circuit diagram.

Figure 7.67

Thing external view. Clockwise from mains switch: Custom low voltage transformer, Admiralty pattern HT transformer, EZ81 rectifier, Admiralty pattern HT choke, heater reservoir capacitor.

Estimating the effect of non-zero leg separation in a balanced system

Although the stage's balanced topology assists in rejecting low frequency interference, rejection inevitably falls as frequency rises because the separation between any given circuit point and its counterpart in the inverted leg can never be zero, and this non-zero separation becomes an increasingly significant proportion of wavelength as the interfering frequency rises, causing the two points to receive slightly different voltages or currents that no longer cancel perfectly.

Fortunately, we can perform some simple calculations to estimate the effect of non-zero separation. The problem is proportional to frequency, so we will investigate 200 kHz because above this frequency penetration depth (see Chapter 1) is sufficiently shallow that

1 mm steel or 3 mm aluminium becomes an effective electromagnetic shield. We first calculate 200 kHz electromagnetic wavelength.

At 200 kHz:

$$\lambda = \frac{c}{f} = \frac{3 \times 10^8}{200,000} = 1500 \ \text{m}$$

where:

λ = wavelength (m)
c = velocity of light in vacuo $\approx 3 \times 10^8$ m/s
f = frequency (Hz).

We will set a limit of -50 dB because this is the CMRR achievable by a practical differential pair (there is no point in achieving perfect matching of induced signals if the differential pair can't reject them). We convert our -50 dB limit into a ratio:

$$\text{Ratio} = 10^{\frac{dB}{20}} = 10^{-\frac{50}{20}} = 3.16 \times 10^{-3}$$

We want to know the phase difference between two sine waves that would result in this ratio as a consequence of imperfect cancellation. This is equivalent to referencing one sine wave to a phase of $0°$ and wanting to know the phase (θ) of the other. Thus, we take the inverse sine of the previous ratio:

$$\theta = \sin^{-1}(3.16 \times 10^{-3}) = 0.18°$$

There are $360°$ in one wavelength, so $0.18°$ is a proportion ($0.18/360$) of one wavelength, and we can find the absolute distance (d) at 200 kHz:

$$d = \frac{0.18}{360} \times 1500 \ \text{m} \times 0.75 \ \text{m} = 750 \ \text{mm} \approx 30''$$

Thus, provided that we can keep the separation between corresponding points of the two legs <750 mm, a perfect differential stage would achieve >50 dB cancellation at all frequencies below 200 kHz; 750 mm within a piece of electronics is a large distance by any standards, and it is more likely that our separation will be 75 mm ($\approx 3''$), suggesting that cancellation should not degrade until beyond 2 MHz. In practice, this theoretical cancellation will be eroded by:

- Unmatched stray capacitances between legs
- Unmatched wire lengths and orientations between legs intercepting the magnetic field and developing different interference signals
- The estimation made the implicit assumption that the magnetic field was homogeneous, which is true for far sources, but not from a local source – another justification for a remote power supply.

Case construction: Shielding and screening

Despite the caveats, the message is clear: balanced construction should enable 50 dB cancellation of magnetic fields <200 kHz, and best rejection is obtained by making the circuitry small (minimising wire length), giving surface-mount devices a useful advantage.

Turning to mechanical considerations, if we know that we can tolerate degraded electromagnetic shielding <200 kHz we don't need to make our case out of mumetal (astonishingly expensive, and requires annealing after working). The high frequency magnetic shielding we need can be provided by 1 mm steel or 3 mm aluminium, and ideal electrostatic screening implies a totally enclosed conductive case having no holes.

Case construction and heat

Totally enclosing the audio circuitry severely impedes heat loss, so we should consider how much heat must be lost. See Figure 7.68.

The audio circuitry requires 48 mA from the +195 V rail, and the Statistical Regulator's Zener chain draws a further 20 mA, resulting in a total HT power dissipation of 16.5 W. (It is permissible to ignore the dissipation in the regulator's CCS because its hot DN2540N5 can be bonded to an external heatsink.) There are four 8416 heaters, each dissipating 1.89 W of heater power, so we have a total internal power dissipation of almost 21 W; 21 W might not sound much, but it must cause a local temperature rise, and the question is whether it is significant.

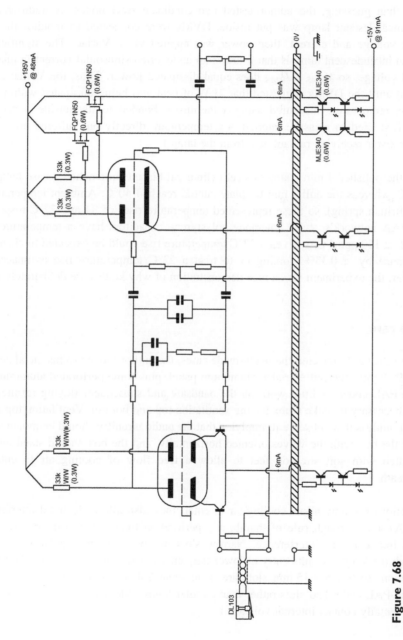

Figure 7.68
Simplified circuit diagram showing where the heat is dissipated.

Rather than guessing, the author tested two candidate steel boxes. A traditional 40 W mains incandescent lamp was put inside, DVMs were connected to monitor the lamp's applied voltage and current, then power was applied via a Variac. The significance of using an incandescent lamp is that it draws an undistorted sinusoidal current in phase with applied voltage, so $V_{RMS} \times I_{RMS}$ **does** equal dissipated power. Thus, the Variac could be adjusted until the DVMs suggested that 21 W of heat was being dissipated within the box and box temperature measured using a thermistor bonded to its outside surface. (The author first tried to measure internal box temperature directly but suffered errors due to the thermistor receiving radiant heat from the lamp.)

Sadly, the polished 1 mm stainless steel cylindrical biscuit tin reached a case temperature of 52°C, whereas the dull 1 mm tin plate cuboid reached 40°C. Ambient temperature was 18°C (British spring), so these represented temperature rises of 34 and 22°C respectively. The RIAA network's ±1% tolerance polystyrene capacitors have a temperature coefficient of ±150 ppm/°C, so even a 22°C temperature rise could be expected to change their capacitance by ±0.33%, forcing us to treat a 22°C temperature rise as unacceptable. However, the experiment was a nice demonstration of why kettles are deliberately shiny!

Making the case

Having reluctantly rejected the ready-made cases, the author looked to his metal and found a 3U 19″ lightly flanged extruded aluminium panel, plus some perforated aluminium sheet from a PAL coder. A little work on the bandsaw and subsequent tidying resulted in the panels necessary to make a box having ventilation top and bottom. Ventilation top and bottom is pointless if we obscure it, implying that the audio circuitry should be mounted on the side of the box with the valves oriented horizontally, and the box should stand on 20 mm legs fitted with soft stick-on feet to allow a free flow of cooling air to enter from underneath.

Perforations not only allow cooling air to enter, they also allow electrical interference to enter. As a very rough rule of thumb, any perforation longer than a quarter wavelength allows free entry of interference. Narrow slots are better than circular holes because although the length permits entry of lower frequencies, such entry is permitted only in one orientation, so 3 mm × 15 mm slots are better than 7.4 mm diameter holes − which is why the PAL coder had slots rather than circular holes. (Slots also make it more difficult to accidentally contact internal voltages.)

In order to ease access, the author screwed the facing side panels together using a 106 mm strut at each corner made of 5 mm × 12 mm cross-section aluminium, then screwed all other panels to these struts. In this way, any panel can be removed simply by undoing four button head socket screws. See Figure 7.69.

Figure 7.69
The case is based on two sections of 3U flanged panel connected by four struts.

M3 button heads are 5.5 mm in diameter, but the fasteners are 2.5 mm from the edge, so the author turned the heads down to 5 mm in diameter, and this didn't weaken the head because button heads use the hex size down from cap heads. Further, those smaller sockets are shallower than in a cap head fastener, so button head fasteners should be reserved for light duties. During the course of Toaster construction, the chassis was assembled and dis-assembled many times and the sockets of eight M3 stainless steel button heads wore out, requiring replacement.

Marking out and drilling holes to the required accuracy is difficult, so the author didn't even try. Instead, he set up a jig on his pillar drill that guaranteed repeatable alignment.

A large needle (with flattened eye cut off) was gripped in the drill chuck, and its tip checked for concentricity by spinning the chuck. Loosening the chuck, rotating the needle, then retightening can often change concentricity, so experiment. The table was adjusted so that its height would allow the quill to drive the tip of a 3 mm drill below the table surface, rotated so that its hole was below the needle, then firmly locked off. That done, the quill was lowered so that the needle was just above a steel rule butted squarely up to an aluminium bar lightly clamped to the table behind the axis of the drill.

The aluminium bar was gently tapped with a soft hammer until the backwards/forwards distance to the needle was 2.5 mm (half the strut thickness), and the clamps locked off. The easiest way to do this is to roughly position the bar so that the spacing is ≈ 0.5 mm oversize, tighten one nut/clamp finger-tight, then ease the other end of the bar from its far end whilst checking needle position above the rule — the leverage from the far end makes accurate positioning much easier. Finally, spin the chuck — it's unlikely that the needle will be perfectly true, so check that the tip deviates symmetrically either side of your intended position. Once you're happy, lightly tighten the clamps with a spanner, taking care not to upset table or clamp positioning — check the setting afterwards with the rule.

A second stop having a square end was then secured lightly by a third clamp in position to act as a left stop, the steel rule placed in position, and the stop's position gently adjusted until there was 10 mm from its end face to the needle, then its clamp gently locked off with a spanner. Check both your distances with the rule after using the spanner to ensure that you haven't disturbed either setting. See Figure 7.70.

Provided that the (square) work is pressed firmly into the (swarf-free) corner provided by the rear bar and left end stop, all holes will be drilled in exactly the same place, and if you can't drill holes to their required position ± 0.1 mm in this way, you're not trying hard enough. The two panels were quickly drilled clearance for M3 (turn each panel over for the other two holes), then the struts were drilled M3 tapping. The author's pillar drill was bought second-hand and (unsurprisingly) came complete with random holes drilled in an arc across its table. Significant pressure (enough to drill) on a strut caused tilting because it wasn't supported directly underneath the drilling point, so each strut was lightly clamped to the table at its far end to prevent this movement. Once drilled, the holes were tapped and the work assembled. Because of the guaranteed repeatability of the jig, everything fitted without fettling, no matter how it was assembled. This does not normally happen with hand-made work, so the technique is recommended not just for its speed, but also for its accuracy and repeatability.

Figure 7.70
Clamped scrap forms a corner into which the work can be located, allowing accurate and repeatable drilling.

The moment all the panels were fitted, the author noted that the case had the proportions and appearance of a bread toaster, so that name has stuck — especially as "Toaster" is considerably less of a mouthful than "balanced hybrid RIAA stage" and most of the construction problems centred around preventing it from toasting its close tolerance components. The front panel overhangs the bottom of the case and almost reaches the surface on which its legs stand. The author is not certain that this is the prettiest solution, but doesn't fancy shortening the front panel to discover that yes, it was. See Figure 7.71.

LED power indicator

Because the author has a lathe (**far** more affordable than they used to be), he turned an aluminium ferrule to an interference fit for an 8 mm red/green bicolour LED, allowing a neater fitting to the front panel and better visibility in bright sunlight (it shades edge light from the surface of the rather weak LED).

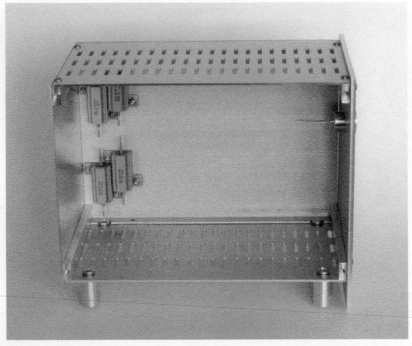

Figure 7.71
To maximise ventilation, the top and bottom plates are perforated, and the Toaster is supported on four 20 mm legs incorporating stick-on soft plastic feet. The LED power indicator in its ferrule can be seen towards the top of the overhanging front panel.

The bicolour LED has a common cathode that connects to 0 V, and at 10 mA its red LED dropped 1.84 V whereas the green dropped 2.03 V. The purpose of the bicolour LED is to simultaneously monitor the heater supply (green) and HT supply (red). At switch-on, the LED immediately glows green, turning to orange when the EZ81 warms up. Red would indicate the fault condition of HT present without the heater supply.

The Statistical Regulator's Zener string could be stood on top of the red LED (passing 20 mA) but even a temporary fault that increased Zener current would destroy the LED. Alternatively, a resistor from the +195 V could supply the 10 mA LED current, but it would dissipate 1.93 W and the Toaster has no need of extra heat. The cool solution was to switch current to the red LED from the heater supply via a transistor. The extra components live on a small PCB screwed to a collar that was an interference fit over the back of the LED's ferrule. The unusual PCB shape was required to avoid it fouling the output coupling capacitors on the main PCB. See Figures 7.72–7.74.

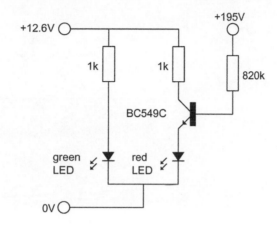

Figure 7.72
LED switching circuit.

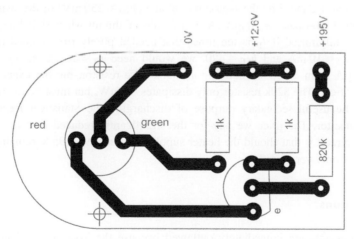

Figure 7.73
PCB layout for LED switching circuit (view from foil side).

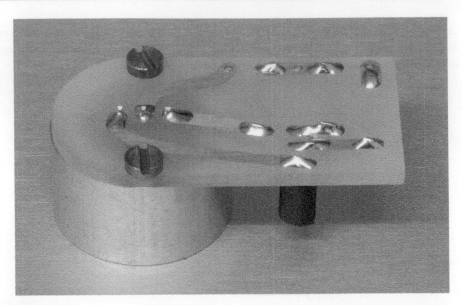

Figure 7.74
LED switching PCB fitted to LED collar.

LED current is limited by the 1k resistors, and the slightly lower forward drop of the red LED is counterbalanced by the saturation voltage ($V_{sat} \approx 250$ mV) of the switching transistor, so both LEDs pass ≈ 10 mA. A useful rule of thumb when switching transistors is that $V_{sat.}$ is minimised if I_B is ten times that needed purely from considerations of h_{FE}. Thus, for $I_C = 10$ mA and $h_{FE} = 420$, we would need $I_B \approx 24$ µA, so for switching we drive 240 µA from $+195$ V, requiring an 820k base resistor, but the exact value isn't in the least critical. The 820k resistor only dissipates 45 mW, but must be rated at 250 V and performs the useful secondary purpose of discharging the Statistical Regulator's 10 µF bypass capacitor. The price we pay for the significantly reduced heat dissipation of the switching circuit is that should the heater supply fail but the $+195$ V remain, the red LED would glow only very faintly.

Audio connections

Phono connectors are immediately outlawed because the Toaster is balanced, making a pair of three-pin male XLRs the obvious output connectors. Logically, the inputs should also be a pair of three-pin female XLRs, or perhaps mini-XLRs, but the author has always used a single five-pin DIN plug to carry stereo balanced signal and earth connections from his arm/cartridge, so he fitted a five-pin DIN socket and did not need a separate binding post for the arm's chassis connection.

The Toaster relies on a balanced connection to the cartridge to minimise the effect of induced interference, but at the time of writing, commercial arms adhere to the unbalanced coaxial cable/phono connector standard. Coaxial cable is far too stiff to pass through a pick-arm bearing, so most arms have a connector at their base, enabling conversion to balanced wiring that can be restored later if necessary. Ideally, you would make your own arm having a fine twisted pair per channel from the cartridge past the bearing (unipivot, string, magnetic, etc.), probably connecting to a thicker (and double-screened) twisted pair leading to the Toaster. Making a pick-up arm is easier than you might think, and the author made his first at school. A double metalwork class sufficed for the machining, and that evening a Shure M95ED in unipivot arm played records. It certainly wasn't pretty, but it made music.

The umbilical cable and EMC

We have previously argued that the umbilical cable between an RIAA stage and its remote power supply should have a single connector at the power supply end, so we simply cut a hole in the Toaster for the cable to exit. To avoid the sharp edges of the metal hole chafing the cable's plastic insulation, we fit a grommet. From an EMC point of view, this is a disaster because:

- Fitting the grommet necessarily requires an even larger hole in the case, leaving a non-conducting annulus between the cable's overall conductive screen and the case through which interference can enter.
- We inevitably connect the cable's screen to the case via a short (but inductive) pig tail, causing the screen to be at a different radio frequency potential to the case.

Fortunately, there is a solution. Compressed gas and vacuum systems require gas-tight connections between flexible hoses and hardware such as pumps, so connectors and adapters are readily available, but make sure that you purchase the plated brass type and not plastic. In the UK, such connectors are specified by their British Standard Pipe (BSP) thread, dimensioned in fractions of an inch. The thread is slightly conical in order to assist in making a gas-tight seal, and the specified dimension relates to the internal pipe bore, not the thread, so a ⅜″ BSP connector has an external thread diameter closer to 14 mm than ⅜″ (9.5 mm). With a little ingenuity, one of these plated brass gas connectors can electrically bond the umbilical cable's screen to the case with no gaps and no pig tails. See Figure 7.75.

Figure 7.75
An air connector avoids EMC issues as the umbilical cable enters the Toaster.

The connector is secured behind the panel by a plated brass nut that started life as an adapter to the next larger pipe size, but was sawn down. Because the gas connector is the incoming earth bond, a serrated washer ought to be between it and the chassis to make a gas-tight electrical contact. Unfortunately, the author didn't have any suitable serrated washers, so he settled for gripping the fixed nut firmly in a bench vice (using soft jaws made of aluminium angle to prevent marking) and tightening the 17 mm A/F free nut as hard as possible using a long ring spanner. The advantage of the vice over another spanner was two-fold:

- Because the vice gripped the nut firmly there was less likelihood of spanner slippage that would chew a nut or gouge the rear panel.
- The fixed nut could be aligned with the rear panel and any panel rotation seen and corrected as the free nut was tightened.

The author used RG174 coaxial cable for the unregulated HT and also for the subsidiary negative supply. Heater power was carried over tightly twisted 0.8 mm² solid-core wire inside a braid screen. These three cables passed snugly through the 7.43 mm internal bore of the ⅜″ BSP pipe connector. A braid overall screen (salvaged from BBC PSF1/2 video cable) went over the pipe connector, and an almost circumferential electrical connection was made by an O-clip of the type used for securing car radiator hoses. See Figure 7.76.

Figure 7.76
The O-clip electrically bonds the braid screen to the chassis.

O-clips must be fastened by the proper tool — not a pair of ordinary pliers. Note that the blunt jaws that do not quite meet are for tightening O-rings, whereas the sharp jaws that meet are for cutting and removing them. See Figure 7.77.

Figure 7.77
O-clip pliers.

If taken under the O-ring, the umbilical cable's outer nylon braid would creep, degrading the earth bond beneath, so it had to go outside. Sadly, any braid has a restricted range of enveloping diameters, and nylon braid that could tighten over the cable would not open over the O-clip, so loose strands were taken over the O-clip and the braid was secured and tidied by large-diameter glue-lined heatshrink sleeving that passed over the O-clip. See Figure 7.78.

The Statistical Regulator

The unregulated HT leaving the Thing's LC filter is reasonably quiet and, more significantly, the +195 V leaving the Statistical Regulator is far too quiet to risk taking it down even a double-screened cable. Thus, the entire Statistical Regulator belongs near its load within the Toaster. Depending on mains voltage, the CCS's outer DN2540 drops 65 V whilst passing 68 mA, implying a typical dissipation of 4.4 W. The Toaster's 3U side panels are 3 mm thick, giving them lower thermal resistance across their surface than the 1.6 mm rear panel, making them better heatsinks, so the Statistical Regulator's CCS should be mounted on a side panel.

Because the Statistical Regulator employs a series CCS, it draws a constant current from its raw supply. The significance of the previous statement is that a constant current through a wire implies an unchanging magnetic field, so we are not concerned with magnetic induction from the raw HT wiring, only electrostatic. Thus, preventing coupling from the raw HT into sensitive circuitry simply requires minimising capacitance between

Figure 7.78
Glue-lined heatshrink sleeving tidies and secures the nylon braid to the chassis. The adjacent DIN socket is secured by 4BA cheese head screws.

them, which implies either distance or an intervening earthed conductive screen. Since raw power enters the Toaster on a screened cable, the only unscreened interference source is the CCS's outer FET. Once the Statistical Regulator's postage stamp CCS PCB had been made, it was obvious that it could be positioned sufficiently remote from sensitive circuitry to render additional screening unnecessary. See Figures 7.79 and 7.80.

The CCS was tested by temporarily thermally bonding its outer FET to a large heatsink and connecting it across a bench supply via an ammeter. Supply voltage was gently increased from zero whilst monitoring current. At 1 V, only a few milliamps passed, so voltage was increased (≈ 7 V) until measured current became constant. Unsurprisingly, the 50 Ω rheostat needed adjustment to set current to 68 mA. Supply voltage was then increased to the 65 V expected across the CCS and measured current increased by ≈ 0.1 mA. The increase in current was due to self-heating changing FET parameters and does not reflect the CCS's dynamic resistance.

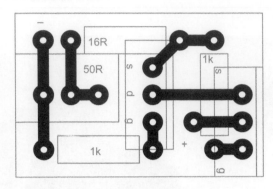

Figure 7.79
PCB layout of Statistical Regulator CCS (view from foil side).

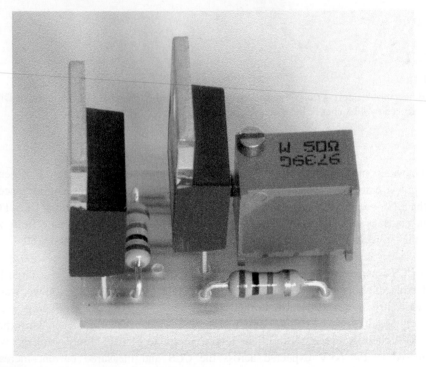

Figure 7.80
The Statistical Regulator's CCS is very small.

The shunt section of the Statistical Regulator needed a tag panel for its Zener string and 10 μF Kelvin capacitor. The inductance of an extended foil film capacitor is minimised by making it short and fat, coincidentally allowing room for more tails at each end. Thus, at each end we connect to:

+ 195 V in (from Zener string)
First stage +195 V (L)
First stage +195 V (R)
Second stage +195 V (L)
Second stage +195 V (R).

This is the reason for the five tails per end that mystified early purchasers expecting only two tails. The opportunity was also taken to increase the voltage rating to 630 V (making it suitable for any non-power stage), so the resulting capacitor body is ≈48 mm diameter and ≈31 mm long. The original two-tailed Kelvin capacitor was flattened during manufacture, making it easy to mount using a foam self-adhesive pad, but the new capacitor is larger, heavier, and requires a little more care.

A 50 mm × 25 mm × ¼″ block of Paxolin was cut with a 24 mm radius to seat the capacitor using a boring attachment in the author's milling machine. This boring process lived up to its name, and a perfectly adequate result could have been achieved in a fraction of the time by careful filing followed by rubbing the Paxolin on abrasive paper wrapped round a 46 mm cylinder. The Paxolin was drilled and tapped M3 to suit the hole size and spacing on the panel but spaced away from it by washers to allow a cable tie to pass freely between the two fastened parts. Holes were drilled in the Paxolin block to allow the cable tie to pass through and secure the capacitor. Pillars allowed the final assembly to be secured to the case. See Figure 7.81.

The M3 screws passing into the Paxolin block ideally use all the available thread, but must not poke into the capacitor, so they needed trimming to length. The author did this in his lathe, but an indistinguishable result can be achieved by sawing and filing slightly oversize, clamping the screw in a split nut (see Chapter 2) in a drill chuck, then running the drill in a stand and bringing the screw's filed end to bear on 240 grade abrasive paper supported on a flat steel block.

The tie-wrap needs to bend 90° through only **one** of the holes in the Paxolin block, so this was countersunk to allow a larger bending radius, and the capacitor was firmly secured.

Figure 7.81
The ¼″ thick Paxolin cross-member has been cut away to locate the 10 μF 630 V Kelvin capacitor.

Flush cutters are best for severing the surplus tail of cable ties because they do not leave a protruding sharp edge that later scratches. See Figure 7.82.

Subsidiary negative supply Zeners and decoupling

The subsidiary negative supply needs a series pair of 5.6 V Zeners decoupled by a 10 μF four-wire capacitor, and these connections were most conveniently provided on a small PCB. See Figures 7.83 and 7.84.

Point-to-point signal wiring or printed circuit board (PCB)?

The Toaster's signal circuitry includes a significant number of semiconductors plus a number of capacitors designed for PCB mounting. It is fashionable to extol the superior leakage resistance and component cooling of point-to-point wiring compared to a PCB, but many of the components required for this design were intended for PCB mounting and

Figure 7.82
The 10 μF 630 V Kelvin capacitor is secured by the tie-wrap passing between tag strip and cross-member. With the pillars and capacitor fitted, this sub-assembly is ready for the 5V6 Zener diodes.

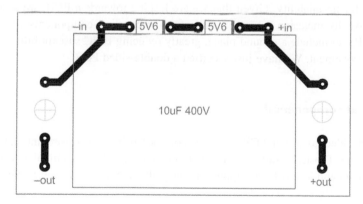

Figure 7.83
PCB layout of subsidiary negative supply's shunt elements (view from foil side).

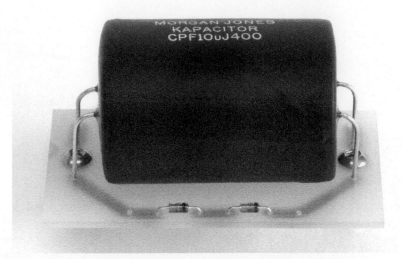

Figure 7.84
This simple PCB supports the shunt section of the subsidiary negative regulator. Note that a 40 V voltage rating would have sufficed even under fault conditions, but 400 V was in stock, so it was used.

supporting their off-centre weight solely on a soldered joint in point-to-point construction is asking for unreliability. Given that we are leaning towards a PCB, we should see if we could make its inclusion a virtue. Aside from supporting components firmly, a double-sided PCB can include a ground plane, greatly reducing EMC susceptibility — and that's a killer improvement. We have just specified a double-sided PCB.

SMD or leaded components?

Having decided to use a PCB, we can consider using surface-mount devices (SMDs). Many of the Toaster's audio components do not have SMD equivalents, but its small-signal silicon and low voltage support circuitry all has SMD equivalents. See Table 7.3.

At the time of writing, SSM2210 was still readily available, so SSM2212 was not enforced. 0805 and SOIC-8 are fiddly for the over-40s to solder, but not impossible, and

Table 7.3

Leaded components and their SMD equivalents

Leaded device	Package	SMD equivalent	Package
SSM2210 (freshly obsolete)	DIP-8	SSM2212 (current)	SOIC-8
BC549C	TO-92	BC847C	SOT23
Cheap red LED	Various	HLMP6000	
MRS25 resistor		Various, thick film	0805

small devices allow a tight layout (remember that EMC susceptibility is proportional to loop area).

There is no device advantage in replacing the BC549C with BC847C (they are electrical equivalents), and LTspice modelling suggests that substituting BFR93A for the outer BC549C would only slightly improve -11 V PSRR beyond 1 MHz whilst leaving input CMRR unchanged. However, not only is the subsidiary negative supply very quiet, but CMRR above 1 kHz is strongly determined by the capacitive imbalance of the incoming arm cable and quickly falls to a value that renders all other considerations immaterial, although it is worth matching the 33k cascode loads to maximise hum rejection.

You might need to search junk shops to find the standard brightness (dim) red LEDs that have a low slope resistance, but when you do, you're likely to pay less for a bag of a hundred than you will for five new HLMP6000s. That's quite an incentive to use leaded LEDs.

General-purpose SMD resistors are generally thick film, which means that they have fine resistive granules mixed into an insulating binder, the composite material being known as a matrix. Conduction requires breakdown of the insulation between resistive granules, and because the exact route of these random breakdown paths is determined by the voltage across the resistor, they suffer excess noise habitually expressed in μV/V. Because achieving the same resistance in a small resistor requires a higher resistivity matrix having fewer or finer granules, the inter-granule paths become more significant and excess noise is inversely proportional to physical size, making the small 0805 device noisier than its leaded equivalent. Excess noise is significant because the Toaster has many constant current sources or, from a noise point of view, many high-gain amplifiers amplifying any excess noise generated by the 1 V forced across their current programming resistor. The most critical component is the 160 Ω CCS current programming resistor at the input stage. Fortunately, the slope resistance seen looking into each of the SSM2210's emitters is so low as to cause the CCS's cascode to have unity voltage gain from its current programming resistor. The resistor's calculated thermal noise (0.23 μV) is similar to what might be expected for the excess noise of a leaded film resistor having 1 V across it (≈ 0.2 μV/V), so we could safely sum these two uncorrelated sources to give ≈ 0.3 μV. This noise is common mode and we can reasonably expect >40 dB CMRR, so the effective noise voltage (**due to this source only**) at the input of the RIAA stage is 3 nV. Competently engineered RIAA stages have an unweighted signal-to-noise ratio of ≈ 68 dB referred to their 5 cm/s sensitivity (0.3 mV for the DL103), so this equates to ≈ 120 nV at the input − 32 dB greater than the speculated 3 nV contributed by the CCS's current programming resistor. Summarising, the CCS's current programming resistor's noise contribution is sufficiently small that even an SMD resistor suffering 1 μV/V excess noise would be fine.

We noted earlier that SMD produces smaller layouts, enabling better cancellation of electromagnetic fields, and this would be a significant advantage if all our components were SMD and small, but valves and many of their associated components are necessarily large, so SMD wouldn't help the Toaster greatly.

There still isn't a firm argument to choose between leaded components or SMD until we consider PCB manufacture. If made commercially, SMD is easier and cheaper (fewer drilled holes), but if the PCB is made at home, producing tracks and pads to the required accuracy is difficult. The author very nearly took the SMD route, but it seemed very slightly more convenient to make the board at home other than generate gerber files and having the board made commercially, so that enforced leaded components.

Standard values, tolerances, and statistics

Some components were broadly specified during the electronic design of the Toaster, such as $\pm 1\%$ tolerance polystyrene capacitors and $\pm 0.1\%$ tolerance resistors for RIAA equalisation, and polypropylene capacitors for inter-stage coupling, but final specification may be determined by other factors.

The author favours LCR's F/Y series of polystyrene capacitors because dissection reveals that LCR have developed a method of making end contacts to all parts of an extended metallised polystyrene film rather than a single point in the manner of an electrolytic capacitor. The downside is that small quantities are only available in E6 values. The design calls for a 3410 pF that could be made up from 3.3 nF and 110 pF in parallel, but the nearest available value to 110 pF in the F/Y series is 100 pF, necessitating an additional 10 pF. However, one possibility might be to select 3.3 nF capacitors having actual values of 3.31 nF (+0.33%).

Although a manufacturer strives to manufacture to nominal value, there is always a spread of values above and below this mean value, so they specify a production tolerance, perhaps ±1%. But such a spread of values always conforms to a statistical curve variously known as the normal distribution, Gaussian distribution, or bell curve, whose characteristics are well known. Thus, if wishing to include 99% of their production run, a manufacturer would specify their tolerance as including three standard distributions (3σ) of the curve. The statistical significance is that if a ±1% tolerance equates to 3σ, ±0.33% tolerance equates to 1σ, and 1σ includes 68% of the population — which is why most ±1% components measure so much better than ±1%. For our present purposes, we could say that 32% of the population is outside 1σ, being either above or below the ±0.33% tolerance. But we require only a value **above** the +0.33% tolerance (known in statistics as a single-tailed probability), so we are interested only in the upper tail of the 32% distribution, and because the curve is symmetrical, this means 16% of the total population. Sixteen per cent is equivalent to saying that the chance of a capacitor falling into the region we require is only one in six, so if we need two capacitors from this region, we must buy at least twelve capacitors, and probably twenty. Out of the author's stock of ten, only one capacitor had the required 3.31 nF capacitance. If these capacitors were cheap, we might cheerfully buy fifty and select, but they aren't, so the approach isn't practical.

Instead, we note that 68% of the population is within ±0.33% of the stated value, so for a 3.3 nF capacitor, that means ±11 pF. Thus, by adding a 2–22 pF trimmer capacitor in parallel we can correct almost 68% of the population to exactly the 3.31 nF value we need (a few won't tolerate the unavoidable 2 pF minimum capacitance, hence the "almost"). Rather than buying twenty capacitors and hoping to find two that are suitable, we now know that we must **reject** 32%, so if we buy three capacitors, one must be rejected to leave two suitable capacitors. In practice, only large populations can be described and predicted accurately, so for stereo we buy five 3.3 nF capacitors and two 20 pF trimmers, then select the two most suitable for trimming at lower cost than selecting the target value from twenty 3.3 nF capacitors.

Trimmer capacitors

We should consider the trimmer capacitor's dielectric. Gut reaction would be to reject ceramic trimmers because leakage equals noise in high-impedance circuits, yet their specified Q (ratio of parallel resistance to reactance) at 1 MHz is generally comparable (≈ 300) to PTFE, suggesting that they use the superior C0G/NP0 dielectric — which would make sense for such small values of capacitance. PTFE trimmers are readily available in the required value and are compact, whereas air trimmers are significantly larger and best obtained at ham radio fairs (where they will be much cheaper than PTFE).

The author couldn't easily test to clear or condemn ceramic trimmers, so he chose PTFE over the marginally superior air because it fitted more easily on his PCB. Be aware that the silver-plated contacts of trimmer capacitors are typically rated for a lifetime of only ten turns — they are not intended for haphazard constant twiddling. Set and forget.

Wirewound resistors and thermocouples

Unlike a common cathode amplifier, the high internal resistance of a cascode cannot shunt the noise of its anode load resistors, so their excess noise must be minimised, which is why wirewound resistors were specified in the original design. Nevertheless, even wirewound resistors are not perfect, and for a 33 kΩ resistor, imperfections centre on the end caps (**not** self-inductance). In the fourth edition of *Valve Amplifiers*, the author speculated that such defects were more likely to be due to uncancelled thermocouple EMFs than magnetic effects, but a measurement would be better.

With the aid of a temperature-controlled hot-air gun and an Agilent 34410A 6½ digit DVM, the author measured the DC EMF of large wirewound resistors under two conditions. The significance of choosing large resistors was that fitting a small nozzle to the gun enabled heat to be applied controllably to only one end termination, preventing cancellation of thermocouple EMFs. See Table 7.4.

The author was surprised to find that not only was a thermocouple EMF easily measurable, but that press-fitted ferrous end caps were measurably inferior. Upon checking in Kaye and Laby [4], the author's surprise abated — the thermal EMF of nickel (at 100°C, relative to platinum) is −1.48 mV and that of iron 1.98 mV, giving a difference of 3.46 mV that entirely explains the observed EMF of 2.4 mV between two nickel and iron alloys of unknown composition.

Table 7.4

Comparison of glazed resistor thermocouple EMFs produced by heating one resistor end termination only

Resistor type, value	Construction/termination	Both ends $\approx 20°C$	One end $\approx 100°C$
Welwyn W23, 33k ± 5%, 25 W	Glazed, press-fitted ferrous end caps, ferrous-cored leads	$\approx 2.6 \pm 25$ μV	2.4 mV ± 25 μV
Welwyn RCSC, 33k ± 5%, 6 W	Glazed, resistive element brazed directly to copper leads	$\approx 2.5 \pm 25$ μV	0.4 mV ± 25 μV

Despite the different power ratings, the two resistors were the same physical size.

The author has always favoured aluminium-clad resistors because conducting heat directly to the chassis minimises internal air temperature, and their tags make convenient wiring points, so further measurement was in order. See Table 7.5.

Table 7.5

Comparison of aluminium-clad resistor thermocouple EMFs produced by heating one resistor end termination only

Resistor type, value	Construction/termination	Both ends $\approx 20°C$	One end $\approx 100°C$
CGS HSA25, 22k ± 5%, 25 W	Aluminium-clad, press-fitted ferrous tags	$\approx 1 \pm 20$ μV	$\approx 40 \pm 20$ μV
Dale RH25, 33k ± 1%, 25 W	Aluminium-clad, press-fitted ferrous tags	$\approx 0.5 \pm 16$ μV	$\approx 10 \pm 20$ μV

The author was astonished by the results in Table 7.5 — he fully expected the brazed copper-leaded Welwyn RCSC to be the best construction, so he queried his experimental technique and repeated all the measurements. The good thermal conductivity of the aluminium cladding tends to equalise end-to-end resistor temperatures, although its thermal mass slows that effect, yet even deliberately placing the hot-air gun's nozzle almost in contact with a tag couldn't produce a significant thermocouple EMF. Despite their ferrous press-fitted tags, thermocouple EMFs for the aluminium-clad resistors were at least an order of magnitude smaller than the best glazed types, and the ± 1% tolerance variant was probably the best.

Having found a measurable difference between constructions, it is tempting to make a connection and state that resistors having a reduced thermocouple effect will sound better, but this is not necessarily true. We can certainly speculate that air currents within the Toaster might cause different end-cap temperatures, leading to DC drift, which we could alternatively describe as being very low frequency noise, and we could also note that RIAA adds 19 dB low frequency gain, but that is probably as far as we could go. For the author, the choice was decided by the fact that his RH25 aluminium-clad resistors were $\pm 1\%$ tolerance whereas the others were $\pm 5\%$, and close load matching improves CMRR in a differential pair, so any improvement due to reduced thermocouple effects would be a bonus.

Wirewound resistors and "annealing"

Although not new, the 33 kΩ wirewound resistors were heated in the oven for a day at 135°C then gently cooled to relieve their winding stresses before being measured and selected for matched resistances to maximise the common-mode rejection of each differential pair. Incidentally, although Scroggie (who originated the wirewound resistor heating process) termed it annealing, it is nothing of the sort, because true annealing heats to incandescence then slowly cools, changing the metal's crystal structure in the process. Remember that a wirewound resistor comprises a coil of wire wound under tension upon a cylindrical insulator, but bear in mind that such a cylinder might not be perfectly centred upon its axis of rotation, causing a varying winding radius and tension, resulting in gentle sinusoidal stretching and consequent necking of the conductor, which would then be clamped in place by its sealant. Deliberate component heating causes all parts to expand and contract in accordance with their temperature coefficients, producing sufficient force to overcome the bond between the wire and its sealant, allowing the necking (which increases resistance) to be relieved. Because the process is primarily dependent on energy rather than temperature, it can also be achieved by more heat cycles at a lower temperature over a longer time (although stiction reduces efficacy), which is why wirewound resistors over four years old show little change from deliberate heating.

Other resistor choices

Each anode load resistor dissipates 0.3 W, which is easily catered for by the 25 W wirewound type specified for the input cascode, but (considering the second stage) is more than can be safely dissipated by a 0.6 W MRS25 resistor randomly positioned on a PCB.

Noise at the second stage is not an issue because r_a is low enough to shunt resistor noise and signal levels are much higher, so 2 W metal film resistors were perfectly adequate, although matching remains worthwhile, so $\pm 1\%$ types would be preferable if only they could be found ($\pm 2\%$ used to be available).

At the other end of the resistor performance spectrum, the source followers need gate-stopper resistors, and as there is no gate current, there can be no voltage drop across these resistors to cause excess noise. Thus, we cheerfully choose carbon film resistors because the increased resistivity of the carbon matrix compared to the nichrome alloys permits reduced inductance and less likelihood of source follower radio frequency oscillation.

The second differential pair needs grid-stopper resistors, and past experience with E88CC has shown that surface-mount types close to the valve pins are the most effective. As mentioned earlier, surface-mount resistors are generally thick film and suffer excess noise, but the same argument applies to the valve as the FET: if there is no grid current there can be no voltage drop and no excess noise.

Having chosen our components, we know how large they are and we can plan our PCB layout.

PCB layout and heat

Having specified a double-sided PCB, with the top foil a ground plane, we know that the lower foil will be for our interconnections, and once the PCB is mounted on the aluminium chassis, those signal connections will be sandwiched between two earthed screens, reducing EMC susceptibility. A commercial design would not only route the DC power on the PCB, but also the heaters. However, this would require breaking the ground plane with many links and vias. For a one-off, we can specify twisted pair heater wiring spaced a little way away from the board (so that it kisses the chassis), thus minimising induced interference. Further, if we use a Kelvin capacitor for the HT with multiple terminations, star connections to it would minimise unwanted voltage drops along PCB tracks (which have a surprisingly high resistance). Having deliberately removed all the long tracks from the PCB, it should be possible to produce a layout that positions components so as to minimise deleterious inter-component coupling capacitance using short tracks themselves having minimum stray capacitance. At the same time, we must consider heat and practicality.

There are eight power transistors each dissipating 0.6 W, or 4.8 W in total. Ideally, these devices should lose their heat to an external heatsink to minimise any rise in internal air temperature that reduces reliability and causes RIAA equalisation component values to drift in value. The best solution would place all eight devices in line so that they can be fastened to a single straight heatsink, and if this heatsink could be at the end of the board, so much the better. One problem with this solution is that we can only route tracks forward of the transistors – we can't easily take them round the back. Far more significantly, this positioning would force the large output coupling capacitors to be close to earlier circuitry.

A better electrical solution would keep the transistors in line across the board but extend the board beyond the heatsink, allowing the output coupling capacitors to be well away from earlier circuitry. See Figure 7.85.

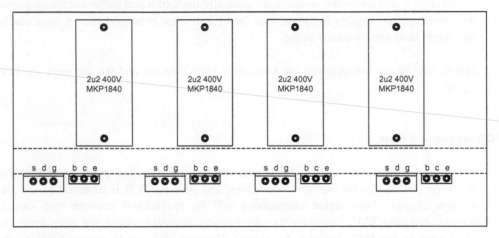

Figure 7.85
The two dashed horizontal lines between the transistors and output coupling capacitors denote the heat transfer bar that cools the transistors.

Insertion or removal of valves flexes the PCB, but the power transistors are necessarily firmly bolted to a heatsink. Thus, changing a valve strains either the soldered joints to the power transistors, or worse, the leads into a transistor package. Straining semiconductor package leads is a capital crime because it breaks the seal to the epoxy, allowing water vapour to enter, and water kills semiconductors. Thus, the PCB has three M3 mounting holes along the $1\frac{1}{2}'' \times \frac{1}{4}''$ cross-section aluminium heat transfer bar whose purpose is not so much to support the negligible weight of the board and its components, but to prevent PCB flexure and consequent straining of the power transistors when a valve is inserted or removed. The substantial cross-section was chosen for the heat transfer bar because calculation using the equations in Chapter 1 suggested that the temperature drops from centre to either end would sum to an entirely negligible 1°C. The heat transfer bar is thermally bonded to a pair of $\frac{1}{2}'' \times 1\frac{1}{4}'' \times 60$ mm aluminium vertical pillars that in turn take the 4.8 W of power transistor heat to the outside chassis, where it can safely be lost to the atmosphere. (The mixed dimensions occur because the author is mostly metric, but smelting aluminium requires cheap electricity, usually hydroelectric, so aluminium stock in the UK comes from Canada, hence its Imperial units.)

PCB layout and power transistor tab insulation

The heat transfer bar solves one problem but creates another. None of the power transistors thermally bonded to it has its tab at earth potential, so each needs an intervening insulator, and therein lies the problem. Although extremely convenient (because they don't need a coating of silicone grease on either side), conformal insulators for semiconductor packages are not necessarily the dielectric that one would choose for an audio capacitor, yet the four MJE340 tabs are all connected to signal, adding a parasitic capacitor to chassis. The author had previously forsworn mica because it is so brittle and easily damaged, but the MJE340 would ideally be insulated from the heat transfer bar by a low-loss dielectric of stable thickness, and that means mica. The FQP1N50P tabs are connected to +195 V rather than audio, so conformal insulators are perfectly acceptable.

PCB layout and earth bonding

The PCB has a 6 mm mounting hole between the first two valves, and this makes the single electrical bond between the 0 V signal earth on the ground plane to the chassis. As usual, the author prefers a large fastener because it can be done up tightly without breaking or stripping — guaranteeing a long-term low-resistance earth bond.

Any earth bond fastener must minimise its voltage drop whilst passing interference currents at random frequencies, but the bond from 0 V signal earth to chassis is critical. When we investigated electrostatic screening in Chapter 1, we saw that screen impedance must be minimised, and the same is true for an enveloping conductive chassis because its large area has capacitance to all points. If the chassis has a non-zero impedance bond to 0 V signal earth, an interference voltage will be developed across that bond, causing the chassis to capacitively couple that interference voltage everywhere. Small signal voltages are proportionately more susceptible to this interference, so this is why we position the bond at the input of the RIAA stage — it avoids adding cable impedance to bond impedance.

Although the large surface area (resistances in parallel) and high contact force (low resistance) of a large fastener guarantees a low-resistance earth bond, it does not necessarily guarantee low inductance. All the earth bond connections from a solder tag to chassis via a serrated washer are coaxial or, to put it another way, they are via a coil having the fastener as its core. Considering the washer, the arc traversed from end to end of each serration is only a fraction of a turn and these inductances are in parallel, but $\mu_r \approx 5500$ for iron, causing the inductance of even a very poor coil to be significantly multiplied by a steel core, so the 0 V signal earth to chassis bond should use a brass screw. The incoming mains earth safety bond can perfectly well use a steel screw because fractional inductance at the end of metres of mains cable is utterly irrelevant.

Because the author's PCB needed to be spaced 18 mm from the chassis (to clear the RH25 resistors), its chassis bond was longer than ideal, so the author rooted out his (very few) large brass screws, drilled and tapped some 15 mm diameter brass to 0BA, then machined it to 18 mm in length. Copper would theoretically have been better (lower resistance) but it's horrible to machine and tap. Having made the spacer, it could be screwed down to the anodised aluminium chassis, thus creating two new problems:

- Anodised aluminium is an insulator.
- There is a significant electro-potential between brass (0.3 V) and extruded aluminium alloy (0.73 V), causing corrosion in the presence of acid.

The first problem was circumvented by choosing a countersunk fastener, necessitating countersinking the 3 mm aluminium panel and exposing bare aluminium. The second problem means that having countersunk the panel, the exposed aluminum must not be touched with fingers (sweat contains acid).

The author is always nervous about countersinking because it is so easy to produce a countersink of incorrect depth, off centre, or misshaped, possibly all three. The solution when precision is essential is to clamp the work firmly to the pillar drill's table. If the chuck axis is aligned by passing a tight drill into the hole, then the work clamped with at least two clamps, the drill can be replaced with a countersink and an accurate countersink can be produced.

In order that the non-magnetic bond be the only electrical bond between ground plane and chassis, the edge of the heat transfer bar was insulated from the PCB ground plane by four narrow strips of 0.25 mm thick clear plastic usually used for the front cover of office reports. The insulator might not have ideal dielectric properties, but that doesn't matter since it is in parallel with a short circuit (the deliberate earth bond), so there will be negligible voltage to cause leakage currents. Making and fixing the strips was easy. The author stuck a 150 mm (6″) length of double-sided adhesive tape near to one edge of the plastic. Using a cheap steel rule as a guide, a scalpel cut was made through the tape and the plastic all along one long edge, roughly 1 mm inside the edge of the tape to leave adhesive tape perfectly aligned with the edge of the plastic. A second cut 5 mm away from this edge produced a 5 mm strip with tape perfectly aligned along its length. Scissors cut the strip into lengths that would cover the edge of the heat transfer bar but not obscure the M3 tapped holes. The tape's paper cover was removed, and handling the strips by their edges, they were stuck to the bar.

The M3 screws that secure the PCB to the heat transfer bar must also be insulated from the ground plane, and this was done by relieving the ground plane around their holes and adding fibre washers. There is no need to take especial care because these insulators are also in parallel with the short circuit of the deliberate earth bond.

PCB layout and accurate RIAA equalisation

The LCR polystyrene capacitors have a dot signifying their outer foil connection so, once positioned, these components were oriented so that their outer foils were connected to the adjacent coupling capacitor — thereby rendering stray capacitance between them irrelevant.

Component values inevitably drift with temperature, and RIAA equalisation is hypercritical of errors. Although the temperature coefficient of the $\pm 0.1\%$ RC55Y resistors is only ± 15 ppm/°C, the $\pm 1\%$ LCR polystyrene capacitors are ten times as sensitive at

± 150 ppm/°C, so they needed to be kept as far away from heat sources as practical. Because the RC55Y resistors had a much lower temperature coefficient and significantly lower surface area than the capacitors (intercepting less radiant heat), they could tolerate moving closer to radiant heat sources such as valves.

General PCB layout considerations

The circuit was broken down into functional modules (CCS, source follower, RIAA equalisation, etc.), and once a good layout for a module had been developed it was placed on the board and copied as necessary. It might be necessary to slightly modify a module's layout once placed in a given position, but the principle of developing a good module and copying it works well. The original circuit diagram showed the push–pull source followers sharing LED voltage references, but this proved awkward to lay out, so each module became self-contained, slightly increasing the current drawn from the subsidiary negative supply, but making future faultfinding much easier (any fault is likely to upset LED current, so a brighter or dimmer LED would indicate the approximate fault location).

Good analogue PCB layout takes time, even if it is a simple circuit – the author required two days of close concentration to develop a PCB layout that would fit his box. Note that in addition to the purely technical requirements, a good layout follows the circuit diagram logically, and if symmetry is possible, this is aesthetically pleasing and aids faultfinding should it be needed. See Figure 7.86.

Any multiple layer PCB must address the problem of registration between layers. It is essential that the pad on one layer aligns perfectly with the drilled and plated via intended to connect it to the corresponding pad on another layer. Conversely, ground planes and tracks must allow leaded components and vias to pass through without shorting. Handmade PCBs derived from paper templates that stretch cannot achieve accurate registration, so the author made this double-sided PCB by marking holes on the signal layer only and immediately drilling all marked holes 0.3 mm, thus providing accurately registered points on the ground plane side. Obviously, drilling before etching risks undercutting of the exposed copper edge at holes, and this is why the author used a 0.3 mm drill – the holes were opened out post-etching to their required size, eliminating the copper undercut. Having made the PCB, the Toaster's electronic complexity is perhaps a little beyond what can be sensibly achieved on a hand-made PCB, so if you can generate gerber files with a minimum of swearing, SMD would be the way to go.

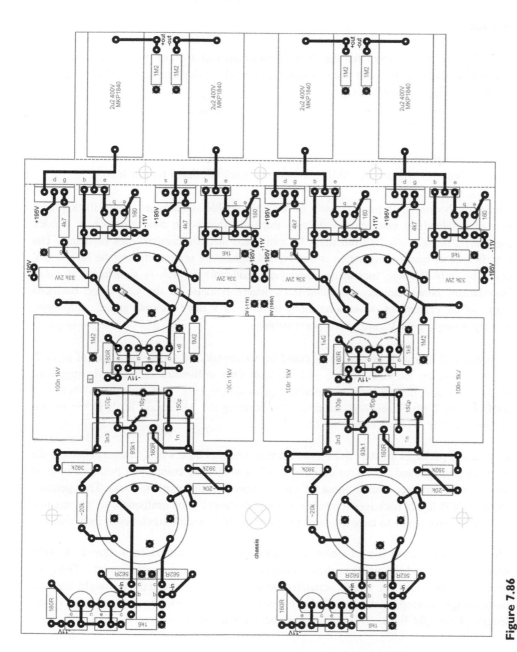

Figure 7.86

Toaster PCB layout (view from foil side). Note that a cross on a pad signifies connection to the ground plane.

SMD layout

If SMD were to be chosen for the PCB, commercial manufacture would be a necessity but good registration between layers would be assured. At that point, it would make sense to use a four-layer board:

Layer 1 (top): 0 V plane
Layer 2: −11 V plane
Layer 3: 0 V plane
Layer 4 (bottom): Signals and SMD components (track width 0.7 mm or less).

Four-layer boards are typically made of an inner double-sided board 1 mm thick plus a pair of single-sided outer layers 0.3 mm thick, so that when the sandwich is bonded together, it forms a standard thickness 1.6 mm board that fits standard rack hardware. The significance of board construction is that layer 4 signals are separated from the layer 3 0 V plane by only 0.3 mm, increasing interlayer capacitance, which is significant at valve impedances, hence the recommended track width. Because of the CCSs and balanced nature of the circuit, the currents returning to the −11 V have minimal signal component, making a single −11 V connection permissible.

If required, 2 mm wide heater tracks could be accommodated round the edges of layer 2, slightly reducing the size of the −11 V plane. One millimetre clearance is suggested around vias and through-hole components to avoid tracking. One unexpected bonus of SMD is that 2 W 1% resistors are available, enabling the 33k anode loads in the second differential pair to be matched.

PCB layout packages all start from a net list that produces a rat's nest of connections in the graphics package, and the layout designer's task is to allocate component positions so as to tidy the rat's nest, then convert those rat's nest connections into explicit tracks. The operating ethos of PCB layout packages forces most layout designers to be draftsmen, so it is not reasonable to expect them to have the same appreciation of a circuit's subtleties as its designer.

The problem begins at the rat's nest stage because as each component is moved, the rat's nest is updated and the package draws straight lines between connected points. There are many ways in which six widely separated points could be connected by straight lines, so the package maximises display clarity by calculating node routing that minimises total length of straight lines. This update can be misleading unless you have a very clear understanding of the circuit because it will cheerfully distribute power from the input device rather than the board's incoming filtering capacitor if that turns out to fulfil its rat's nest criteria. Move a single component and a node could be routed entirely differently.

Like all drawings, the nearer to completion you are, the harder a change becomes, so inappropriate component positioning can be very difficult to fix later on. The best layouts are produced by having the circuit designer present during the entire layout process to advise as to which components or tracks can move close and which must be separated. This is expensive, but not as expensive as having to scrap a layout and start again.

Some tracks around the RIAA equalisation are long because they guard sensitive nodes; conversely, tracks at the input must be kept short. Separation between equalisation and the adjacent CCS is deliberately generous to prevent coupling. Series resistor combinations are in a specific order to maintain a low impedance and reduced sensitivity to coupling at their junction. To summarise, if converting the author's main PCB layout to SMD, maintain component orientation and track routing, just shrink it as necessary.

Populating the PCB

The Toaster's PCB contains sufficient components that they needed to be fitted in the right order (otherwise, later components could be difficult to fit). Tall components were fitted last to avoid them obscuring lower parts, so from first to last:

Resistors
Valve sockets
LEDs
Small transistors (but not the SSM2210).

The resistors and valve sockets must be soldered first because some pins need connections to the ground plane, requiring explicit soldering to the top of the board in the absence of plated-through holes, and these soldered joints were promptly defluxed. Note that the four 1k surface-mount grid-stoppers on the PCB's signal side **must** be soldered using solder containing at least 2% silver. Following the chastening experience of mislabelled resistors whilst constructing the Bulwer-Lytton, the author checked all components before soldering, particularly LED polarity and transistor lead-outs. Beware in a cascode CCS that although the upper LED can be tested in situ, the lower LED'c current will be stolen by the transistor's base—emitter junction and emitter resistor, so it only lights when the full circuit is in correct operation.

With the resistors, small transistors, and LEDs in place, the four differential pair CCSs could be tested individually by connecting their −11 V to a bench supply and sequentially connecting their outputs via an ammeter to 0 V. For each CCS, the bench supply voltage was gently increased from zero and at ≈3 V the LEDs lit dimly whilst the ammeter registered ≈5 mA, which rose to 6.2 mA at the design voltage of −11 V, at which point the LEDs glowed quite brightly. See Figure 7.87.

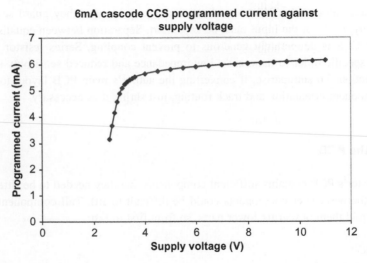

Figure 7.87
The cascode CCS starts operating at ≈3 V.

The switch-on knee is quite soft at ≈3 V and in an ideal CCS would lead to a horizontal plateau. Although the cascode CCS has a high output resistance with low capacitance, the crude DC reference (red LED plus series resistor) that confers its low noise causes its DC characteristics and rejection of power supply interference to be comparatively poor, which is why constant current diodes are so often seen in the LED chain. Constant current diodes are expensive, and the Toaster would need eight, so it was cheaper to take care over the negative subsidiary supply.

The heat transfer bar was temporarily fitted, and with all power transistors inserted and carefully positioned, but **not** soldered, a scriber marked their fixing holes. The bar was removed and strips of masking tape temporarily applied to the face edges of the bar to prevent subsequent swarf contacting and sticking to the edges of the insulating plastic strips. All the transistor fixing holes were drilled and tapped M3, whilst the TO220 holes were counterbored to a depth of ≈ 1 mm (using the pillar drill's depth stop) using a 3.7 mm drill on the device side only, and all holes were carefully deburred on both sides. Careful deburring is essential to prevent damage to insulating pads. The masking tape was removed when the bar was free of swarf.

Fitting a single TO220 device to a heatsink with an insulating kit and persuading its floppy conformal pad to align squarely is fiddly, but fitting four at once is really fiddly. Flat tweezers (Ideal-Tek 2A.SA) were immensely helpful for hanging conformal pads from overlong plastic bushes that protruded from the back of the TO220 transistors. The bushes then slipped into the 3.7 mm counterbores, and everything fitted together surprisingly easily. With the bar secured to the TO220 transistors, it was tightened up to the board using the three M3 screws.

Each MJE340 was insulated by a circular mica washer, but because these fouled the adjacent FQP1N50P's conformal pads, a chordal segment needed to be cut away using clean sharp scissors. Each washer was lightly but thoroughly smeared both sides with silicone grease and positioned on the bar using flat tweezers — the grease held it in place while the transistor leads were slipped through their PCB holes with the body held slightly away from the washer. Once the transistor was located, it was eased back onto the washer and secured. There wasn't a great deal of room, so the head of a hex key was sawn short to allow it to access the screw head more easily.

Resistance was measured between the bar and the PCB ground plane, and (much more importantly) between collector or drain of each transistor to the bar. Upon final tightening, one MJE340's collector was found to have a short circuit to the bar and the subsequent post mortem showed this to have been an entirely preventable failure. With the faulty mica butted up to the adjacent conformal washer, its central hole very slightly obscured the tapped hole in the heat transfer bar. On fitting the screw, the author had felt a slight binding, but persevered and his reward was to create mica grinding paste that produced a fine conductive swarf that was carried around the mica's hole to bridge the gap between transistor collector tab and heat transfer bar. See Figure 7.88.

Figure 7.88
Conductive black swarf can be clearly seen bridging the hole of this mica washer at four o'clock. The scattering across the face is presumed to have occurred on removal.

The contaminated transistor and mica were discarded, a bottoming tap run through the heat transfer bar from the other side, and the hole deburred again. The whole area was squirted with copious quantities of isopropyl alcohol, closely inspected for swarf on bar and PCB, and a fresh mica was cut and greased before inserting a new transistor. It probably wouldn't have helped in this instance, but hindsight suggests that fractionally counterboring the MJE340 holes might have been a good idea. Once all the insulation tests had been passed, soldering resumed in the following order:

Power transistors
SSM2210
Fixed capacitors
2–22 pF trimmer.

Once the semiconductors were in place, the board's foil side was defluxed. The 3.3 nF and 100 pF capacitors were measured and the trimmer capacitors adjusted appropriately so that total capacitance was the required 3.31 nF. The capacitors were fitted and the board carefully defluxed – taking care to minimise the amount of defluxer that made its way to the component side of the board as the (open) trimmers, tape-wrapped LCR polypropylene capacitors, and plastic insulating strips on the heat transfer bar were presumed fragile. The main PCB was now complete. See Figure 7.89.

Figure 7.89
The two channels are side by side and PCB layout follows the circuit diagram from top to bottom, so the SSM2210 can be seen at the top and the output coupling capacitors below the heat transfer bar.

The main PCB sets the position of the 8416, and once its positioning was known for certain, the +195 and −11 V regulators could be fitted to the opposite panel such that the tag panel and large 10 μF 630 V capacitor didn't foul the valves – there was sufficient space, but little excess. With hindsight, if the +195 V panel's support pillars had been secured to the main PCB, the Zeners would have been much further from the (hot) valves, but this disadvantage was not apparent at the time of PCB layout, so the pillars had to be secured to the case's opposite panel. Mounting the −11 V PCB vertically on the same panel near the front panel kept it cool and maximised air flow past the valves. The Statistical Regulator's CCS needs to be kept cool, so mounting it in the corner near the front panel with the rheostat's adjusting screw facing the (easily removable) top of the box seemed ideal. See Figure 7.90.

Figure 7.90
This panel faces the main PCB, so the cross formed by the tag panel and 10 μF 630 V capacitor sits between the four 8416s. The PCB carrying the shunt element of the −11 V subsidiary negative supply can be seen to the left (Zeners obscured by capacitor), and the CCS for the +195 V at the top left is well away from sensitive audio.

Mounting the Statistical Regulator's CCS by its DN2540N5 tab turned out to be astonishingly fiddly, ideally needing three hands. An M3 screw and nut secured the DN2540N5, and an overlong insulating bush was used as before to locate the conformal pad, requiring the panel's inside face to be counterbored to a depth of ≈1 mm using a 3.7 mm drill.

Eventually, everything was persuaded to stay in place whilst the screw was tightened. Insulation between chassis and tab was immediately tested using an insulation tester applying first 50 V and finally 1000 V. No measurable leakage or breakdown occurred and the author heaved a sigh of relief. Note that we need to be very careful when applying high voltage insulation tests, not just from the obvious safety viewpoint, but also that we stress only the intended insulation. The CCS was utterly unconnected to other circuitry, so connecting from one point on its board to an external point could only stress the conformal pad's intervening insulation.

The four 33 kΩ RH25 cascode load resistors on the main panel had been removed for drilling the earth bond hole and each dissipates only 0.3 W, but there's no virtue in allowing them to become warmer than necessary so they were refitted with a smear of silicone grease to their undersides.

Wiring

The PCBs having been populated, and mechanical work completed, it was time to add the interconnecting wiring. The main PCB is quite complex, and although the low voltage CCSs had been tested successfully, its remaining functionality was untested. The author isn't fond of nasty surprises, so he wanted to test as much as possible before final installation, and that meant fitting the twisted pair heater wiring first, then tails for the numerous +195 and −11 V connections, which could be temporarily connected to bench supplies for testing.

It's much easier to work on parts that don't wobble, so the author added temporary spacers to the top of the main PCB. See Figure 7.91.

Heater wiring

The heaters are in parallel, and hum must be minimised, so an individual tightly twisted pair of solid-core wire from each of the four valves needed to connect to the incoming twisted pair. Adding a small tag strip having a central chassis bond underneath the PCB made the heater wiring easy, enabled the addition of radio frequency bypass capacitors, and made pre-installation testing easier. This is just as well because cutting that short tag strip neatly from a much longer strip was difficult − the phenolic delaminated at the slightest provocation. The central tag has a plain washer above and star washer between it

Figure 7.91
Temporary pillars at the front of the PCB make it stable for heater wiring. Behind, the RH25 resistors and earth bond pillar are visible inside the case, and the 20 mm legs with stick-on feet outside. Note that work is carried out on a large sheet of paper — this protects both work and bench from scratches due to wire fragments, etc.

and chassis, and has been tightened firmly with an M4 cap head screw to ensure a low-resistance bond. See Figure 7.92.

Good heater wiring is fiddly and time-consuming — so anyone who tells you different is either unduly modest or possessed of superhuman skills. Starting from valve pins, points to note are:

- Black wire was 0 V (and presumed quiet), blue was +12.6 V (and presumed dirty).
- Black went to pin 4 because pin 3 is cathode (and sensitive), blue to pin 5 because pin 6 is an anode and less sensitive than the cathode.
- Each twisted pair immediately rose from the PCB to minimise capacitance to adjacent tracks.
- Each twisted pair bent sideways at 18 mm so that when the PCB was mounted, the twisted pairs would just kiss their chassis panel.

Figure 7.92
The heater wiring is in place and terminates on the tag strip at bottom right. Note the 1k SMD grid-stopper resistors touching pins 2 and 7 of the nearer valves.

- Each long twisted pair was routed into the edges of the chassis panel.
- The twisted pairs from the left-hand valves are laced together to ensure that they hold their form and because tight proximity assists cancellation. Plastic spring-loaded clips were useful for holding them in position as they were formed and laced.
- It was permissible for the laced loom to cross the PCB at foil level because the PCB has continuous ground plane at that point. (Crossing at foil level made wiring to the tag strip easier.)
- Knots in lacing cord can come undone, but a drop of cyanoacrylate glue on the knot solves that problem.
- There is a heater tag adjacent to each end of an output coupling capacitor. The end of the output coupling capacitor connected to the source follower has a lower impedance to ground and is more able to short-circuit hostile interference, so its adjacent tag carries blue wires.

617

- Normally vilified, a pair of 100 nF high-K ceramic capacitors on the tag strip are ideal for ensuring that incoming heater wiring is earthed at radio frequencies.

A potentially nasty little gremlin greeted the author when he tested power to the 8416 heaters. A Variac powered the Thing, enabling heater power to be applied gently and backed off at the first sign of trouble, such as excessive current. The Variac was gently increased from zero, and as the ammeter showed nothing untoward, quickly increased to full mains voltage. Meanwhile, the 8416 heaters began to glow and, as the structure warmed and filament resistance increased, current gradually fell and stabilised at 750 mA; 750 mA is 25% more than the author had been expecting (600 mA), but he can't blame the data sheet because he's never seen a manufacturer's data sheet for the 8416. There's no doubt that the 8416 **is** a premium quality version of the ECC88 because measurement shows that the anode characteristics are identical, but the only other known fact is that the 8416's natural habitat is inside a Tektronix oscilloscope.

Increased heater current increases both reservoir capacitor ripple voltage and regulator drop-out voltage. The author promptly backed off the Variac until ripple appeared across the heaters at an applied mains voltage of 235 V_{RMS}. This was marginally too high compared to the author's observed mean mains voltage of 245 V_{RMS}. Fortunately, the Thing's HT transformer had a mains primary tapped at 10, 0, 200, 220, and 240 V, so he had mechanically routed his mains wiring in such a way that moving a single wire could accommodate a different mains voltage (perhaps 220 V), and a happy corollary was that the neutral wire to the low voltage transformer could be moved from the 0 V to the 10 V tapping, increasing its primary voltage by 10 V. Increasing the primary voltage applied to the low voltage transformer was acceptable because part of its specification was a higher number of turns per volt than usual, so there was no danger of core saturation at a higher voltage — an off-the-shelf transformer might not be so tolerant.

Following the modification, the heater supply dropped out at an entirely acceptable mains voltage of 228 V_{RMS} (245 V − 7%). This might seem a lot of fuss about a small problem, but remember that the Toaster's input stage is a cascode, so C_{hk} would couple regulator drop-out spikes to the audio on the cathode.

Connecting the umbilical cable

The umbilical cable contains two RG174 coaxial cables plus the heater twisted pair in a braid screen. Another braid screen holds these cables together and the two braid screens form the chassis bond from Toaster to Thing, and thence to mains safety earth, so we need to be certain that they have reliable low-resistance connections to the chassis at either end.

The umbilical connector was plugged into the Thing (to hold it stable), and its rear cover unscrewed and pushed far down the cable, followed by a 50 mm length of glue-lined heatshrink sleeving. The nylon braid and outer conductive braid were pushed backwards to concertina along the cable and reveal their inners. The braid surrounding the twisted pair was trimmed back to expose 30 mm of inners, and a 20 mm length of PTFE sleeving passed over the twisted pair but under the braid to protect the inners from heat. The braid was then tinned all round using the largest possible tip on the iron so that all strands were soldered. When the braid is soldering properly, rather than there being a globule of solder at the applied point, capillary action suddenly draws the solder into the braid. The iron and solder need to be moved all round the braid and quite a lot of heat is needed, hence the protective PTFE sleeve.

A 20 mm long PTFE sleeve was passed over both RG174 cables and the outer conductive braid returned towards the connector and trimmed to the twisted pair's braid length and the braid encircling soldering process repeated, making sure that the outer and twisted pair braids soldered together, and a pig tail was added to the braid. When the screening braid had cooled down, the outer nylon braid was returned to the connector, trimmed to length and covered with the 50 mm length of heatshrink sleeving already on the cable. (Try slipping sleeving over the exposed ends of nylon braid one day and you will discover exactly why the author puts it on the braid very early on.)

The umbilical cable now had a twisted pair, two RG174 coaxial cables and a screen pig tail protruding beyond the heatshrink sleeving, and these cables could be soldered to the appropriate connector pins. The connector's rear cover was dropped back down the cable, screwed on, and its clamp tightened onto the heatshrink.

Inside the Toaster, 300 mm (12″) of inners protruded from the gas connector, and exposed braid around the heater twisted pair would have caused short circuits if unrestrained, so it was cut back. Low-interference audio needs low resistance earth bonds in the right place and although the outer braid screen was explicitly connected to the chassis by an O-clip, the author couldn't resist unravelling the remaining inner braid from its twisted pair and tightly twisting it to form a 2.5 mm diameter wire 50 mm long. The wire was tinned, sleeved with heatshrink sleeving and a long 4BA solder tag curled circumferentially around its end, lightly crimped with pliers, then soldered. The solder tag was bonded to the chassis by one of the 4BA screws holding the five-pin DIN input socket in place.

Testing the umbilical cable

The umbilical cable was 1 m long, yet a four-wire resistance measurement from Thing to Toaster chassis registered only 15 mΩ. To put the previous figure into context, the allowable earth loop resistance for an appliance requiring a 13 A mains fuse is 100 mΩ.

The author hadn't bothered to discriminate between the two RG174 cables, so he bared their ends at the Toaster, removed the EZ81 rectifier from the Thing and switched it on with the umbilical cable plugged in. With a meter connected between inner and outer, one RG174 cable had -29 V on its inner, and this identified it as the subsidiary negative supply. A 5 mm wide ring of red insulating tape was promptly put round the **other** RG174 cable as it exited the gas connector.

Interconnecting wiring

With the heater wiring done, interconnecting wires could be added to the main PCB prior to it being fitted in place. Bent nose pliers turned out to be ideal for pushing 0.7 mm diameter solid-core silver wire through 0.7 mm holes surrounded by components. The author uses silver wire in PTFE sleeving partly because he once compared it and thought it sounded better, but mainly because it produces visibly excellent soldered joints (perhaps why it sounded better).

It quickly became apparent just how many wires needed to be taken to the $+195$ and -11 V Kelvin capacitors. The PCB ground plane meant that 0 V from each stage couldn't be star connected to the capacitors, so the 0 V end of the $+195$ V capacitor had three of its five wires cut away to leave one wire for incoming power and one for outgoing (to the ground plane). More significantly, with the Kelvin capacitor assemblies in place on their facing channel panel, access was restricted. The solution was to remove the assemblies from the panel, then use it as a template to mark the fixing holes through onto a piece of scrap metal. The scrap was then cut away to form a skeletal jig that allowed easy access for wiring. See Figure 7.93.

Figure 7.93
The skeletal jig plate supports the +195 and −11 V supplies for wiring but allows easy access. Note the tightly twisted pairs for the input signals and lighter twisted pairs for the output signals.

The 35 off 5V6 Zener chain of the Statistical Regulator was tested on a bench supply set to limit current to 20 mA. Voltage and current were monitored as supply voltage was increased from zero, current lifted from zero as 195 V approached, and when it reached 20 mA, the voltage stabilised at the expected ≈ 195 V. The Kelvin capacitor was connected across the Zener chain, using red PTFE sleeving to denote positive.

The skeletal jig worked really well, and although dressing wires neatly took time, accessing them for soldering was easy.

Testing

The EZ81 was removed from the Thing and mains applied via a Variac whilst the subsidiary negative supply voltage was monitored. As usual, the Variac was advanced gently from zero, but noting that the monitoring meter immediately responded with the correct polarity, it was quickly swung to full mains voltage, whereupon the subsidiary negative supply registered -11.2 V, exactly as it should, and a few LEDs glowed.

It was no use prevaricating about the bush; the moment of truth had arrived. Probes were removed and the front panel fitted. Mains was switched off, Variac removed, EZ81 plugged in, further test instruments switched on, and one output of the Toaster plugged to the audio test set. Mains power was switched on.

LEDs at the input stage lit immediately, the 8416 began glowing, quickly followed by the remainder of the LEDs. Anode voltages were as expected (≈ 95 V), but reality returned when 50 Hz hum was discovered at -48 dBu.

Fettling

The first issue when investigating hum is to distinguish between the three ways in which it can couple into the audio:

- Power supply ripple
- Electrostatic
- Electromagnetic.

The best test equipment for hum diagnosis is a digital oscilloscope preceded by an audio test set. The audio test set has variable gain and filters to restrict the measurement bandwidth to audio frequencies, whilst the oscilloscope allows any change in waveform shape to be seen instantly. Further, oscilloscope automated measurements (frequency, V_{RMS}) are useful, and the ability to save traces enables before and after comparisons. Triggering the oscilloscope from line and invoking averaging allows misleading noise to be removed from hum waveforms. If you aren't fortunate enough to have access to an audio test set, then a low noise amplifier having gain switchable from 0 to 60 dB in 20 dB steps plus a switchable 22 kHz, 24 dB/octave filter is very nearly as good (and a lot cheaper).

Fifty hertz hum is invariably magnetic hum, and this was confirmed when the bench fan was rotated and hum amplitude changed — the fan's motor was very leaky, so pointing its coil at the Toaster made hum worse, whereas rotating by a further $90°$ minimised hum.

Moving the fan towards the Toaster case made hum worse, but draping the umbilical cable over the fan's motor made no difference. These quick tests suggested that the umbilical cable was innocent and that it was the Toaster's internal wiring that was picking up hum. The two channels were the same, so that suggested that a planning error was more likely than a wiring error.

The traditional way of curing hum depends heavily on the attending engineer's past experience of "things that have caused hum", so various "solutions" are tried until either the hum or the engineer goes away. The author decided to try being a little more scientific. Knowing that the cause was magnetic coupling, a search coil would be driven from a function generator to deliberately induce hum. Even driven by the generator's maximum output (7 V_{RMS}), a coil the size of a fingertip could not create a very intense magnetic field, but setting the generator to 70 Hz and triggering the oscilloscope from the generator's sync output should make the waveform distinguishable from mains hum. More significantly, because the search coil was so small, its magnetic field would be very localised, so it should be possible to identify the most sensitive part of the Toaster, and thereby determine the problem. Pleasingly, the scheme worked very well, and each channel of the Toaster was found to be most sensitive near the main PCB, on a vertical line between the input valve and the SSM2210, and the author promptly realised that instead of the two 562 Ω input resistors sharing a common earth connection (as in Figure 7.86), their connections to the ground plane were separated by 38 mm. See Figure 7.94.

In theory, a ground plane has zero impedance between all points, but practical ground planes have non-zero impedance. Bear in mind that the DL103 is specified to produce 0.3 mV at 1 kHz, 5 cm/s, and RIAA equalisation increases gain by 17 dB at 50 Hz, implying 42 μV_{RMS} sensitivity. But we expect a signal-to-noise ratio of ≈ 70 dB, so we would like hum to be better than 70 dB down on 42 $\mu V_{RMS} = 13$ nV_{RMS}, so even a very small resistance passing a small current could develop a voltage of this order, and this is why earthing is crucial.

Returning to the practical problem, the 38 mm separation between the earth points of the two resistors constituted a wire having non-zero resistance that could develop a hum voltage, and because an induced current would flow from one end to the other, it would produce a positive voltage at one end and negative at the other — a differential mode signal to which the Toaster is sensitive.

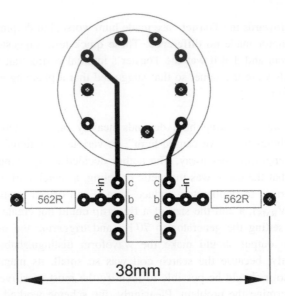

Figure 7.94

Input ground plane connections. Note that although this layout minimised track length for the sensitive input signals, it separated the ground plane connections (denoted by a cross) by 38 mm.

Theory is useful, but a fault has only been diagnosed when it has been fixed. The author lifted the resistors and connected their earth points to a common point on the ground plane to eliminate the 38 mm wire.

Playing the 1 kHz, 5 cm/s track from a test record produced +8.8 dBu (2.13 V_{RMS}) from the Toaster. With the arm in its rest, the audio test set measured noise at −62 dBu over a 22 Hz–22 kHz bandwidth, so referring that to the 5 cm/s test level gave an unweighted signal-to-noise ratio of 71 dB − very much as expected. What the author wanted to know was whether hum was discernible within that noise. See Figure 7.95.

The audio test set was set to 60 dB gain and its output analysed by a Rohde & Schwarz RTO1012 oscilloscope to produce the Toaster's noise spectrum from 0 to 200 Hz, using a linear scaling of 20 Hz per division. Vertical scaling is referenced to 0 dBV and is 14 dB per division in order to encompass programme peaks at +18 dBV (2.13 V_{RMS} = +6.6 dBV, and we expect programme peaks 12 dB higher). The noise rises towards low frequencies

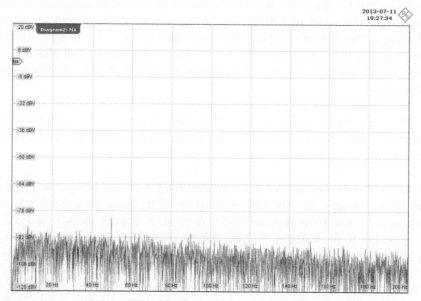

Figure 7.95

Toaster 0–200 Hz noise spectrum. FFT parameters: 20 s capture time, 125 kS/s sample rate, 2.5 MSa record length, rectangular window. FFT preceded by 30 kHz low-pass filter and 60 dB analogue gain.

primarily because of RIAA equalisation, but also because of base current noise in the SSM2210. A clear hum peak at 50 Hz can be seen poking out of the noise at a level of −82 dBV, and since the 1 kHz, 5 cm/s line-up level is +6.6 dBV, this means that hum is at −89 dB referred to 1 kHz, 5 cm/s.

Note that although the noise spectrum broke the frequency scale into bins 0.1 Hz wide and showed typical bin amplitudes of −99 dBV at 20 Hz and −108 dBV at 200 Hz, we (and the audio test set) hear the **sum** of all bins, which is why the true signal-to-noise ratio referred to 5 cm/s is only 71 dB.

A more accurate (and easier) way of measuring hum amplitude is to remain in the time domain, trigger the oscilloscope from line, and average over as many waveforms as possible to remove the random component (the noise) and leave the hum. Setting the Tektronix TDS3032 to average over 512 traces, an automated oscilloscope measurement read

95 mV$_{RMS}$, and because it was preceded by the 60 dB gain of the audio test set, equated to 95 μV$_{RMS}$ at the output of the Toaster. Given that the 1 kHz, 5 cm/s signal produced 2.13 V$_{RMS}$ at the Toaster's output, this meant that hum was at −87 dB. This waveform measurement is the more truthful because it included the fundamental plus all harmonics of the hum waveform, whereas the spectral measurement only included the 50 Hz fundamental.

Note that these hum and noise measurements were made not with a substitute resistance connected directly across the Toaster's input, but connected to an arm and cartridge on a turntable.

The author powers his Garrard 301 turntable from a Martin Bastin "Wave Mechanic" 60 Hz power supply, so this was switched on. Fifty hertz peak amplitude remained unchanged, but a 60 Hz peak appeared 6.5 dB below it. Starting the 301 loaded the power supply, so the 50 Hz peak rose by 2 dB, but the 60 Hz peak rose by 20 dB. Swinging the arm from its rest towards the end of the record only increased the 60 Hz peak by 2.7 dB, so this suggested that the Toaster was picking up hum directly from the 301's (not terribly well shielded) motor. Given that the Toaster was to the left of the turntable (very poor positioning), this seemed highly likely. Placing a 75 × 75 × 10 slab of mild steel close to the known sensitive region but between it and the motor only attenuated the 60 Hz peak by 2 dB. It was time to position equipment properly.

The Toaster was placed to the right of the turntable, the Thing and Wave Mechanic to the left. The 50 Hz peak fell to −86 dBV, and 60 Hz to −85 dBV. However, swinging the arm towards the centre of the record increased the 60 Hz peak to −70 dBV, indicating that the Thing was now hum-free and that something needed to be done about the Garrard 301 (fitting a 401 motor).

RIAA response

For a variety of reasons, very few twentieth century RIAA stages had accurate equalisation, but computer modelling means that there is no longer any excuse for errors. The convenience of being able to hang a known accurate inverse RIAA network onto the front of a function generator and enter frequencies via a keypad tempted the author into ignoring the fact that such a system produced an unbalanced signal. After all, an unbalanced signal into a balanced input could hardly change the frequency response, could it? See Figure 7.96.

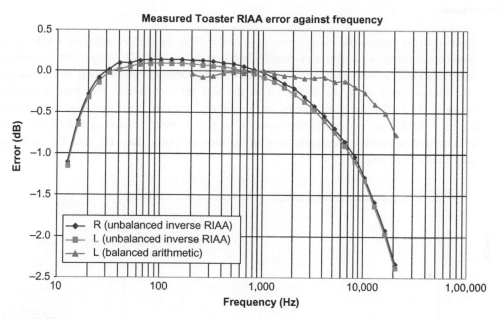

Figure 7.96

Frequency against amplitude response, but see text.

As can be seen, the two channels match very closely, dominated by a 0.06 dB gain error. It's just a pity that the measurement technique is invalid, resulting in a poor high-frequency response. Measuring the harder way, by connecting the audio test set's proper balanced output directly to Toaster input, measuring Toaster output, then subtracting the theoretical RIAA curve from that figure produced the correct result (balanced arithmetic). Be warned, taking liberties with balanced to unbalanced connections can sometimes produce misleading results.

Returning to the valid measurement, no measurements were taken <200 Hz because they necessitated either a very small signal (making it difficult to keep it hum-free), or risking destroying the audio test set's input stage (again) by applying a large signal. Turning to the high frequencies, if the 3.18 μs time constant is neglected, we should expect the response to be −0.64 dB at 20 kHz, whereas the Toaster is actually −0.77 dB. That's good enough, and close to measurement error.

Distortion

Remembering that 1 kHz 5 cm/s produced +8.8 dBu, the author adjusted 1 kHz input level to achieve +8.8 dBu and found distortion to be 0.013%. When the harmonics were analysed, it was possible to state that second harmonic dominated, but remaining amplitudes could easily be attributed to the test set's oscillator. We expect musical peaks at +12 dB referred to 5 cm/s, so the author increased level until the Toaster produced +18.8 dBu, and distortion rose to a rather more measurable 0.062%, with a distortion spectrum pleasingly free of harmonics other than H2 and H3 at equal amplitudes. See Figure 7.97.

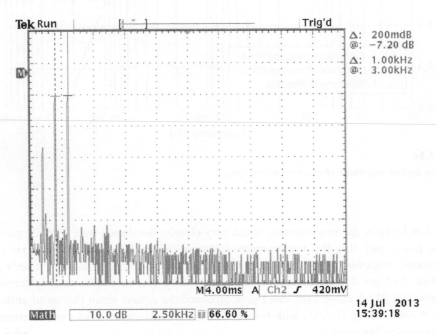

Figure 7.97
1 kHz distortion structure at +18.8 dBu (6.75 V$_{RMS}$).

Overload

Disconnecting the Toaster from the input of his audio test set, the author monitored Toaster output with an oscilloscope connected via a high voltage differential probe, and increased 1 kHz level until clipping occurred at an output of 150 $V_{pk\text{-}pk}$ differential (+ 36.7 dBu). See Figure 7.98.

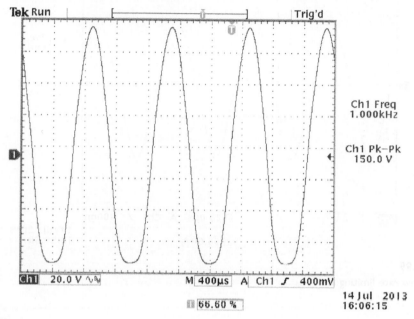

Figure 7.98
Clipping at 150 $V_{pk\text{-}pk}$.

Increasing frequency to 20 kHz, and oscillator level to compensate for RIAA, slewing distortion became evident at 89 $V_{pk\text{-}pk}$, but at this point any following amplifier will be in serious trouble. See Figure 7.99.

Heat

The author's remaining concern was operating temperature, so a Type K thermocouple was poked through the ventilation slots into the general area of the equalisation components (carefully avoiding high voltages). The thermometer stabilised at 54.0°C, but when the thermocouple was removed to measure external temperature, it stabilised at 28.2°C

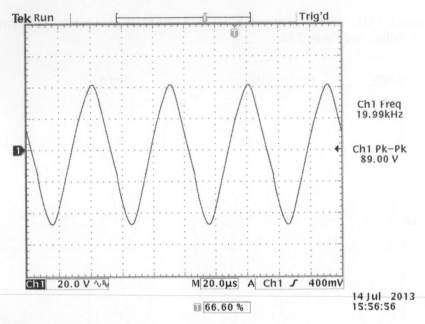

Figure 7.99
20 kHz slew rate limiting at 89 V$_{pk-pk}$.

(British heatwave), implying an internal temperature rise of 26°C. This was more than the author had hoped for, but not atypical for valve electronics cooled by convection. External chassis temperature rises varied a little, 9°C at the bottom, 12°C at the top, with mild hot-spots where heat was conducted directly to the chassis by the heat transfer bar headers, cascode anode loads, and +195 V CCS.

Toaster summary

The Toaster lives up to its name but pleasingly achieves its design objective of providing hum-free amplification from a low output moving coil cartridge without needing expensive input transformers. At 157 mm × 118 mm × 190 mm ($h \times w \times d$), it's quite small. See Figure 7.100.

Figure 7.100
Toaster external view. The single slotted screw is the 0 V to chassis bond, and the M5 button head screws near the front panel secure the heat transfer bar headers. The umbilical cable can be seen exiting to the right

Construction summary

The construction of each of these projects was punctuated by many simple tests and, when necessary, modifications, and such interruptions are inevitable for all but a production run using tried and tested components. Until formally documenting these three projects, the author was not consciously aware of just how many tests he makes en passant, but such tests find and fix minor problems before they become major disasters and are thoroughly recommended. Never ignore an opportunity to check your work — especially measurements.

The tyro alternative is to assemble the entire project at one fell swoop, probably finishing late at night, certainly when tired, cross fingers, and apply power. When it comes to excitement and disappointment, the author prefers it to come from the music, not from gambling the survival of expensive components and hours of work.

References

1. M. Jones, Rectifier snubbing — background and best practices, *Linear Audio*, April 2013, 5, pp. 7—26.
2. J. Millman and H. Taub, *Pulse and Digital Circuits*. McGraw-Hill (1956).
3. J. Mortimer, *The Best of Rumpole*. Penguin (1993).
4. G.W.C. Kaye and T.H. Laby, *Tables of Physical and Chemical Constants*, 16th edn. Longman (1995).

APPENDIX

Two-terminal modelling equations and spreadsheets

Whether we measure a capacitor, explicit inductor or a transformer winding's leakage inductance, we are simply measuring the impedance of a combination of components between two terminals. We imply a model the moment we assign LCR component values to that impedance. Modelling accuracy invariably breeds model complexity, so this section aims to show you how to create and test spreadsheet models ranging from simple to complicated with an absolute minimum of pain.

Once we have a mathematical model of a component's impedance, a spreadsheet can very easily calculate impedance at many different frequencies, plot the results as a graph of impedance against frequency, and overlay measured impedance against frequency. Iteratively adjusting model values until the two curves are perfectly overlaid allows a component's parasitic values to be determined, allowing more realistic and reliable modelling of entire circuits. But first we need that model. . .

Spreadsheets were developed for accountancy, not engineering, so they don't handle complex algebra particularly well, and the way that equations have to be entered makes mistakes likely. However, breaking the problem into small steps allows simple equations that can be entered and checked easily. Obviously, LTspice can analyse and plot the response of a complex circuit with far less effort, but it **can't** overlay your measured result for comparison, and that's why we tolerate the clunkiness of a spreadsheet for generating component models. Thankfully, we only have to enter the equations for a particular model once. Thereafter, we rename our spreadsheet and enter new measured data for comparison.

We know we will produce a complex spreadsheet having many rows and columns, so rather than adjusting model values in their individual columns, we list our components at the top left of the spreadsheet and use absolute rather than relative addressing (B5, rather than B5) in the calculations needing to pick up component values. This makes it easy to adjust model values once the graph has been plotted.

Each reactive component needs its own column to calculate reactance at each frequency of interest. Thus:

$$X_L = 2\pi f L$$

$$X_C = \frac{1}{2\pi f C}$$

Rather than neatly collecting these columns to the left of the spreadsheet, reactance columns are best added as each new component is introduced, keeping them close to the equations that use them.

No matter how many parasitic components there may be in total, a two-terminal problem can always be resolved into series and parallel combinations, but with the added difficulty that each step adds its single component (R, jX_L, or $-jX_C$) to a complex impedance ($A + jB$).

There are six possible situations:

Adding a series resistance:

$$A' + jB' = (A + R) + jB \tag{A.1}$$

Adding a series inductance:

$$A' + jB' = A + j(B + X_L) \tag{A.2}$$

Adding a series capacitance:

$$A' + jB' = A + j(B - X_C) \tag{A.3}$$

Adding a parallel resistance:

$$A' + jB' = \frac{R(A^2 + B^2 + AR)}{(A+R)^2 + B^2} + j\frac{BR^2}{(A+R)^2 + B^2} \tag{A.4}$$

Adding a parallel inductance:

$$A' + jB' = \frac{AX_L^2}{A^2 + (B + X_L)^2} + j\frac{X_L(A^2 + B^2 + BX_L)}{A^2 + (B + X_L)^2} \qquad \text{(A.5)}$$

Adding a parallel capacitance:

$$A' + jB' = \frac{AX_C^2}{A^2 + (B - X_C)^2} - j\frac{X_C(A^2 + B^2 - BX_C)}{A^2 + (B - X_C)^2} \qquad \text{(A.6)}$$

Note that each of the six preceding equations is in $A + jB$ form, so one column is required to calculate the A (real) term and another for the B (imaginary) term. If you have never understood complex numbers and treasure that ignorance, don't panic. Quietly ignore the significance of "j", but be aware that a complex number is composed of an A and jB term, that they are different, and may not be mixed. Think of them as apples and oranges, if you wish.

You will notice that the denominator (lower line) for the parallel equations is the same for the A and jB terms, so equation entry can be simplified by adding a third column for the denominator, then calculating the A and jB terms by entering the equation for the top line (numerator) and dividing it by the result already calculated in the denominator column. Further, you may notice that the parallel equations have been expressed in a way that highlights their very similar shape, making them easier to enter or check. Additionally, it is moderately easy to convert an inductance cell into a capacitance cell and vice versa, enabling copying and pasting followed by small edits, rather than entering everything individually in full.

Just like the DC resistive analysis described in Chapter 1 of *Valve Amplifiers*, we start analysis from as far away from the output terminals as possible, picking a point where we can combine just two components. We then work outwards towards the output terminals one component at a time, adding columns at each point to the spreadsheet. If we need to add both resistance and reactance in parallel, it does not matter in which order we perform the calculations, we can either enter the $A + jB$ values from the resistance addition into the capacitance or inductance addition, or vice versa. If we later find that we want to modify the model and insert an extra component or two, we simply insert columns at the appropriate point and modify the next set of columns to pick up the appropriate result.

Once we have introduced all our components, the far right of our spreadsheet has a pair of columns describing the network's impedance in $A + jB$ form, but we need to convert into Z, Φ form because that is what our test jig measures:

$$Z = \sqrt{A^2 + B^2}$$

And if we wanted the phase angle:

$$\Phi = \operatorname{atan}\left(\frac{B}{A}\right)$$

Once we have the magnitude of impedance (Z), we copy our cells down for as many frequencies as required, enter each individual frequency, and out will pop a calculated impedance that we plot against frequency (double-click on each axis and select a logarithmic scale). The graph of Z against frequency is very useful for checking equation entry, so if this is plotted early on, its $A + jB$ to Z conversion columns can be temporarily set to pick up the $A + jB$ result of each component addition and the graph inspected to see if it looks plausible. Thus, we might start with a series combination of inductance and resistance and plot their resulting impedance, expecting to see a graph with a horizontal line equal to the resistance, then a line rising with frequency for the inductance. If we then added parallel capacitance, we would expect to see a resonant peak, after which impedance would fall with frequency. We could even check the resonant frequency using:

$$f = \frac{1}{2\pi\sqrt{LC}}$$

Bear in mind that this simple resonance equation does not take account of the frequency pulling effect of resistive components, so provided the model equations suggest a resonant frequency within a few per cent of the simple resonance equation, they are right and the simple equation is wrong.

If we add a shunt resistance, we should expect to be able to adjust peak amplitude.

Once we have a plot of modelled impedance against frequency, we add the measured impedance against frequency to the graph and adjust model values until the two overlay. We use the measurement frequencies as input frequencies to our model because this allows perfect overlay for a perfect model. Useful tip: If you have a frequency counter (many DVMs include one) use this when you make the impedance measurement and enter the actual frequency (perhaps 1.0245 kHz) rather than the intended nominal frequency of 1 kHz; in this way, the model will calculate using the applied frequency, and compare like with like.

If we finally add a column to our calculations that divides modelled impedance by measured impedance, subtracts 1, and expresses the result as a percentage, we have the percentage discrepancy, and this can be linearly plotted against frequency (logarithmic scale), allowing model values to be fine-tuned for minimum discrepancy. If this graph uses the same frequency scale as the impedance graph and is positioned directly below it, glancing between the two makes it easier to identify which component value needs adjusting to minimise discrepancy.

Note that LTspice includes the four-component model as standard for its inductors and capacitors (although not resistors), and calculates more efficiently if this is used than if parasitic components are added externally. Obviously, if a six (or more) component model is required to describe a component, it will have to be entered as a collection of individual elements.

Once you have modelled a few components (inductances, especially), you will discover that it is possible to generate two models whose discrepancy graphs have very similar maximum variation (but entirely different shape), so which one is **right**? The answer is neither. There is no absolute **right**. What you have is a collection of components that behaves in a similar fashion **over the tested frequency range**. It is very dangerous to attempt to extrapolate over a wider frequency range unless you have an excellent physical understanding of the modelled component. Even then, it is far better to take extra measurements so that the comparison encompasses the working frequency range.

Assuming that the comparison covers a sufficiently wide frequency range, it can generally be said that a poor model produces many sharp variations in the discrepancy graph whereas a good model tends to have a smoothly varying discrepancy.

Six-component worked example

We want to model a 10 H HT choke using the split inductance model. See Figure A.1.

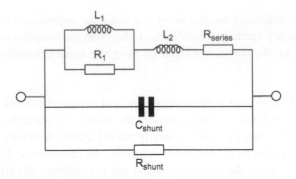

Figure A.1
This split inductance six-component model can describe a practical inductor quite accurately.

We have six individual component values, so we use cells 1–6 in column A to label them in order (L_1, R_1, L_2, R_{series}, C_{shunt}, R_{shunt}), and enter tentative values in column B (4, 5000, 4, 300, 150, 800,000). We expect to change the values in these cells later on, so the author changes their fount to bold to remind himself that only cells in bold may be changed. The equations require base units, so although the choke inductance (and probably series resistance) can be entered directly, its shunt capacitance C_{shunt} is likely to be only 150 pF or so. We will enter the shunt capacitance in cell B5 in pF, and to the cell immediately to the right (cell C5), take that value and perform a division by 1E12 (10^{12}) so that the capacitance is expressed in farads by typing = B5/1E12. Similarly, R_{shunt} would be more conveniently entered in kΩ, so we change cell B6 to 800 and multiply by 1E3 in cell C6 by typing = B6*1E3. If we later modify the spreadsheet to model leakage inductances, we might well use the same technique to allow us to enter values in mH (divide by 1E3) or μH (divide by 1E6). Remember to amend the cell labels by adding (pF) and (kΩ) as appropriate.

We leave a row blank, and use cell A8 to label the Frequency (Hz) column, then enter some hypothetical measured frequencies below it, perhaps IEC third-octave frequencies extended to 100,000, one frequency per cell. We leave columns B and C blank (we will need them later). We start our calculation by noting that L_1 is in parallel with R_1, so we need a reactance column (D) for L_1, and type XL1 in cell D8 as a label. In cell D9, we type = 2*pi()*A9*B1 to calculate reactance at each frequency. We are starting from the simplest point where we only have a resistance and are adding an inductance to it, so A = R and B = 0, simplifying some of the next equations.

Because we are calculating a parallel combination using equation (A.5), we need a denominator, so we use column E, type Denominator in cell E8, and in cell E9 we type $= \$B\$2\^2 + D9\^2$. Columns F and G will be our new A and B terms, so we type A in cell F8 and B in cell G8. In cell F9 we type $= \$B\$2*D9\^2/E9$ and in G9 we type $= D9*\$B\$2\^2/E9$. So that we know which columns are dealing with which combination, we type XL1 in cell D7.

We add the series resistance R_{series}, using equation (A.1). We type A in cell H8 and B in cell I8. In cell H9 we type $= \$B\$4 + F9$ and in I9 we type $= G9$. We type R_{series} in cell H7.

We want to add the series inductance L_2, so we create a reactance column and type XL2 in cell J8. In cell J9 we type $= 2*pi()*A9*\$B\$3$. We type A in cell K8 and B in cell L8. We use equation (A.2), and in cell K9 we type $= H9$ and in cell L9 we type $= I9 + J9$. We type XL2 in cell J7.

We want to add the parallel capacitance C_{shunt}, so we create a reactance column, and type XC in cell M8. In cell M9, we type $= 1/(2*pi()*A9*\$C\$5)$. Note that we pick up the value in the base units of farads, not the everyday value in pF. We use equation (A.6), and a denominator column makes life easier, so we type Denominator in cell N8, and in cell N9 we type $= K9\^2 + (L9-M9)\^2$. O and P will become our new A and B, so we type A in cell O8 and B in cell P8, then in cell O9 we type $= K9*M9\^2/N9$, and in cell P9 we type $= -M9*(K9\^2 + L9\^2-L9*M9)/N9$. We type C_{shunt} in cell M7.

We want to add parallel resistance, and use equation (A.4), so we type Denominator in cell Q8, and in cell Q9 we type $= (O9 + \$C\$6)\^2 + P9\^2$. We type A in cell R8 and B in cell S8. In cell R9 we type $= \$C\$6*(O9\^2 + P9\^2 + O9*\$C\$6)/Q9$, and in cell S9 we type $= P9*\$C\$6\^2/Q9$. Note again that we pick up the value in base units, not the value in kΩ. We type R_{shunt} in cell Q7.

Columns R and S represent the impedance of our circuit in $A + jB$ form, so we convert this into the modulus of impedance Z. We type Z in cell T8, and in cell T9 we type $=$ sqrt $(R9\^2 + S9\^2)$.

We now copy all our calculation cells down for as many frequencies as we need (to row 49 in this example), and if we plot column T against column A with logarithmic scales, a curve having the following shape should result. See Figure A.2.

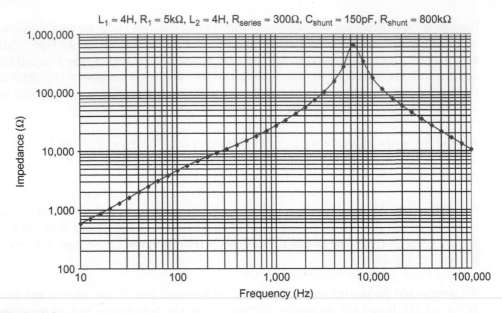

$L_1 \approx 4H, R_1 \approx 5k\Omega, L_2 \approx 4H, R_{series} \approx 300\Omega, C_{shunt} \approx 150pF, R_{shunt} \approx 800k\Omega$

Figure A.2
If entered correctly, the impedance curve produced by the six-component model should look like this.

We can now play with model values. Presently, the circuit resonates at ≈ 6.3 kHz. Change C_{shunt} to 600 pF and it resonates at ≈ 3.15 kHz. Restore C_{shunt} to 150 pF. The circuit is quite undamped, and increasing R_{shunt} from 800k to 10,000k increases peak amplitude slightly, but reducing it to 10k flattens the curve. Restore R_{shunt} to 800k. Reduce L_1 and L_2 to 0.1 H, and note that the peak moves to 50 kHz. Reduce R_{series} to 200 Ω, and note that the left-hand flat portion of the curve drops from 300 to 200 Ω (the new R_{series}). Restore L_1 and L_2 to 4 H, but leave R_{series} at 200 Ω. Toggle R_1 between 5000 and 2000 Ω, and note that the kink in the rising curve below resonance moves from left to right (it's harder to see this kink with only a few plotted points, but it's there).

Having proven a theoretical model, we can now compare it to practical measurements. We type V_{out} in cell E1, set our function generator to its maximum sinusoidal output, measure V_{CycRMS} using our digital oscilloscope, and enter that voltage, perhaps 5 V, in cell F1. We type r_{out} in cell E2 and enter the function generator's output resistance (probably 50 Ω) in cell F2.

We type R_{series} in cell E3 and enter its value, perhaps 51 Ω, in cell F3 — we need to choose R_{series} so that we **never** see more than $V_{out}/2$ in column B. 51 Ω is a minimum value, but we might need more. Be prepared to do a quick frequency sweep to estimate a suitable value — it isn't critical. If you're using this technique to measure from 10 Hz to 10 MHz, you might need to break the measurement into frequency bands and use different R_{series} in different bands. Just add more cells below F3, label them, and refer to them at the appropriate frequencies.

We type Voltage in cell B8 and Impedance in cell C8. In cell C9 we type = (F2 + F3)/(F1/B9-1) to calculate impedance from measured voltage, and we copy this cell down as necessary. We add a plot of column C against column A to our graph, and add the legend Measurement. We add the legend Model to our original curve.

We drive our test inductor from the function generator via R_{series}, we invoke oscilloscope averaging over 32 traces, and use the oscilloscope's automated measurements to measure V_{CycRMS} directly across its terminals (thus making a four wire measurement), enter that value in cell B9, and its frequency in A9. We take as many measurements as necessary to give detail wherever we need it, then fit the model to the measurement. The "Sort" function in the "Data" menu becomes useful when extra measurements are added.

UK to US glossary

The author has been made aware that the UK and the US are two countries divided by a common language and that the innocent rugby term "hooker" has a different connotation in the US. In an effort to minimise the problem, the following non-inclusive glossary was kindly provided by Stuart Yaniger:

UK	*US*
Allen key	hex wrench, Allen wrench
auto punch	generically referred to as center punch
bodging	improvising
dial gauge	runout gauge
dividers	compass
earth	ground

(Continued)

641

(Continued)

UK	US
folding machine	sheet metal brake
G-clamp	C-clamp
Goddard's Silver Dip	Tarn-X
haberdasher	specialist supplier of sewing materials
HT	high tension, high voltage, B+
jigsaw	not the same as an American jigsaw — saber saw
knicker elastic	elastic band material, available at sewing supply shops
loft	attic
loo roll	toilet tissue
LT	low tension (invariably heater supply)
methylated spirits	denatured alcohol
mole wrench	vise grip
nutrunner	nut driver
Paxolin	Garolite
rugby	contact sport similar to American football, but without armour
semi-flush cutters	diagonal cutters
serrated/shakeproof washers	lockwashers
sheet metal punches	chassis punches
skip	dumpster
solder tags	solder lugs
spanner	wrench
tag strips	terminal strips
tension file	rod saw
tin snips	tin shears
valve	tube
washing-up liquid	detergent

INDEX

Note: Page numbers followed by "*f*" refer to figures.

Printed and bound by CPI Group (UK) Ltd, Croydon, CR0 4YY

03/10/2024

01040310-0013